THE NEW NATURALIST LIBRARY

A SURVEY OF BRITISH NATURAL HISTORY

THE BURREN

THE BURREN

DAVID CABOT

and

ROGER GOODWILLIE

With photographs by Fiona Guinness

WILLIAM
COLLINS

This edition published in 2018 by William Collins,
an imprint of HarperCollins Publishers

HarperCollins Publishers
1 London Bridge Street
London SE1 9GF

WilliamCollinsBooks.com

First published 2018

A CIP catalogue record for this book is available
from the British Library.

Set in FF Nexus

Edited and designed by
D & N Publishing
Baydon, Wiltshire

Illustrations by Martin Brown

Printed in China by RR Donnelley APS

Hardback
ISBN 978-0-00-818378-3

Paperback
ISBN 978-0-00-818379-0

Contents

Editors' Preface

WITH *THE BURREN*, the New Naturalist Library achieves a significant milestone. Regional volumes have been a distinctive part of the collection ever since the beginning; *London's Natural History* was the third of the series. However, the Burren is the first Irish natural region to receive the full treatment, although it is no surprise that this is not its first appearance in the series. This extraordinary area of Carboniferous limestone, situated to the south of Galway Bay in Co. Clare, has long been known to generations of botanists for its remarkable flora. David Cabot, in his major New Naturalist volume *Ireland* (NN 84), gives it almost 40 pages. It first appears in *Wild Flowers of Chalk and Limestone* (NN 16), in which J. E. Lousley draws attention to the area's unique association of species: 'there is nowhere in Europe where Mediterranean and arctic–alpine plants grow together in a similar way'. And Michael Proctor, in his now classic *Vegetation of Britain and Ireland* (NN 122), describes the Burren as supporting 'the richest limestone grassland in Ireland'.

It has to be said that all this eulogising over the area's exciting botany is in marked contrast to some of the earliest reports commissioned by officialdom. David Cabot, in his previous volume, describes how one of Oliver Cromwell's commissioners entering the Barony of Burren describes it as 'a country where there is not water enough to drown a man, wood enough to hang one, nor earth enough to bury him'.

It is true that for the visitor unfamiliar with the area, the first impression is often of a bleak, inhospitable vista of bare limestone. However, closer acquaintance reveals characteristic shallow transitory lakes, known locally as turloughs, fens and Hazel scrub, all with as much interest for the zoologist as for the botanist. It is also easy to overlook that the Burren has an Atlantic coastline, including the Aran Islands, with the spectacular bird cliffs of Inishmore and Inishmaan.

David Cabot and Roger Goodwillie have collaborated to produce a masterly account of this remarkable area, integrating all aspects of its physical geography, landscape history and wildlife. The final chapter looks to the future and considers the problems of harmonising the competing pressures of farming and tourism. Both authors are among Ireland's most distinguished field naturalists. David Cabot is both an environmental policy advisor and eminent ornithologist, and already an experienced contributor to the New Naturalist Library with his earlier volumes *Ireland, Wildfowl* (NN 110) and, with Ian Nisbet, *Terns* (NN 123). Roger Goodwillie is a botanist and ecologist with a special interest in biogeography as well as in turloughs and woodlands. He has been a consultant on habitat evaluation and environmental impact for many years. The text is enhanced by the photography of Fiona Guinness, which was specially commissioned for the book.

Authors' Foreword and Acknowledgements

THIS IS NOT A GUIDEBOOK TO THE BURREN. It does not set out to tell the reader where to go to see the greatest spectacle or the rarest plant, although it does give some of this information in passing. Rather, it is an attempt to give the background story that will enrich a visit and hopefully bring to light something the reader is not aware of, whether it is the diversity of habitat in this small area, the history of its discovery or the relationship of a particular species to its environment.

The Burren attracts any naturalist with an eye for beauty, but it is the intricacies of the species' ecology, their links to the soil or to a particular insect, that is really fascinating. All the time it invites questions about the past, the present and the future of the area, which the following chapters seek to address. Not only do plants here seem to grow on next to nothing, but all the organisms have survived the comings and goings of woodland, the multiple mouths of grazing animals and the passage of several civilisations over 6,000 years. How they have persisted in such exuberance and diversity is a testament to their past evolution and to the gene complement they have accumulated over several million years, allowing them to adapt to a multitude of different conditions.

In the chapters that follow, the botanical scientific and common names follow those in the third edition of Clive Stace's *New Flora of the British Isles* (2010). After the first mention of the scientific name in each chapter, the common name is employed thereafter. Place-names follow those in the Ordnance Survey Ireland Discovery Series 1:50,000 maps (Sheets 51 and 52). The notation BCE means 'Before the Common Era' (before BC) and CE means 'Common Era' (AD).

DC FOREWORD

I knew little about the Burren until 13 June 1961, when Bill Watts, one of my botany lecturers at Trinity College Dublin, took our class for a week's fieldwork there. We stayed at Ballynalacken Castle Hotel, which I recall was rather spartan, but now, some 57 years on, has surely become somewhat luxurious. We were a serious group – there were no student pranks or high jinks. Bill, a thoughtful and sometimes solemn character, introduced us to all the major habitats, where we carried out descriptive analyses of plant communities, paying homage to Braun-Blanquet's analytical methods. We even 'stung' the fen at the base of Mullagh More with a long auger, driven down to the 'plastic clay' at the very bottom, 630 cm down. Those were five halcyon days in a magical landscape with a display of bewildering vegetation to confound phytogeographical botanists. What I observed and recorded on that precious trip resides in a brown standard exercise book I still sometimes dip into.

David Webb, Professor of Botany at Trinity, was my main mentor at college and for several years afterwards. He took me on several personal field trips to the Burren. We camped in Fanore dunes and stayed in a rented house near Kinvarra, and we roamed all over the Burren. He was gathering records for his future publications and my job was to record the birds of the Burren as part of the British Ecological Society Burren Survey Project. The records were gathered, but I lacked confidence with my methods and the data shamefully remained unused in my field note-books. I have, however, quarried some records for this book.

I well remember David's impatience when, after barking out some 20 scientific names in rapid fire, expecting me to retain all perfectly, I failed miserably. We also made an amateurish film about the Burren, now lurking in attic detritus but shown at several public lectures. Later, in 1969, I returned more fully plumaged with the BBC Natural History Unit and my producer friend Richard Brock to make an hour-long documentary on Ireland – *The Green Island* – in which we featured the Burren. It was shown several times on BBC and RTE television. We also made a radio programme on the Burren with Tony Soper. In the 1970s, the late Mary Gillham organised two tours of the Burren for her Cardiff adult education programme. I was fortunate enough to lead them, posing as an expert on the Burren.

Just over 20 years later, it was my turn, together with my film colleague and writer friend Michael Viney, to get behind the camera and produce a documentary about the eminent cartographer Tim Robinson. We spent many long days in the Burren filming Tim going about his work, mapping the archaeology, landscape features and other items of interest. Tim opened our eyes in a way nobody else could have done. The film, *Folding Landscapes* (1991), was shown several times on RTE television and also on Channel 4.

In order to freshen up my knowledge of the Burren for this book, I spent four wonderful weeks with Fiona Guinness, with visitations by Roger Goodwillie, in perfect weather in May and June 2016, followed by another two weeks during June 2017.

It has been a great pleasure working again with Roger Goodwillie, renewing our collaboration and friendship from the 17 years we worked together in An Foras Forbartha (the National Institute for Physical Planning and Construction Research) in Dublin and, latterly, in 1987, on a three-month expedition to north-east Greenland.

In this book I hope we have been able portray the magic of the Burren while scratching a little deeper than usual into the fields of interpretation, providing a bit of speculation and hopefully some explanation of this extraordinary area. One of my overriding concerns is that the Burren should never become a plaything or be overpromoted in the interests of tourism. It is the jewel in the ecological crown of Ireland.

RG FOREWORD

I first visited the Burren on a school field trip in the 1960s with my botanical mentor Richard McMullan, and I still remember the excitement of finding plants we had only read about in David Webb's (unillustrated) *An Irish Flora*. I returned on a field course with the Botany Department of Trinity College, which was led by Colin Dickinson, who had previously published an account of microclimate on the limestone pavement. He stressed the importance of the environment to plant life – the conditions that each plant has to master before it can become a permanent member of the vegetation. Thus began a lifelong interest in ecology that has led me to most of the habitats of the country, and to work on bogland, woods and turloughs. During this time I have benefitted hugely from the expertise of members of the Botanical Society of Britain and Ireland, from ecological discussions with many in the National Parks and Wildlife Service, and with Daniel Kelly and Micheline Sheehy Skeffington, who between them have supervised numerous recent Ph.D. projects on ecology in the Burren and elsewhere.

In the preparation of this book we have been blessed with the facility of electronic searching, whereby the world of research is open to all. I am humbled by the volume and quality of such research and fully acknowledge its creators.

This book is not a scientific treatise, so does not continually quote references as a basis for going forward. But I would not like anyone to feel that their work was not appreciated – even if it is not cited. I salute them all without hesitation.

In writing the habitat sections I am indebted to my wife, Olivia, for her constructive criticisms, which have improved the clarity and explanation of ecology. She was also part of the university class of 1970 and, more recently, organiser of a number of field trips we took to the Burren over 20 years, bringing outside groups of naturalists to favourite places.

My recent fieldwork in the Burren and Aran Islands has built on this base. In particular, during the spring photography sessions in 2016 and 2017 I enjoyed being part of the team, and am deeply grateful to Fiona Guinness for her hospitality and dedication. For me, this project was a return to working with David Cabot after years in a government planning institute and a few months in Greenland. I would hope that neither my editing of his words nor his of mine will change our opinion of, and friendship with, each other.

ACKNOWLEDGEMENTS

There are many people we would like to thank for their help with the preparation and writing of this book.

Helen and Enda Healy, managers of Deelin Mór Lodge, not only made us very welcome during our fieldwork and on our trips in 2016 and 2017, but also facilitated local contacts, while the lodge owners, James and Diana Moores, allowed us to stay at their wonderful house. Also at the lodge, Yvonne Naughton looked after us almost too well.

Liz Griffith assisted with trying to locate Pine Martens (*Martes martes*) we might capture with a camera trap, while Emma Glanville, National Parks and Wildlife Service (NPWS) Conservation Ranger, Burren National Park, guided us through the intriguing Rockforest woodland. Ann Bingham in the NPWS Information Centre, Corrofin, provided general information. Both Tony Kirby and Mary Howard, professional Burren guides, directed us to some of the more elusive Burren flowers. Mary Angela Keane, a friend of David's from the 1970s, took us to the Large-flowered Butterwort (*Pinguicula grandiflora*) site at the Spa in Lisdoonvarna. Cahill Murray at Aillwee Caves gave us access to photograph the Brown Bear (*Ursus arctos*) bones.

Carl Wright at Caher Bridge Gardens allowed us to set up a camera trap for Pine Martens; he also provided many photographs, several of which we have used. Steve O'Reilly arrived with his drone for some aerial photographs. Other photographers were also generous with their images – in particular, we thank Philip Strickland for his series of moth and butterfly shots, Colin Stanley, Dave Allen, John Fox, Ken Kinsella, John Breen, Gavin O'Sé, Cilian Roden and Jenny Seawright. Máire Caffrey, Head Librarian at Teagasc, and John Finn, also of Teagasc, provided a high-resolution image of the soils of Co. Clare. Kevin Walker kindly gave us permission to use the map from his paper on the rediscovery of Arctic Sandwort (*Arenaria norvegica*). Carol Gleeson of the Burren and Cliffs of Moher UNESCO Global Geopark allowed us to use some of their diagrams, as did Mike Simms for his geological profile of the Burren. One of David's best friends, Michael Longley, a devotee of the Burren, kindly gave us permission to reproduce his poem 'Burren Prayer', from *The Weather in Japan* (Jonathan Cape, 2000).

The portrait of Robert Lloyd Praeger (Fig. 6, top) is reproduced with the permission of the National Museums of Northern Ireland. The map of Black Head (Fig. 38) is reproduced with permission of the Ordnance Survey of Ireland: permit no. 9152 © Ordnance Survey Ireland/Government of Ireland. We would also like to thank Peter Snowball for allowing us to use his Woolly Mammoth painting on p. 127.

Ken Bond read the sections on moths, butterflies and other invertebrates in Chapters 3 and 5, and made improvements, as did Kieran Craven with regard to the Pleistocene and other geological issues in the same chapters. Brian Nelson pointed out some errors concerning invertebrates and other matters in those chapters, while Brendan Dunford provided constructive comments on Chapter 12. Jim O'Conner helped unravel a difficult taxonomic conundrum. Veronika Reven in Ljubljana assisted with information on Slovenian karst, as well as commenting on Chapter 5. In answering individual queries, we are very grateful to Maria Long, Una Fitzpatrick, Julian Reynolds, Cilian Roden and Emma Glanville. Whatever their advice, responsibility for the exactness of the text remains solely with us. Louise and Liam Cabot helped with computer technical issues.

We were extraordinarily lucky to have been joined on this project by Fiona Guinness, whose dedication and determination are apparent in her wonderful photographs. She gave up some of her valuable time to work with us, while also facilitating our stay at Deelin Mór Lodge.

Finally, we would like to thank David Streeter, assigned to the project from the New Naturalist Editorial Board for his advice and guidance. Also at HarperCollins, the indefatigable Myles Archibald hovered above us with encouragement while Julia Koppitz held everything together as an outstanding *chef d'orchestre*. Susi Bailey was our skilled copy editor, while David and Namrita Price-Goodfellow oversaw the copy-edit and designed the layout. Robert Gillmor has honoured us with a magnificent book jacket design that perfectly captures the essence of the Burren.

BURREN PRAYER

Gentians and lady's bedstraw embroider her frock.
Her pockets are full of sloes and juniper berries.

Quaking-grass panicles monitor her heartbeat.
Her reflection blooms like mudwort in a puddle.

Sea lavender and Irish eyebright at Poll Salach,
On Black Head saxifrage and mountain-everlasting.

Our Lady of the Fertile Rocks, protect the Burren.
Protect the Burren, Our Lady of the Fertile Rocks.

Michael Longley

From *The Weather in Japan* (Jonathan Cape, 2000)

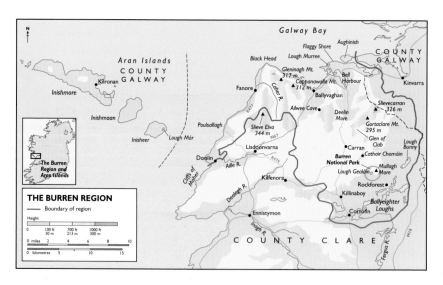

Location of the Burren, Co. Clare. Official place names and heights are those on the Ordnance Survey Ireland Discovery Map 51, 1:50,000 First Edition.

CHAPTER 1

Perspective

V IEWED FROM THE NORTH ACROSS Galway Bay, the Burren hills
appear like smooth grey beached whales, with a small pod of three
others offshore – the Aran Islands. Burren is the name of an ancient
barony – a collection of many townlands – that was under the control of the

FIG 1. Eastern flank of Gortaclare Mountain, looking towards the eastern hills of the Burren.
(Steve O'Reilly)

FIG 2. The earliest detailed map of north-west Co. Clare. The broken line marks the boundary of the Barony of Burren, while the solid lines mark the parish boundaries. From Petty (1685).

FIG 3. Deelin More – a typical Burren landscape, with terraced limestone hills overlooking a grassy valley based on glacial sedimentary deposits. Hazel and Ash woodlands grow at the junction of rock and soil, and extend onto the limestone in places. (Fiona Guinness)

O'Loughlins, the most powerful family in the north-western part of Co. Clare in the seventeenth century. There were 331 baronies in Ireland, which were created by the English as administrative areas. In 1672, the barony of Burren consisted of some 30,092 hectares, a predominantly limestone area on the Atlantic coast, stretching from the south-western part of Galway Bay around Black Head and south to Doolin. Eastwards, it skirted Lisdoonvarna, Kilfenora and Killinaboy towards Lough Bunny, then extended northwards to Kinvarra. Although baronies are now administratively obsolete, their boundaries are still used for land registry and planning purposes (Fig. 2).

The name Burren is derived from the Irish *boireann*, meaning 'big rock', 'place of stone' or 'rocky district'. It cannot be far removed also from 'barren', which comes from old French (Norman). At first sight, the Burren resembles the dry, rocky skeleton of a landscape, but it is a place that has a startling, powerful and lasting impact on many visitors. Close to, it has a dramatic physical appearance, but it also houses an extraordinary profusion and mixture of wild flowers. To many people, the Burren is a very special place, and in the words of writer and cartographer Tim Robinson (1999), it is 'one of the world's most precious and delicate terrains'.

EARLY DESCRIPTIONS OF THE BURREN

One of the earliest descriptions of the area comes from Lieutenant General Edmund Ludlow (c.1617–92). He was second in command to Henry Ireton, Lord Deputy of Ireland and son-in-law of Oliver Cromwell, and was famous otherwise for signing the warrant for the execution of Charles I. While on military action in Ireland, Ludlow reconnoitred the Barony of Burren in November 1651. In his *Memoirs* (1722), he reported:

> *After two days' marching, without anything remarkable but bad quarters, we entered into the Barony of Burren, of which it is said, that it is a country where there is not water enough to drown a man, wood enough to hang one, nor earth enough to bury him; which last is so scarce, that the inhabitants steal it from one another, and yet their cattle are very fat; for the grass growing in turfs of earth, of two or three foot square, that lie between the rocks, which are of limestone, is very sweet and nourishing.*

One of the early naturalists known to have explored the area was Frederick Foot (1830–67), who worked for the Geological Survey of Ireland. He was an able geologist and an excellent botanist who produced 13 explanatory geological

MEMOIRS

OF

Edmund Ludlow Efq;

Lieutenant-General of the Horfe,

Commander in Chief of the FORCES in *Ireland*, one of the Council of State, and a Member of the Parliament which began on *November* 3. 1640.

In THREE VOLUMES.

The SECOND EDITION.

To which is prefix'd,
Some Account of his LIFE and WRITINGS, Collected from the Earl of *Clarendon*, Biſhop *Kennett*, and Mr. Archdeacon *Echard's* Hiſtories.

WITH

An Account of his Conduct during his Baniſhment in *Switzerland*, and a Copy of the Inſcription upon his Monument. From Mr. ADDISON.

LONDON:

Printed for W. MEARS, and F. CLAY, without *Temple-Bar*; and J. HOOKE, and T. WOODWARD in *Fleetſtreet*. M.DCC.XXII.

FIG 4. Left: Lieutenant General Edmund Ludlow (c.1617–92). From Anon. (1755). Right: title page of Ludlow's *Memoirs* (1722).

memoirs, either solely or jointly, for 30 sheets of geological maps. He explored the Burren extensively in the early 1860s while mapping the geology of the region. It would be hard to better his classical description of the Burren, written nearly 150 years ago:

Almost all the barony of Burren is composed of the upper portion of the carboniferous limestone, with the coal measure shales occupying its S.W. corner, and stretching away south, higher beds of grits and shales appearing to the S.W. and S. The limestone rises into hills, upwards of 1,000 feet in height above the sea, intersected by valleys and deep ravines, their sides being in a step-like succession of bold bluffs and steep perpendicular cliffs with broad terraces of bare rock at their feet, which present to the geologist all the appearance of sea-beaches, elevated from time to time. The rock is traversed by different systems of joints, which form innumerable fissures in the flat beds, suggesting the idea of the surface of a glacier. At a distance these bare rocky hills seem thoroughly devoid of vegetation, and the desert-like aspect thus imparted to the landscape has been compared to that of parts of Arabia Petraea. But on closer inspection, it will be found that all the chinks and crevices, caused by the above mentioned joints, and the action of rain, are the nurseries of plants innumerable, the disintegration of the rock producing a soil, than which none is more productive. So rich and fattening is the pasture in the valleys, and often in the barest-looking crags, that high rents are paid for tracts of grazing, which a stranger en passant would hardly value at two pence per acre. (Foot, 1864)

THE MAGIC OF THE BURREN

The Burren is one of those rare and magical places where geology, glacial history, botany, zoology and millennia of cultural history converge to create a unique landscape of extraordinary natural history interest. It is without equal to any other area in Ireland or Britain. This veritable paradise for naturalists does, however, present unresolved questions concerning the exuberance of the wild flowers and the juxtaposition of plants from widely different origins – Mediterranean, Atlantic, Arctic and Alpine regions. As David Webb (1912–94) and Mary 'Maura' Scannell (1924–2011) stated in their *Flora of Connemara and the Burren* (1983), 'We must confess that a general explanation of the Burren flora is still to seek.' Some

FIG 5. Mullagh More, with Lough Gealáin in the foreground. Most of this area is included within the Burren National Park. (Steve O'Reilly)

FIG 6. Top: Robert Lloyd Praeger (1865–1953), painted in 1950 by Sara Cecilia Harrison. Above: cover of Praeger's *A Tourist's Flora of the West of Ireland* (1909).

attempts to answer these questions will be found later in this book, in Chapter 5.

For many years the Burren attracted the attention of only a few travelling naturalists. At first, these included just a trickle of mostly theologians, gentlemen travellers and a few trained specialists, but by the mid-twentieth century there were hundreds of visitors a year and today, in 2018, literally tens of thousands of tourists, university students and schoolchildren travel here. There are so many, in fact, that the coastal road from Ballyvaghan to Black Head and Poulsallagh can become clogged with traffic in summer, and often backed up with coaches and parked cars. Fortunately, however, most of the interior parts of the Burren remain relatively free from visitors; here, one can enjoy a feeling of solitude and consider the landscape and ecology without interruption. We discuss the procession of naturalists and their publications in Chapter 2.

It was not until the doyen of Irish botany, Robert Lloyd Praeger (1865–1953), presented the first popular account of the remarkable flora of the Burren in *A Tourist's Flora of the West of Ireland* (1909) and the slightly updated version in *The Botanist in Ireland* (1934) that other naturalists woke up to the richness on their doorstep. But Praeger also realised that such natural areas should be protected and was a prime mover in the establishment of the National Trust for Ireland (An Taisce) in 1948. So, while on the one hand he increased the flow of visitors to the Burren, on the other, he had a hand in limiting their impact.

DEFINITION OF THE BURREN AREA

The area ecologists and geologists include in their definition of the Burren differs from the barony boundary. Moreover, to complicate matters further different ecologists have their own definitions of the Burren. Where does the Burren start and finish, and what are its boundaries? The boundary of the Burren we follow in this book is that of David Webb, who, in his seminal 1962 paper 'Noteworthy plants of the Burren: a catalogue raisonné', states that the area is

> bounded on the north and west by the sea, on the south-west by the shales which overlie the limestone and bear an entirely different flora, with a sudden and dramatic contrast at the geological boundary, and on the south-east and east by arbitrary lines which separate the mainly karst-like country of the Burren area from the adjoining country which, though it contains patches of karst, is mainly drift covered and similar to the rest of the central plain. Small areas, which

FIG 7. Much of the Burren is a mosaic of limestone rock and thin, stony soils. Deeper soils derived from glacial sediments are rare except in the valleys. (David Cabot)

although limestone, are excluded near Kilfenora and north-east of Lisdoonvarna, are mainly farmed and have no floristic interest.

Taking Webb's definition of the Burren (Fig. 9), the mainland portion amounts to 419 square kilometres. Adding the Aran Islands (46 square kilometres), which are geologically and ecologically the same, the total area is 465 square kilometres, or 46,500 hectares. This amounts to 13.5 per cent of the total area of Co. Clare, although the Aran Islands are technically in Co. Galway.

A much smaller area for the Burren – 360 square kilometres – is often quoted in the general and tourist literature; this corresponds to the hills or 'high' Burren. This upland area rises to about 300 metres in the north and 100–150 metres in the south, and is made up of carboniferous limestone. The significant areas below

FIG 8. The western coastline looking south towards Poulsallagh, one of the most interesting botanical areas in the Burren. On the rocks here is the most extensive growth of the Hoary Rock-rose (*Helianthemum oelandicum* ssp. *piloselloides*) to be found in Ireland or Britain. (Fiona Guinness)

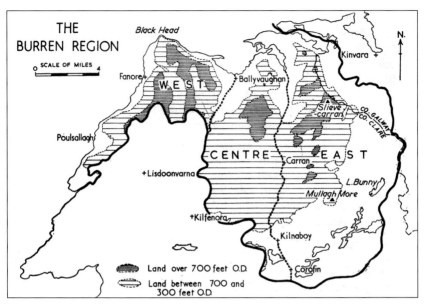

THE BURREN REGION

SCALE OF MILES

Black Head

Fanore

Poulsallagh

W E S T

Ballyvaughan

Kinvara

Slieve carran

CO. GALWAY
CO. CLARE

C E N T R E
Carron

E A S T

+Lisdoonvarna

L. Bunny

Mullagh More

+Kilfenora

Kilnaboy

Corofin

N.

Land over 700 feet O.D.
Land between 700 and 300 feet O.D.

FIG 9. Definition of the area of the Burren (marked by the heavy black line), according to Webb. Note that the three Aran Islands are not included on this map. From Webb (1962).

50 metres are the relatively narrow coastal fringes on the northern and western edges of the area, and the eastern lowlands inside a ring formed by the towns of Gort, Corrofin and Kinvarra. Two major valleys, also below 50 metres in altitude, cut into the hills south of Ballyvaghan and Bell Harbour. These are ancient valleys where rivers no longer flow.

The northern and western boundaries of the Burren are defined by the coastline as far as the low-water mark. The southern boundary follows the junction between the shales that overlie the characteristic limestone of the area and is extremely erratic, stretching from Doolin in the west, moving northeast towards Slieve Elva and Knockauns Mountain, then diving south-east to Kilfenora and Corrofin. This part of Co. Clare represents a complete contrast to the Burren, with rushy fields, conifer plantations and peat bogs. The eastern boundary is much less clear-cut, as the limestone terrain extends into the rest of Co. Galway to the north-east. There are many lakes here, some but not all of which are included in the Burren. The boundary then curves northward to meet the sea in Aughinish Bay. The Aran Islands are essentially a western extension of the karst limestones of the Burren, and are therefore included in Webb's and our definition.

A GLACIO-KARST LANDSCAPE WITH A UNIQUE BIODIVERSITY

The Burren is the best-known glacio-karst landscape in Ireland and Britain, and has a unique flora and fauna. It is described as such because its rocks were exposed to the atmosphere long enough to become permeable and develop underground rather than surface drainage, and were then covered and eroded by ice sheets during the last glacial period. Although its celebrity status overshadows other glacio-karst areas in these islands, the Burren does not have a monopoly on this landscape type, and especially one of its key features, limestone pavement. There are also limestone pavements in Britain, smaller in extent but still with the main features. These are refuges for many rare plants and have some notable similarities with the Burren.

In *Wild Flowers of Chalk and Limestone* (*New Naturalist* 16, 1950), Edward Lousley (1907–76) discusses the flora of Carboniferous limestone areas in Britain. Some limestone pavement occurs in the Derbyshire Dales, in the

FIG 10. South-eastern part of the Burren, with Mullagh More in the distance and Hawthorn (*Crataegus monogyna*) in flower, May. (Fiona Guinness)

FIG 11. The green road from Fanore to Ballyvaghan is a popular walking track today. (David Cabot)

southern parts of the Pennines, which Lousley describes as 'roughly a great tableland' with 'deep sided winding valleys with limestone exposed on their cliffs and slopes'. The area here is about 970 hectares, a figure exceeded only by the Great Scar Limestone pavement, on the northern side of Ingleborough in Yorkshire, which occupies some 1,050 hectares, more than a third of the estimated total in Britain (Lee, 2013). There are close similarities between the geology and ecology of these pavements and those of the Burren, including the common presence of several species of rare plants, mostly arctic–alpine and montane species.

Mountain Avens (*Dryas octopetala*), Spring Gentian (*Gentiana verna*), Hoary Rock-rose (*Helianthemum oelandicum*) and Shrubby Cinquefoil (*Potentilla fruticosa*) all occur on the Carboniferous limestone of Upper Teesdale, as they do in the Burren. However, the Shrubby Cinquefoil and Hoary Rock-rose found there are genetically distinct from those in the Burren; in fact, the Burren Shrubby Cinquefoil resembles the Continental form and must have had a different origin. Other Upper Teesdale plants in common with those found in the Burren include northern species such as Limestone Bedstraw (*Galium sterneri*), Spring Sandwort (*Minuartia verna*), Mossy Saxifrage (*Saxifraga hypnoides*) and Dark-red Helleborine

(*Epipactis atrorubens*). It is notable that there are no Mediterranean–Atlantic species in Upper Teesdale, such as Maidenhair Fern (*Adiantum capillus-veneris*), Dense-flowered Orchid (*Neotinea maculata*) and Lusitanian Large-flowered Butterwort (*Pinguicula grandiflora*), which do occur in the Burren.

CHAPTER 2

Burren Explorers

EARLY EXPLORERS, AND THEIR REPORTS AND BOOKS

The procession of explorers to the Burren, both botanical and otherwise, started as a trickle during the mid-seventeenth century, with the first plant records published in 1650. Throughout the eighteenth century a few more visitors arrived, recording their discoveries, and either publishing the records themselves or passing the information on to others for publication. A major turning point in the unfolding of the Burren's unique flora and fauna was the arrival of the geologist Frederick Foot in the early 1860s. Although he was ostensibly working for the Geological Survey of Ireland, Foot was no mean amateur botanist. The Royal Irish Academy published his remarkable account of the Burren's flora in 1864. Surprisingly, Foot's paper was to stand as the only comprehensive view of the Burren's flora for exactly 98 years, until David Allardice Webb's 'Noteworthy plants of the Burren: a catalogue raisonné' was published by the Royal Irish Academy in 1962.

During the interval between Foot's and Webb's accounts, an increasing number of botanists and other scientists were drawn to the Burren. Since 1953, the Botanical Society of Britain and Ireland (BSBI) had been collecting records of flowering plants and ferns in the Burren and across the rest of Ireland, and some of these contributed to Webb's 'Catalogue raisonné'. Then, with the encouragement of the Irish Regional Branch of the BSBI, Webb teamed up with Maura Scannell of the National Botanic Gardens, Dublin, to amalgamate the Burren and Connemara records. They added their own extensive knowledge to produce, in 1983, the first comprehensive book on the flora of the Burren, *Flora of Connemara and the Burren*.

June, 1962

62 B 9

PROCEEDINGS

OF THE

ROYAL IRISH ACADEMY

VOLUME 62, SECTION B, No. 9

D. A. WEBB.

NOTEWORTHY PLANTS OF THE BURREN:
A CATALOGUE RAISONNÉ

DUBLIN:
HODGES, FIGGIS, & CO., LTD.

1962

FIG 12. Left: David Webb (1912–94). 'He was, the *éminence grise* (and, in later years, *éminence blanc*) of Irish botany and a colossus with one foot firmly placed in his native country but the other planted in Britain and Europe. For two generations he was not only the leading taxonomic botanist in Ireland but the best known, and respected, Irish botanist in international circles, with his major contributions to *Flora Europaea* and the genus *Saxifraga*' (Perring, 1995). Right: Webb's 'Noteworthy plants of the Burren' (1962).

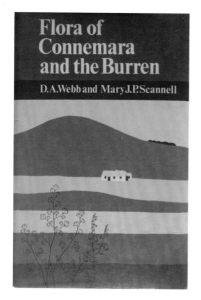

BRITISH ECOLOGICAL
SOCIETY (BES) BURREN
SURVEY PROJECT

An important event occurred in 1957, when Webb encouraged the BES to establish the Burren Survey Project. This was launched in 1958, with the aim of exploring the ecology of the Burren. Funds were made available to support visiting Irish and British botanists, zoologists and others to carry out various ecological studies.

FIG 13. David Webb and Mary Scannell's *The Flora of Connemara and the Burren* (1983). Although published some 35 years ago, it remains the most complete account of the Burren's flora.

The project was a success, and led to a notable vegetation analysis of Burren plant communities in 1966 by Robert Brian Ivimey-Cook (1932–) and Michael Proctor (1929–2017) of Exeter University. Since this seminal work, a plethora of publications – including more books, accounts, guides, and many scientific papers and notes – have appeared, culminating in Charles Nelson's book *The Burren*, first published 1991 and reprinted in 1997, and the most recent book on the area, *The Breathing Burren*, written by Gordon D'Arcy and published in 2016.

The following review of Burren explorers is necessarily selective, based on published works and deliberately focused on the earliest visitors to the area. It has not been possible to include every account, but most major publications have been mentioned. The Burrenbeo Trust has compiled a flora and fauna listing that contains many references to other studies not mentioned in this chapter (Burrenbeo Trust, 2005). In addition, A. J. G. Malloch published a partial annotated bibliography of the Burren in 1976. Studies concerning areas of archaeology or historic buildings have been omitted for the most part, as we

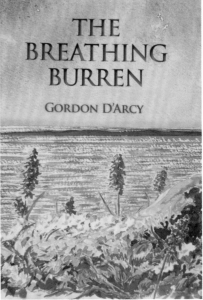

FIG 14. Two recent books on the Burren. Left: *The Burren* by Charles Nelson and Wendy Walsh, 'a companion to the wildflowers of an Irish limestone wilderness', published 1991 with a second edition in 1997. Right: Gordon d'Arcy's *The Breathing Burren* (2016) is a personal account of 30 years living in the Burren and exploring its landscape, flora and fauna.

focus primarily on the natural history of the area. There are many botanists and others who have gathered valuable information about the natural history of the Burren. Some of these records have been published but many remain frozen in notebooks or held as fading memories. One exceptional visitor who falls into this category was the Belfast-based artist Raymond Piper (1923–2007), a fanatical orchid specialist whose knowledge of the Burren's orchids was remarkable. Fortunately, he was persuaded to commit some of his records to paper before he died (now in David Cabot's personal library).

EARLIEST PUBLISHED RECORDS

The first published records of wild flowers in the Burren appeared in William How's *Phytologia Britannica* (1650). Three species were listed: Mountain Avens (*Dryas octopetala*), which 'makes a pretty shew in the winter with his rough heads like Virona'; Spring Gentian (*Gentiana verna*), reported growing 'abundantly' from 'the rocks betwixt Gort and Galloway'; and Dwarf Juniper (*Juniperus communis* ssp. *nana*), found 'upon the rocks at neer Kilmadough'. The author of these records, who had passed the information on to How, was the Reverend Richard Heaton (1601–66), an amateur botanist who has been described by Robert Lloyd Praeger as the source of the earliest records of Irish flowering plants. He was a Dublin clergyman, later Dean of Clonfert, who explored the country near at hand and pushed his researches as far west as Co. Galway (Praeger, 1949). Heaton's records probably dated from the late 1630s while he was travelling between his two parishes at Birr, Co. Offaly, and Iniscattery, south Co. Clare, according to Charles Nelson (1979).

The next naturalist who *probably* visited the Burren was the Welshman Edward Lhwyd (1660–1709), a naturalist and botanist, linguist and geographer. He was the first to describe and name scientifically what we would now recognise

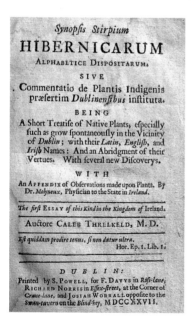

FIG 15. The first Irish Flora, by Caleb Threlkeld, was published in Dublin 1727. Many of the Burren species are listed.

FIG 16. Spring Gentian (*Gentiana verna*), one of the first species reported from the Burren, in 1650. Black Head, May. (Fiona Guinness)

as a dinosaur, and also the earliest botanist to make a collection of Irish plants that still survives today. He reported three species from the Burren. On the Aran Islands he found 'great plenty' of the Maidenhair Fern (*Adiantum capillus-veneris*), probably in June 1700 (Lhwyd, 1707). He also found there a 'sort of matted Campion with a white flower which I bewail the loss of it for an imperfect sprig was only brought to me'. This could have been the Field Mouse-ear (*Cerastium arvense*). He found '*Pentaphylloides fruticosa*' (Shrubby Cinquefoil, *Potentilla fruticosa*) 'plentifully amongst Lime-Stone Rocks on the Banks of Loch Crib [Corrib] in the County of Galloway [Galway], and '*Vaccinia rubra foliis Myrtinis crispis* (a very beautiful plant) to be no rarity in this Kingdom', which was possibly Bilberry (*Vaccinium myrtillus*). Lhwyd may have been in the Burren when he found Juniper (*Juniperus communis*) – 'I observ'd this plant to be so called in the County of Clare' – and Bearberry (*Arctostaphylos uva-ursi*). However, there is some doubt if he was *actually* in the Burren, otherwise why did

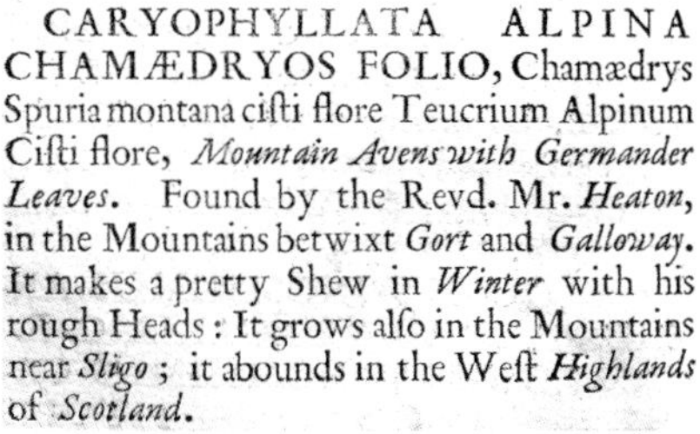

CARYOPHYLLATA ALPINA CHAMÆDRYOS FOLIO, Chamædrys Spuria montana cifti flore Teucrium Alpinum Cifti flore, *Mountain Avens with Germander Leaves*. Found by the Revd. Mr. *Heaton*, in the Mountains betwixt *Gort* and *Galloway*. It makes a pretty Shew in *Winter* with his rough Heads: It grows alfo in the Mountains near *Sligo*; it abounds in the Weft *Highlands* of *Scotland*.

FIG 17. The entry for Mountain Avens (*Dryas octopetala*) in Threlkeld's *Synopsis Stirpium Hibernicarum* (1727).

FIG 18. Maidenhair Fern (*Adiantum capillus-veneris*), first reported from the Aran Islands, probably in 1700. Black Head. (Fiona Guinness)

he fail to mention some of the more obvious species there that were not well known at the time and therefore of interest? Lhwyd travelled widely, recording little-known species in the mountains of Co. Kerry, Co. Sligo, Connacht and Co. Donegal (Lhwyd, 1712).

Charles Lucas (1713–71) was born in Ballingaddy, Co. Clare, where his family had been granted lands following the Cromwellian conquest of Ireland in the 1650s. He was trained as an apothecary and later as a medical doctor, and ultimately became a parliamentarian in Dublin. As an apothecary, he became familiar with herbs and wild flowers in the Burren. His contribution to charting the Burren flora was contained in a letter – as was the practice at the time, when junior scientists and others sent missives to more established experts. The recipient was a fellow physician, Sir Hans Sloane (1660–1753), an Irishman who was also a botanist and great collector of natural history objects; the bequest of his natural history collections laid the foundation of the Natural History Museum in London. Sloane published Lucas's letter in *Philosophical Transactions* (1739), and in it were three new Burren records: Yew (*Taxus baccata*), Goldenrod (*Solidago virgaurea*) and Vervain (*Verbena officinalis*), the last probably introduced to the Burren. Following his move to Dublin, Lucas continued to visit the Burren periodically, and in 1740 he collected three or possibly four species that were previously unrecorded for the area: Bearberry, Stone Bramble (*Rubus saxatilis*) and Rue-leaved Saxifrage (*Saxifraga tridactylites*), and possibly Sea Wormwood (*Artemisia maritima*) (Nelson, 1997).

NINETEENTH-CENTURY VISITORS

Joseph Woods (1776–1864)

Next to record visits to the Burren was Joseph Woods, an architect and botanist born in London of Quaker stock, who was quite an explorer and a frequent traveller throughout Europe. He arrived in Waterford in 1809 by packet steamer from Milford Haven, Wales, and kept a detailed diary of his observations during his travels (Lyne & Mitchell, 2014). His entries between July and October of that year contain important observations on the Irish landscape and its antiquities, as well as some botanical records. Woods was, however, disenchanted with the Irish landscape, especially with the lack of woodland, and complained that there was nothing there that reminded him of the England he knew: 'a mixture of woods, cultivation and habitation, hedgerows with trees in them, sheltered villages and farmhouses with their orchards'.

Woods spent some days in the Burren on his way to Connemara. His records and observations from the area are, however, disappointing for a man with such a critical eye and botanical knowledge. He observed some turloughs at Kilfenora, noting that:

> *These are among the singularities of this Country. They are pieces of flat land, generally very fine and rich pastures. After continued rain the water rises through some holes or under some rocks on the borders and presently converts them into lakes. After a time the water runs off again generally by the same openings. The Subterraneous Rivers whose overflowings supply the Turloughs are the only streams in the Limestone country – there is no continued stream of water and no constantly descending valley in which such a stream could run. (Lyne & Mitchell, 2014)*

It is unlikely Woods would have had prior knowledge of turloughs – only one is recorded for the whole of Britain. And he was not well informed, because the Caher is a 'real' river, the only permanent one flowing above ground in the Burren, descending westwards to the sea at Fanore.

After visiting the Cliffs of Moher and returning to Doolin, Woods 'hunted a long while in vain for the Maiden hair on the Shore and returned from a long and dreary walk without meeting any botanical recompense' (Lyne & Mitchell, 2014). While travelling near Corcomroe Abbey, he recorded the following:

> *Even the valley here is rock with scarcely any covering but straggling patches of the Dryas octopetala – called here the Burrin rose & I think it might almost be called the Burrin grass, it constitutes so large a proportion of the whole vegetation. A*

variety of Saxifraga hypnoides or rather perhaps a distinct species [In the electronic version of Woods's diary, the editors noted: 'presumably Saxifraga rosacea'] is abundant thro most of the district. (Lyne & Mitchell, 2014)

On his way back to Cork from Connemara on 20 September, Woods passed through Gort and visited Kilmacduagh:

Taylor's Map of Ireland had induced me to believe that Kilmacduagh was situated immediately at the foot of the Burrin mountains and I had flattered myself with the hope of examining their vegetable productions at a considerable distance from my former track and especially of renewing my search for Potentilla fruticosa but the weather and the distance united to prevent the execution of this plan. There are however one or two Turloughs or pieces of ground of that nature in the neighbourhood where I hunted without success. (Lyne & Mitchell, 2014)

FIG 19. Shrubby Cinquefoil (*Potentilla fruticosa*). Mullagh More can be seen in the background, May. (Fiona Guinness)

Although Shrubby Cinquefoil has two flowering seasons – late spring and early autumn – Woods would have seen it in flower, but he was well off track because the species does not extend as far east as Kilmacduagh.

Before departing for England, Woods made a general, but double-edged, observation:

> We picture to ourselves in England the Irish as a very hospitable people but we are apt to imagine that a considerable part of that hospitality consists of making their guests drunk, or at least in pressing upon them abundance of wine. I have no doubt that where an inclination of that sort exists, a person will find it very easy to gratify it in Ireland, & that even where reason struggles against inclination the latter will find a powerful advocate in the host. But for my own part I can truly say that I found it no more difficult to avoid intemperance in Ireland than in England but that I experienced a hospitality whose generous object seemed only to be to gratify every wish & to forward me in every object that I had in view. (Lyne & Mitchell, 2014)

Charles Carter (dates unknown)

A somewhat enigmatic Englishman of whom we know little, Charles Carter was possibly in a business that took him to the west of Ireland from time to time. He clearly had an interest in botany, although he is not listed in the *Dictionary of British and Irish Botanists and Horticulturalists* (Desmond, 1994). He also alludes to excursions made to ornithological sites at Ailsa Craig, Bass Rock and the Mull of Kintyre (Carter, 1846). Carter visited the Burren in mid-July 1846, spending some time around Ballyvaghan and Black Head. His paper, 'Botanical ramble in Ireland', in which his address is given as 'Oranmore, Galway' and which was published in *The Phytologist* that year, demonstrates his competence in botany. Carter expanded the list of Burren plant records, adding Mossy Saxifrage (*Saxifraga hypnoides*) and the small shrub Tutsan (*Hypericum androsaemum*), together with several species of fern. He was, however, somewhat disparaging about his Irish experiences:

> From New Quay I walked a distance of five miles round the creek to Ballyvaughan, a wretched little village. Such fare, and such a bed for a poor weary naturalist, who is only enlivened by a far too intimate acquaintance with hosts of the order Siphonaptera.
>
> Getting clear at daylight of these loving friends, I trudged up a long and weary ascent to the top of Beal-na-thulloch, and then what dreariness is seen around! Towards Black-head were piles upon piles of large lime-stone blocks, and mass upon mass extended until lost in the thick mist that covered the headland.

> Beyond the hill I saw the mountain avens (Dryas octopetala) abundantly in
> flower, the rock bramble with its bright red berries, and a few plants of Cistopteris
> and Grammitis Ceterach, with frequent tufts of Saxifraga hypnoides – a wretched
> tract, however, for the botanist. (Carter, 1846)

In an interesting aside to Carter's Burren visit, he also reported that he had
been in Connemara that September and that William Ogilby (1808–73), 'an
accomplished naturalist and botanist', had discovered another station for 'Erica
Mackkaii' (Erica mackaiana). Carter then received a strong but restrained rebuke
from Charles Farran in the next issue of The Phytologist for ascribing the discovery
to Ogilby, Farran pointing out that it was William McAlla (now correctly spelt
McCalla) (1814–49) who had originally found the species, in the summer of 1835
(Farran, 1847).

By now, the middle classes had more leisure time and resources to peruse
their hobbies and interests. The rise of natural history societies such as the
Dublin Natural History Society and the Belfast Natural History Society, along
with further explorations by naturalists, led to greater interest in the Burren and
natural history generally. Forays to the Burren were initially led by Dublin-based
naturalists, but they were soon followed by naturalists from Northern Ireland and
further visitors from Britain.

Francis Whitla (1783–1855) and Thaddeus O'Mahony (1821–1903)

In late July 1851, Dublin solicitor Francis Whitla was accompanied by Thaddeus
O'Mahony to the Burren on a three-day botanising trip. O'Mahony was an
undergraduate of Trinity College Dublin at the time, later becoming Professor
of Irish. He published an account of their joint visit in the Back Proceedings of
the Dublin Natural History Society in June 1858, which was the longest account
of the Burren flora to that point. Whitla had reported to the Dublin Natural
History Society in March 1851 that he had found Thyme Broomrape (Orobanche
alba) parasitising Wild Thyme (Thymus polytrichus). Like earlier accounts of
Burren flora, O'Mahony's paper was simply a narration of botanising, and did
not question the curious juxtaposition of alpine–montane and Mediterranean
species. However, in the mid-nineteenth century the European distribution of
many species was not well known.

Rather than providing a straight listing of species, O'Mahony remarked on
the habitats of flowers observed as well as commenting on how some species,
such as Common Valerian (Valeriana officinalis) and Hemp-agrimony (Eupatorium
cannabinum), could thrive in such dry conditions in the Burren when he had seen
them in moist locations in Co. Kerry. He recorded Dark-red Helleborine (Epipactis

FIG 20. A carpet of Hoary Rock-rose (*Helianthemum oelandicum*) with a clump of Spring Gentian (*Gentiana verna*). Ridge top looking south towards Poulsallagh, May. (David Cabot)

atrorubens) growing among broken limestone slabs close to Bearberry and Juniper on the summits of the highest hills. There were Spring Gentians near Fanore, and at Glencolumbkille and Rockforest, Alder Buckthorn (*Frangula alnus*) and Shrubby Cinquefoil were both in flower.

Like many early explorers, Whitla and O'Mahony were eager to find species new to Ireland – this was a way of making their mark as botanists among their peers. They thought they had a 'first' in finding Dropwort (*Filipendula vulgaris*) beside a small lake near Glencolumbkille. Leaves and roots were presented to an expert for identification. When the species was confirmed, O'Mahony wrote, thinking he had a new Irish record: 'I do not think it possible for man to enjoy a pleasure more innocent, more unalloyed' (O'Mahony, 1858). Alas, he was unaware that it had already been reported as occurring in Ireland, growing wild in the Burren, by the Reverend John K'Eogh (1681–1754) in his *Botanalogia Universalis Hibernica*, published in 1735. It is unclear where K'Eogh obtained this record.

Whitla and O'Mahony did, however, add some new records to the Burren flora, such as the Hoary Rock-rose – it had previously been recorded only from the southern Aran Islands and the southern coast of Cape Clear Island,

True MAIDEN HAIR, Hib. *Choſſa duh.* Lat. Adjanthum vulgare, ſeu Capillus veneris. The leaves are ſmall, round, and ſerrated, the ſtalks are black, ſhining, and ſlender, near a foot high, it grows on ſtone walls, and Rocks, the beſt in this Kingdom, is brought from the rocky mountains of *Burrin* in the *County* of *Clare*, where it grows plentifully, from thence it is brought in ſacks to *Dublin*, and ſold there, It is *Pulmonic*, *Lithontriptic*, *Emmenagogic*, *Hepatic*, *Splenetic*, *Nephritic*, *Pectoral* *Dieuretic*, *Styptic*, and *Deopulative*, * it wonderfully helps thoſe, that are afflicted with *Aſthmas*, ſhortneſs of breath, and *Coughs*, occaſioning a free Expectoration, it is alſo good againſt the *Jaundiſe*, *Dropſy*, *Diarrhæa*, *Hæmoptyſis*, and the bitings of mad dogs.

FIG 21. Excerpt from Reverend K'Eogh's *Botanalogia Universalis Hibernica* (1735) on the uses of Maidenhair Fern (*Adiantum capillus-veneris*).

BOTANALOGIA
UNIVERSALIS
Hibernica,
Or, A General IRISH
HERBAL
Calculated for this KINGDOM,
GIVING AN
ACCOUNT
Of the Herbs Shrubs, and Trees, Naturally Produced therein, in Engliſh, Iriſh, and Latin; with a true Deſcription of them, and their Medicinal Virtues and Qualities.
To which are added, Two Short
TREATISES.
One Concerning the Chalybeat, Waters, Shewing their Origin, Situation, Medicinal Virtues, &c.
Another of the Prophylactic, Or, Hygiaſtic Part of Medicine, Shewing how Health may be preſerved, and Diſtempers which human Bodies are ſubject to, prevented.
Authore Job, K'Eogh, A. B. Chaplain to the Rt. Hon. the Lord KINGSTON.
He cauſeth the Graſs to grow for the Cattle, and the Herb for the Service of Man. Pſal. 104. 14. V.
CORKE, Printed and ſold by GEORGE HARRISON at the Corner of Meeting houſe Lane, 1735.

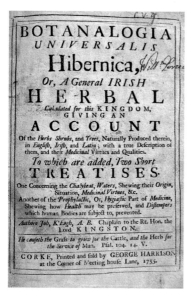

FIG 22. Reverend John K'Eogh's *Botanalogia Universalis Hibernica* (1735), the first Irish herbal, with several records of the Burren's flora.

The Roots of *Avens* are only uſed, which give a pleaſant Taſte, and ſmell to Wine, and chears the Spirits. Pain ariſing from Cold, or Wind in the Bowels, is aſſwaged by it; being of a binding Nature, it is uſeful in Fluxes.

FIG 23. Another 'prescription' from Reverend K'Eogh's *Botanalogia Universalis Hibernica* (1735) for the use of Mountain Avens (*Dryas octopetala*).

Co. Cork, according to James Mackay's (1775–1862) *Flora Hibernica* (1836). Also, the Maidenhair Fern was a new addition (apart from records on the Aran Islands and at Roundstone, Co. Galway). The pair remarked that large amounts of the fern used to be taken to Dublin for the manufacture of capillaire syrup, supposed to have medical properties as an astringent and aromatic (O'Mahony, 1858). Boiled Maidenhair Fern was sweetened with honey or sugar, then mixed with milk or water, and was said to be emetic when taken in large quantities. Another fern record new to the Burren was the Hard Shield-fern (*Polystichum aculeatum*).

Among the Mountain Avens plants the two botanists came across, they found polypetalous and sessile flowers, and remarked that the species was the most 'widely diffused of all flowers' in the Burren (O'Mahony, 1858). They based this on the fact that the plants were so plentiful, the locals bringing them down from the hills in loads to use as a fuel.

Alexander Goodman More (1830–95)

In the same year that Whitla and O'Mahony were exploring the Burren, a young keen naturalist, London-born Alexander Goodman More, arrived on holiday with

FIG 24. Alexander Goodman More (1830–95), a towering figure in all branches of Irish natural history.

his parents to stay with the Shawe-Taylors, acquaintances of the family who lived at Castle Taylor near Gort. More took to the area, returning there to botanise with the Shawe-Taylors several times and eventually settling in Ireland. He subsequently became one of Ireland's outstanding all-round naturalists, spanning the whole gamut of Irish flora and fauna. With David Moore (1807–79) he wrote *Contributions Towards a Cybele Hibernica* (1866), which was a big advance on Mackay's *Flora Hibernica*.

More was initially appointed as an assistant in the Dublin Natural History Museum in 1867, later being promoted to Keeper of the Natural History Division in 1881. He retired, due to ill health, in

FIG 25. Profusion of Mountain Avens (*Dryas octopetala*) on the southern side of Cappanawalla, May. (David Cabot)

1887. The naturalist made a large contribution to the expanding list of the Burren's flora, including the discovery of the Pyramidal Bugle (*Ajuga pyramidalis*) on the Aran Islands and the Brittle Bladder-fern (*Cystopteris fragilis*) at Black Head and between Ballyvaghan and Kinvarra. While botanising with Walter Shawe-Taylor at Poulsallagh, one of the favoured Burren botanical localities, More encountered Bloody Crane's-bill (*Geranium sanguineum*), Mountain Avens, Mountain Everlasting (*Antennaria dioica*), Spring Sandwort (*Minuartia verna*), Blue Moor-grass (*Sesleria caerulea*), Sea Spleenwort (*Asplenium marinum*), Rock Samphire (*Crithmum maritimum*) and the sea-lavender *Limonium transwallianum*, the last originally reported as *Statice occidentalis* by More in 1860. The sea-lavender plants found by More at Poulsallagh were only about half the size of *L. binervosum* agg. specimens growing elsewhere and had a less branched panicle. Writing in their *Flora of Connemara and the Burren* (1983), Webb and Scannell were convinced that the plants More had found, and those growing today near Black Head, are *L. transwallianum*.

More and his sister Alice made the remarkable discovery of the Dense-flowered Orchid (*Neotinea maculata*), one of the Burren's iconic species, early in the month of April 1864 while botanising at Castle Taylor. Although this is outside the Burren area as defined by Webb (Chapter 1), the orchid nevertheless occurs in many locations throughout the Burren and northwards in limestone areas to Lough Carra in east Co. Mayo and beyond.

The Mores first noticed this unfamiliar orchid as 'appearing above ground at a singularly early date', growing in a locality where they knew the Early-purple Orchid (*Orchis mascula*) was the only early-flowering species. Specimens were collected and dried, and then sent to Dublin, where the species' identity was

FIG 26. Pyramidal Bugle (*Ajuga pyramidalis*) with one flower remaining. The species was first discovered on the Aran Islands by Alexander More in 1854. Poulsallagh, May. (Fiona Guinness)

confirmed by More's botanical collaborator, David Moore (1864). Specimens were also sent to Heinrich Gustav Reichenbach (1823–89), one of Europe's leading orchid experts at the time, who confirmed the orchid's identity and provided a full description of it in the *Journal of Botany* (Reichenbach, 1865). It was a species new to Ireland and Britain. More was overjoyed, commenting, 'we thus had the great pleasure of adding one more to the list of southern plants which grow in Ireland without reaching Britain proper' (More, 1865).

While O'Mahony had noted that various Burren wild flowers were associated with environments atypical of their habitats elsewhere in Ireland, he was not concerned with the wider problems of phytogeography. More, however, was perplexed as to which geographical plant grouping the Dense-flowered Orchid should be assigned, while also intrigued that it was found growing with so-called alpine and arctic–alpine species such as Spring Gentian and Mountain Avens. He dismissed its grouping with the Lusitanian species of Irish Heath (*Erica Erigena*), St Dabeoc's Heath (*Daboecia cantabrica*), Large-flowered Butterwort (*Pinguicula*

FIG 27. Transparent Burnet (*Zygaena purpuralis sabulosa*) on Burnet Rose (*Rosa spinosissima*). This subspecies is confined to Ireland and found principally in the Burren. May. (Fiona Guinness)

grandiflora) and the 'Spanish saxifrages', and thought it was perhaps better classified with the Mediterranean Strawberry-tree (*Arbutus unedo*), but was disturbed by the disjunct distribution of *Arbutus* in Ireland. Webb and Scannell (1983) state that the distribution of the Dense-flowered Orchid is entirely Mediterranean, while the *New Atlas of the British and Irish Flora* (Preston *et al.*, 2002) ascribes it to the Mediterranean–Atlantic element (Chapter 4). In pondering the phytogeography of the Dense-flowered Orchid, More was the first to begin the discussion about the origins of some of the more unusual Burren species.

More was also struck by the occurrence of a very distinct species of a day-flying moth at Castle Taylor, the Transparent Burnet (*Anthrocera minos*, now *Zygaena purpuralis sabulosa*; Fig. 27), found in the same field as the orchid. He surmised from this association that he would expect also to find the orchid in the Burren in locations where the Transparent Burnet was well known (More, 1865). He was right – today, we know that the distribution of the Transparent Burnet closely mirrors that of the Dense-flowered Orchid, although there is no known ecological connection. Both species require limestone, so the moth is found in parts of south-east Co. Galway, around Lough Corrib and Lough Mask, and on the Aran Islands. The other moth location is at Barrigone, a limestone area near Askeaton, in north Co. Limerick (*MothsIreland*, 2017).

Frederick Foot (1830–67)

Frederick Foot also visited the Burren in the early 1860s, and was more concerned with recording where the various species grew. An excellent field geologist working for the Geological Survey of Ireland, Foot was – like More and many Victorians – a skilled all-round naturalist with a keen interest in botany. During the months he spent working in Co. Clare, he investigated every hill and corner of the Burren, noting aspects of the surface geology and detailing the distribution of the flora. He was able to appreciate the annual cycles of the flowers and ferns, as his work extended over 12 months.

With the information he gathered, Foot produced the first detailed botanical map ever published in Ireland or Britain (Fig. 28), all the more remarkable for how accurate it remains today, 150 years on (Foot, 1862). He marked the exact location of some species on a six-inch to the mile Ordnance Survey map, using a number code for each one. He then transferred the data to a one-inch map, from which he constructed his botanical map. Foot commented that the map might help direct visiting botanists to the interesting species and their locations. The information also allowed botanists in the years to come to establish whether some species were declining or increasing. According to Webb

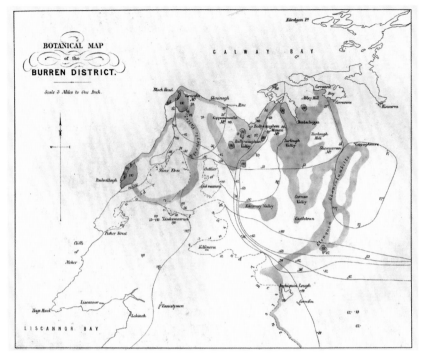

FIG 28. Frederick Foot's remarkable botanical map (1862) of the distribution of rare flowers and interesting botanical areas in the Burren.

and Scannell (1983), Foot added 17 previously unrecorded flowers and ferns to the Burren list.

Foot (1862) divided the Burren flowers and ferns into four groups:

- Plants abundantly and more or less equally distributed over the whole area, with only a few of the best stations marked on the map – 73 species.
- Plants confined to an area of the Burren and abundantly distributed in that area, the boundary line limiting the range of the plant, and the number (according to Foot's species list in his paper) of the plant written on it and a small arrow indicating the side of the boundary on which the plant occurs – 15 species.
- Plants occurring locally but growing in colonies marked on the map and coloured with the number of the plant marked – 2 species.
- Plants occurring locally, their positions marked on the map by a dot and the number of the plant – 25 species.

FIG 29. Hazel (*Corylus avellana*) woodland, near Cathair Chomáin in May. It is interesting that Frederick Foot omits any mention of Hazel woods or scrub in his 1864 paper. (Fiona Guinness)

Foot listed 114 species of flowering plants and ferns, and provided detailed notes on their distribution and abundance, sometimes in extraordinary detail. For example, he itemised the different stations where Maidenhair Fern was found, giving particulars of their elevation and distance from the sea. His precise detailing of the distribution and abundance of species provided an early baseline, allowing scientists today to assess roughly how the flora has changed over the past 150 years.

The most startling change since Foot's time has been the spread of Hazel (*Corylus avellana*) – the naturalist did not mention Hazel or any woodland in his account. Landscape photographs from the 1880s also confirm the absence of Hazel, but today it is dominant as scrub over wide areas and has spread into new districts within the past 30 years. Foot also wrote that Shrubby Cinquefoil occurred near Ballyvaghan, extending over an area of more than 2 acres (0.8 hectares) of low-lying ground or turlough. This was a famous – indeed, almost mythical – site for many years. The area was photographed by Robert Welch and featured on p. 351 of Praeger's *The Botanist in Ireland* (1934), accompanied by the caption: 'An almost continuous low growth, about two feet in height, extending over several acres of flat heathy limestone ground occasionally flooded'. Since then, a procession of botanists has searched for the site, including Charles Nelson, who in 1989 discovered a few 'woebegone' shrubs using Welch's photograph (Nelson, 1997). A better stand was found in the midst of scrub in 2004, but drainage and other agricultural work has put its future in doubt and it has largely disappeared. The impact of agricultural improvement on the flora, landscape and general ecology of the Burren are issues we shall return to in later chapters.

The status of Bearberry is confusing. Foot (1862) described it as 'spreading over the Hills'. Webb and Scannell (1983) stated that nineteenth-century records suggest a far greater abundance of the species than has been observed recently and that it is apparently diminishing, although they were unable to offer an explanation for the decline. Bugle (*Ajuga reptans*) was reported by Foot as 'by no means common' and observed only at Corkscrew Hill, whereas Webb and Scannell reported that by the 1960s it was occasional and frequent in many other sites, especially in Hazel scrub. Yew, noted by Foot as equally abundant as Juniper, has declined since the early 1860s due to lack of regeneration, severe grazing and other unknown factors, according to Webb and Scannell. Foot found only one Burren location for the Bee Orchid (*Ophrys apifera*), at Acres near Ballyvaghan, whereas Webb and Scannell reported it more widely.

As mentioned above, Foot was the first person to establish a rudimentary botanical baseline of the Burren flora. The unusual thing is that, in the best tradition of Victorian naturalists, his botanical work was a sideline to his professional work. The geologist modestly ended his paper:

> *I have now arrived at the end of my list, and must state that this list and map must not be taken as a perfect botanical survey of Burren. To complete such, it would be necessary to examine accurately every part of the district at different times of the year, according as the plants are in flower or leaf, and my other duties did not admit to this. (Foot, 1864)*

Thomas Corry (1869–83)

The link between Foot and the next Burren explorer, Thomas Hughes Corry from Belfast, is a tragic one. As recounted in his obituary notice in the *Geological Magazine* of 1867, Foot died at the age of 36:

> *On the evening of January 17th, a number of people were skating upon the ice of Lough Kay, near Boyle, in Ireland. Two of them having ventured upon a weak portion of the ice, it gave way, and they fell into the lake. Seeing their extreme danger, Mr. Foot came to their assistance, and in a noble effort to save their lives lost his own. They were both rescued, but he was drowned. (Anon., 1867)*

Corry also died by drowning, but at the earlier age of 22 and in Lough Gill, Co. Sligo. His obituary appeared in the *Journal of Botany, British and Foreign*:

It is happily but seldom that the pursuit of botany is attended with results so disastrous as those which we have had lately to deplore in the death of two Irish botanists – one of them already eminent in more than one branch of the science, the other, a comparatively recent student, but bidding fair to become a leading local authority. But the investigation upon which Mr. T. H. Corry and Mr. Charles Dickson were engaged terminated fatally and abruptly on Thursday, the 4th of August. On the morning of that day the two gentlemen went out in a boat upon Lough Gill, for the purpose of examining the botany of the islands in the lake. The weather was very boisterous, and the wind rough; the boat was upset, and the two botanists were drowned. (Anon., 1885)

Four years before his death, in July 1879, Corry visited the Burren with D. J. Calder, describing his reasons as follows:

Vivid and glowing were the accounts that reached me of the flowers of Clare, nurtured amid the balmy breezes of the green western land, and especially those which came from eyewitnesses of their beauty, and who described the celebrated Burren district as a spot meet to be called 'the botanist's happy hunting ground' for more rare and local plants flourish in that limited space together than can be met with anywhere else in the British Islands, under the same conditions. Accordingly my friend Mr. D. J. Calder and myself determined to profit by such counsels, and if possible judge for ourselves as to their accuracy. (Corry, 1880)

Corry and Calder left Belfast on 1 July 1879, travelling first to Athenry, Co. Galway. They then explored the adjoining districts, followed by the Burren area. Their paper to the Belfast Natural History and Philosophical Society in January 1880 is lyrical, full of information on historical buildings, descriptions of sunsets and the appearance of the sky, and details of all the species they encountered, with information on flowers, mosses and ferns and their habitats. Corry's 40-page paper contrasts with previous reviews written by Burren explorers in that it encapsulates the best of these but also highlights the Burren's uniqueness and its special atmosphere (Corry, 1880).

At first, Corry and Calder spent time botanising in the Hunting Course, a large, open rocky field at Castle Taylor where Alexander and Alice More had discovered the Dense-flowered Orchid. Here, Corry found the Transparent Burnet again, clinging on to the stems of Spring Gentian and Bloody Crane's-bill.

FIG 30. Clooncoose, a sculptured valley in the eastern part of the Burren. This was the location of the discovery of the Pearl-bordered Fritillary (*Boloria euphrosyne*) in 1922. (Fiona Guinness)

Corry (1880) noted that the moth formed its cocoon under stones on the ground and not on flower stems. The pair then visited 'the remote district' of Glenquin, on the eastern side of the Burren, the first part of the Burren proper. Among the limestone crags here they saw large quantities of Limestone Bedstraw (*Galium sterneri*), Squinancywort (*Asperula cynanchica*) and Wild Madder (*Rubia peregrina*). Between Killinaboy and Glenquin they spotted Elecampane (*Inula helenium*), along with a (then) rare grass, Barren Brome (*Anisantha sterilis*), and Common Calamint (*Clinopodium ascendens*). At the base of the Glenquin Hills were Fly Orchids (*Ophrys insectifera*), together with Mountain Avens, Spring Gentian and Bloody Crane's-bill. 'But by far the best of all the denizens of this spot were the fruit stems' of the Dense-flowered Orchid, growing quite sparingly, but hitherto recorded only in the Hunting Course at Castle Taylor (Corry, 1880). The orchid, together with the Irish or Erin Hawkweed (*Hieracium iricum*) and Fly Orchid, found here and at Castle Taylor, were considered by Corry to provide evidence that the Hunting Course should be judged as forming part of the Burren district.

Climbing the terraced hills above Glenquin, the two botanists encountered Hoary Rock-rose, Stone Bramble, Common Cow-wheat (*Melampyrum pratense*) and Goldenrod. Mountain Avens, many with double flowers, dominated the high terraces, providing soft cushions for weary feet, while Yew grew in an extremely dwarfed form. Corry (1880) described the landscape: 'The view from the summit of these hills… a country almost entirely covered with grey crags so as to have quite a strange and weird appearance, and amid which neither road nor inhabitants can be distinguished, the desolate character of the country being only relieved here and there by little lakelets.' On their descent, they came across Alder Buckthorn and, at the side of a small lake, Northern Bedstraw (*Galium boreale*) and Shrubby Cinquefoil. Local legend had it that Shrubby Cinquefoil was once a tall shrub with spines, which Roman soldiers used to form the crown on the head of Christ. The spines disfigured Christ's head and, ever since, the plant has reduced in size and lost its spines. Further north, at the base of Slievecarran, the pair found Dropwort, and another hawkweed, *Hieracium hypochaeroides*.

The naturalists then turned their attention to the western part of the Burren, starting at Poulsallagh and working their way up the west coast:

> *This point, in fact, is the southern termination of a bold escarpment, very striking in character… which stretch[es] up from the bay into the hills, [and] forms the promontory of Blackhead, rising to a height of 1000 feet within a mile of the shore… This escarpment is the steep slope of a great plateau, the summit of which, inclining gently southwards, is intersected by many deep valleys and gorges, which cut it up into numerous hills. These hills are composed of pale grey limestone rock, with scarce a trace of grass, and their sides are, as it were, hewn into giant steps formed by a succession of small cliffs, with nearly flat terraces of rock at their bases and summits, and these lines of cliff and terrace may be traced by the eye along the hills for miles together… they produce a most singular and striking effect, some of the hills, with their flat tops of bare rock, looking like great citadels surrounded by fortification walls. The breeze that flows in from the bosom of the great sea over these hills is warm, yet balmy and delicious, imparting a freshness peculiar to itself. (Corry, 1880)*

On some low cliffs by the roadside, Corry and Calder found *Limonium recurvum* ssp. *pseudotranswallianum*. This sea-lavender was only about half the size of the one they had encountered on Howth Head, Co. Dublin, and had a less branched panicle. Close by was Sea Spleenwort, with dark green fronds measuring a full metre in length. Rock Samphire, with its fleshy, edible stems, was also here. Further along the coast towards Ballyvaghan they encountered Squinancywort, Limestone Bedstraw, Wild Madder and the vanilla-smelling Fragrant Orchid (*Gymnadenia conopsea*). Webb and Scannell (1983) state that most, and perhaps all, of the Burren specimens of this orchid can be assigned to the subspecies *densiflora*, which has achieved specific rank in modern floras.

Corry and Calder found the surface of the cliffs matted with Mountain Avens. Maidenhair Fern was growing deep down in grikes, the frond tops almost reaching up to the surface, as was Lesser Meadow-rue (*Thalictrum minus*). There were Hoary Rock-rose and Spring Gentians, the latter's 'starry flowers' described as 'Deep, deep pure blue, the colour which Greeks gave to the eyes of Athena... associated as it is in our minds with the colour of Alpine lakes, which contrast so vividly with the fairer green of Alpine pastures' (Corry, 1880). Bloody Crane's-bill was 'the prevailing glory and beauty of these limestone rock gardens. Lovely indeed is the tricolour flag waving on the hills of Clare, formed of the Geranium, Dryad and Gentian'. Corry also came across Thyme Broomrape parasitising Wild Thyme at a location far to the south-west of that recorded by Foot. Irish Saxifrage 'formed little cushions on the rocks'. Higher up was more Maidenhair Fern and some Hart's-tongue Fern (*Asplenium scolopendrium*). Rustyback Fern (*Asplenium ceterach*) 'also delighted and greeted us among the rocks with its dark green fronds, and their almost golden or aureolin under-sides, attaining the extraordinary dimensions that I have spoken of elsewhere'. Common Dogwood (*Cornus sanguinea*) was found in the rocky valley of Ballynalackan.

Taking a ride with a jaunting car down Corkscrew Hill was 'a sensation of being jolted to pieces.' At the foot of the hill, beside the road leading to Ballyvaghan, they found the grey stems and flowers of Vervain. Another alien was Swine-cress (*Lepidium coronopus*), introduced from either South America or the Mediterranean region. In a turlough south of the old Ballyvaghan workhouse, the pair found Northern Bedstraw, Shrubby Cinquefoil and an unusual dwarf and prostrate form of buckthorn (*Rhamnus* sp.), along with Bloody Crane's-bill and Squinancywort. They also found Salad Burnet (*Sanquisorba minor*), with flower heads of a reddish-purple hue, a species new to Co. Clare. They walked to Black Head along the main road:

> *The full sun of noonday poured his rays down upon us as we worked, and we experienced what it is to hunt amid grey limestone rocks with an atmosphere flashing with light all around, and very glad we were of the slight relief afforded by shades of neutral tint glass. The fields appeared one blaze of flowers, some cornfields especially reminding me, by the brilliancy and varied hues of their colouring, of Gustave Dore's admirable picture entitled 'The Prairie', one especially so, a field of golden brown wheat, amid which rose one blaze of silk and scarlet flame, due to the splendid flowers of the Poppy Papaver rhoeas. (Corry, 1880)*

The Common Poppy (*Papaver rhoeas*) was a new record for the Burren and certainly introduced. Mingling with the red poppies were the light blue flowers of Field Scabious (*Knautia arvensis*) and Common Cornsalad (*Valerianella locusta*). Two introduced plants the botanists recorded that have since nearly died out were the Dwarf Spurge (*Euphorbia exigua*) and Rough Poppy (*Papaver hybridum*), seen near Cusheen. These last four species were previously unrecorded from the Burren. Near Gleninagh Castle, Corry and Calder found the Brittle Bladder-fern 'peeping forth from under the interstices in the fences of grey stone by the wayside' (Corry, 1880). Mugwort (*Artemisia vulgaris*) and Common Mallow (*Malva sylvestris*) were here also; the latter is now commonest on the Aran Islands.

On some short grassy pastures near Gleninagh the naturalists found Lesser Clubmoss (*Selaginella selaginoides*), growing near its southern limit in Ireland, along with Spring Gentians, which were common on the hillsides. Corry then made 'the acquaintance of some fine specimens of the goat, known technically here as "The Burren deer"' (Corry, 1880). On the higher rocks Mountain Avens formed 'immense cushions', while Bloody Crane's-bill also abounded in the area. Lesser Meadow-rue was found where Foot had reported it, along with Mossy Saxifrage.

FIG 31. Looking north from the Flaggy Shore over Galway Bay, with the mountains of Connemara in the distance – a view Thomas Corry may well have admired. (Fiona Guinness)

At the village of Murroogh, north of Fanore, Corry found a colony of White Horehound (*Marrubium vulgare*), a medicinal herb that is now extinct in the Burren and, indeed, most of Ireland (Webb & Scannell, 1983). Between the village and Black Head, Corry reported that his friend Thomas Johnson Westropp (1860–1922) had found the rare Ivy Broomrape (*Orobanche hederae*) parasitising Common Ivy (*Hedera helix*) (Corry, 1880). After returning to Ballyvaghan in an open donkey cart – 'a most disagreeable shaking' – Corry reported that they 'were charmed by a lovely view of the distant Connemara mountains... lying bathed in the evening light, the light blue colour of the near hills across the bay fading into intense purple in the distance'.

Corry summed up the results of his and Calder's botanical ramble, noting the extension of the range of the Dense-flowered Orchid into the Burren (as predicted by More) and adding 23 new species to the previous records from District 6 (Babington's (1859) botanical district covering Co. Limerick, Co. Clare and east Co. Galway), most of them within the Burren. Corry was an outstanding botanist but he was also a poet – a volume of his poems, *A Wreath of Wildflowers*, was privately published in 1882, a year before his death.

TURN OF THE CENTURY

By the end of the nineteenth century, naturalists had woken up to the attractions of the Burren, and naturalist field clubs were already well established in Belfast, Dublin, Limerick, Cork and Galway. A joint field club excursion to the Burren took place in July 1895, with the naturalists travelling by boat from Galway to Ballyvaghan. It was organised by the future doyen of Irish botany, Robert Lloyd Praeger, who was then honorary secretary of the Irish Field Club Union. The visit was a success, and from then onwards botanising in the Burren was not solely restricted to intrepid botanists and naturalists. One would have thought that after visiting the Burren for the first time, Praeger would have championed the area for its outstanding botanical features. He did include accounts of the district in his various popular books, but apart from his classic 1932 'The flora of the turloughs: a preliminary note', he never applied his energy into exploring the Burren. Instead, he organised other projects, such as the Clare Island Survey and surveys of the flora of under-recorded parts of the country.

The eminent British botanist George Claridge Druce (1850–1932) visited the Burren and other western areas in June 1905, turning up a few interesting species and varieties (Druce, 1909). He found Field Madder (*Sherardia arvensis*) at Black Head, which he considered unmistakably native. At Ballyvaghan, he was

FIG 32. O'Kelly's Spotted Orchid
(*Dactylorchis fuchsii* ssp. *okellyi*). Deelin
More, May. (Helen Healy)

the first to record the eyebright *Euphrasia rostkoviana* (now *E. officinalis* according to Stace (2010)), which is still local. Druce's main finding, however, was O'Kelly's Spotted-orchid (*Dactylorhiza fuchsii* ssp. *okellyi*), originally recorded by the botanist as *Orchis maculata* var. *okellyi*. Druce described this variety as having long, narrow, spotted, pale green leaves and pure white flowers that are in a dense, blunt oblong-cylindric (not tapering) spike. He wrote, 'From its being known for so long by its finder [P. B. O'Kelly, a Burren farmer], and who has done so much to investigate the flora of this rich neighbourhood, and who had in fact called it *immacutata*, I have, with his permission, connected his name with this interesting plant.' There is still some discussion as to whether this orchid is a subspecies or just a form.

The farmer acknowledged by Druce was Patrick Bernard Kelly (1852–1937), known locally as 'Dr' P. B. O'Kelly, who acted as guide to the botanist in 1905. O'Kelly was born in New Quay, between Kinvarra and Ballyvaghan, of a local Burren farming family. He was a most unusual person, a good botanist, farmer and nurseryman, who advertised for sale Burren wild flowers. He had clients throughout Ireland and Britain, and even some in Europe. O'Kelly settled in an attractive large house outside Ballyvaghan, which, when we visited it in May 2017, had almost been consumed by the surrounding vegetation.

O'Kelly's knowledge of the Burren flora was extraordinary. He is credited with the discovery of a rare *Potamogeton* hybrid, *Potamogeton* × *lanceolatus*, in the Caher River in 1891, and also found Water Germander (*Teucrium scordium*) and Mudwort (*Limosella aquatica*). In addition, he claimed that he had found a hybrid between the Pyramidal Bugle and Bugle (*A.* × *pseudopyramidalis*, according to Stace (2010)), although this was not fully accepted in later years. In 1988, five years after the publication of Webb and Scannell's *Flora of Connemara and the Burren*, Maura Scannell stated that Druce had reported the hybrid from Sutherland, commenting 'To this may also probably be referred plants gathered by Mr. P. B. O'Kelly, near the coast of Clare'. Scannell presented additional evidence to support the successful hybridisation and claimed that O'Kelly had been vindicated.

FIG 33. Patrick O'Kelly's abandoned and overgrown house and garden at Glenarva, on the road from Ballyvaghan to Corkscrew Hill. It seems ironic that the domain of such a famous collector and seller of Burren plants is being swallowed up by the local vegetation. (Fiona Guinness)

O'Kelly's lasting claim to fame was the discovery of the white-flowered O'Kelly's Spotted-orchid. Although he did not publish his plant records, he did issue two catalogues of plants and shrubs for sale, the first – *A Complete List of the Rare Perennial Plants and Shrubs of The Burren Mountains of Ballyvaughan, Co. Clare*, reproduced in Charles Nelson's *The Burren* (1997) – listing species dug up from the Burren that were available for sale. These included Pyramidal Bugle, 'very rare, 2 shillings each' (about £12 at 2017 prices); Large-flowered Butterwort; Mountain Avens; and Maidenhair Fern. Modern botanists would have viewed his depredations of the Burren species with alarm, but Charles Nelson (1997) does not believe he caused any long-term decline to the hallowed species.

TWENTIETH CENTURY

Druce's 1909 paper was one of the last original contributions on the botany of the Burren, and brought the early exploration of the area to a close at the start of the twentieth century.

In August 1908, the British Vegetation Committee undertook an excursion to the west of Ireland. It was organised and led by Praeger, and its members included Arthur George Tansley (1871–1955) (Praeger, 1909). They visited Ardrahan, the summit of Slievecarran, the woodland at the base of Slievecarran and Garryland Wood. A full inventory of the flora was reported, and Tansley later published accounts of limestone pavement, limestone heath, closed Hazel scrub and Ash (*Fraxinus excelsior*) wood in his monumental *The British Islands and Their Vegetation* (1939).

The first ecological study of the Burren came in 1932, when Praeger published his preliminary work on the flora of turloughs (Chapter 9). He studied three different locations: a turlough near Tierneevin Chapel, which is small and deep; an unnamed turlough south of Garryland Wood, with some permanent small pools; and Caherglassaun, a low-lying area of water surrounded by a turlough that is subject to tidal influence even though it is kilometres from the coast. Praeger noted that the Caherglassaun water levels fluctuated by about 30 cm twice a day, although the water was not brackish. The tidal waters acted like a hydraulic ram, pushing the fresh underground water back into the turlough at high tides. Praeger never returned to complete his preliminary study.

The Ninth International Phytogeographic Excursion took place in Ireland in July 1949. Twenty-three international botanists spent 16 days exploring the botany of Ireland, including a day in the Burren, where stops were made at Ballyvaghan, Lisdoonvarna, Poulsallagh and Black Head. All the classic Burren species were found (Webb, 1952).

Job Edward Lousley (1907–76) probably visited the Burren in the late 1940s, prior to the publication of his *Wild Flowers of Chalk and Limestone* (1950). On his visit, in mid-June, he found Maidenhair Fern growing extremely luxuriantly in a grike and reaching up more than 30 cm. Near Ballyvaghan, he noted Shrubby Cinquefoil with two different types of flowers on different bushes. The smaller were 'three eighths of an inch across, with deep chrome-orange petals much shorter than the sepals. The larger and commoner kind were an inch across and had much paler petals longer than the sepals.' The significance of this observation is not entirely clear, and no morphological differences were noted in variations in the European populations of Shrubby Cinquefoil by Trevor Elkington in his detailed examination of the species (Elkington, 1989). Lousley also noted Spring Gentian, Mountain Avens and Irish Eyebright (*Euphrasia salisburgensis*) (Lousley, 1950). He found Dense-flowered Orchids, dried up and in fruit, and photographed Pyramidal Bugle, showing the dense, shaggy bracts in which the flowers are half-hidden. Around Black Head and at Ballyvaghan, Lousley saw a saxifrage that 'grew in extremely beautiful patches on the limestone'. This is now known as the Irish Saxifrage (*Saxifraga rosacea*).

Visits to the Burren by naturalists continued, but it was not until the publication of Webb's 'Noteworthy plants of the Burren: a catalogue raisonné' in 1962 that interest in the area livened up again. Webb remarked that it was curious how little had been written on the flora of that 'remarkable district of western Ireland'. His systematic survey of the area attempted to define more precisely the nature of the flora and it peculiarities. For the purposes of his study, he set limits on the Burren area that are followed in this book (Fig. 9). He provided more information on the distribution and status of 151 selected species, including varieties, than anyone before.

A few years earlier, in 1957, as mentioned at the start of this chapter, Webb had suggested to the BES that they set up the Burren Survey Project to encourage naturalists to visit the area. The most impressive study that arose from this was the 1966 work by Robert Ivimey-Cook and Michael Proctor, which described and

FIG 34. Two rare species: Miriam Rothschild (1908–2005) and David Webb (1912–94). Black Head, July 1970. (David Cabot)

classified the vegetation communities. During eight weeks of fieldwork in the summer of 1959, the two botanists recorded the cover or abundance of plant species over several hundred random quadrats, usually measuring 2 × 2 metres. This classic phytosociological method, devised by Josias Braun-Blanquet (1884–1980), allows plant communities to be identified by floristic composition (Braun-Blanquet, 1932). Twenty groups of communities were identified by the pair, corresponding to broad habitats with upwards of 50 separate associations. Grasslands, heaths and fens were the predominant vegetation types, each with many subdivisions.

In late May 1959, during a period of exceptionally dry and warm weather, John 'Jack' Heslop-Harrison (1920–98) carried out a series of temperature and vapour deficit measurements in microhabitats in the Burren (Heslop-Harrison, 1960). He took readings on inland and coastal limestone pavements, in scrub woodland, and at cliffs and escarpments in an attempt to improve understanding of the ecology of the area. He found striking effects of Hazel scrub on the microclimate, which reduced the extremes found on the open pavement. The grikes provided a habitat that is constantly cool and humid, insulating them from the heating and drying effects of prolonged sunshine. This work was followed up in August 1961 by Colin Dickinson, M. C. Pearson and Webb, who compared the flora in the field layer of a well-developed dense Hazel scrub with that of grikes in limestone pavement in relation to micro-environmental factors (Dickinson et al., 1964). They commented on the similarities between grike and woodland floras, and the high relative humidity of each.

Ten years earlier, Heslop-Harrison had come across a 'considerable colony' of Large-flowered Butterwort on the south-eastern slopes of Cappanawalla, beside Ballyvaghan (Heslop-Harrison, 1949). The plants were growing mainly in an area about 400 metres across and were in full flower in May. They occurred at the base of low limestone cliffs and above the wall that marked the upper limit of farming. Several tiny streams from the cliff base kept the site moist. Heslop-Harrison was convinced that this was a naturally occurring population, separated from its main centre in south-west Ireland. The species was still there in May 2017, and we measured the population as covering an area of 107 × 50 metres, with upwards of 1,000 plants. In their 1983 *Flora*, Webb and Scannell reported that some plants appeared to be referable to the hybrid *Pinguicula* × *scullyi* (*P. grandiflora* × *vulgaris*), with flowers displaying a very pale lilac colour. Stace (2010) accepts this as a valid hybrid.

The Large-flowered Butterwort is a member of the 'oceanic-temperate' flora (Preston et al., 2002), occurring in north Spain and throughout the Pyrenees and western Alps to the Jura. It is thus both a Lusitanian and a Mediterranean species. Webb and Scannell (1983) noted that the plant is regarded by many as the most

FIG 35. Large-flowered Butterwort (*Pinguicula grandiflora*) at its Cappanawalla site in May. (Fiona Guinness)

beautiful in the Irish flora. It is absent from Britain as a native, although there have been several introductions.

Praeger had previously reported the Large-flowered Butterwort from a wet shale cliff a few hundred metres from the pump house at the spa in Lisdoonvarna, a specimen of which was sent to him by A. Birmingham in 1903 (Praeger, 1903). It was the first record for the Burren area, but the atypical site suggested that it might have been introduced. It was still growing recently on the vertical shale cliffs on the north side of the river running through the spa (Fig. 36). According to Webb and Scannell (1983), however, the discovery of the species at Cappanawalla rendered any reservations about the spa location irrelevant or unjustified.

FIG 36. South-facing shale cliff at the Spa, Lisdoonvarna, another site of the Large-flowered Butterwort (*Pinguicula grandiflora*). Note the prolific growth of the introduced Montbretia (*Crocosmia × crocosmiiflora*) on the opposite bank. May. (Fiona Guinness)

THE ARAN ISLANDS

Islands hold a special fascination for people, and nowhere more so than the Aran Islands. They are relatively small, their limits are well defined, and their flora and fauna raise many questions. When were the islands separated from the mainland? How and when did the plants and animals arrive? Are there any endemic species, varieties or subspecies? Can we see evolution in progress? How do the island flora and fauna compare with those on the mainland?

The three Aran Islands – Inishmore (3,000 hectares), Inishmaan (900 hectares) and Inisheer (400 hectares) – form a sort of gateway across Galway Bay, with Inisheer only 2 kilometres from the Burren mainland. The islands are a geological extension of the Burren and consist of horizontally bedded limestone without any covering of shale. There is a minimal cover of glacial till – less than 10 per cent – to provide a basis for farming, so plaggen soils have been created from sand and seaweed brought up from the beaches. In former years, these were mostly used for cultivation (potatoes, along with rye for thatching); nowadays, many have reverted to grass. Otherwise, the islands are bare limestone pavement with a scattering of granite erratics. There are many small fields, separated by high stone walls (Fig. 37). A few brackish coastal lakes occur, and in places small areas of machair back the sand dunes. The islands are virtually treeless apart from some patches of Hazel scrub on Inishmore.

As with the Burren, a long procession of botanists and other naturalists have visited the islands since the first record there of Maidenhair Fern was published by Lhwyd in 1707, following his probable visit in June 1700. Twelve publications recounted explorations in the nineteenth century, the most notable being that of Henry Chichester Hart (1847–1908), who published the fullest list of species, doubling the number recorded on the islands to 372 (Hart, 1875). Praeger visited in 1895, bringing the number of plant species recorded on the islands up to 408 (Praeger, 1895). Following this there was very little activity until the period between the mid-1960s and 1970s, when many more visits were made by university staff and students, and others, as summarised by Webb in his 1980 paper 'The flora of the Aran Islands'. At the time, the Aran Islands' flora had numerous inconsistencies, depending on different recorders. Webb reconciled many of the disparities and compared all the previous and present botanical records. His paper was essentially a corrected listing of the 437 plant species recorded on the islands. He did not discuss, apart from a brief comment on the absence of Mountain Avens, the differences between the island and Burren floras.

FIG 37. Enclosed fields with man-made plaggen soil on Inishmore. Limestone pavement is visible in the distance. (David Cabot)

Burren plant species absent (up to 2018) from the Aran Islands are: Mountain Avens, Brittle Bladder-fern, Grass of Parnassus (*Parnassia palustris*), Water Avens (*Geum rivale*), Bitter-vetch (*Lathyrus linifolius*), Autumn Gentian (*Gentianella amarella*), Sea Rush (*Juncus maritimus*), Reed Canary-grass (*Phalaris arundinacea*), Hairy Sedge (*Carex hirta*), Dark-red Helleborine, Greater Butterfly-orchid (*Platanthera chlorantha*), Shrubby Cinquefoil, Fen Violet (*Viola persicifolia*) and Mossy Saxifrage. What are the reasons for these absences? The list has been shortened in the last 35 years as five of the formerly 'missing' species have since been found on the islands, so perhaps it is just a matter of time before others are also found growing there.

ECOLOGICAL INVESTIGATION

In the mid-1960s, horticultural scientist J. G. D. 'Keith' Lamb (1919–2011) examined the occurrence of Spring Gentian on three different soils in the Burren to determine any relationship between soil type and the growth of the plant (Lamb, 1966). He found that the openness of the vegetation was an important factor favouring good growth, whether that was on a stony rendzina soil, a deeper loam or the sand dunes at Fanore. Later, in 1972, Elkington related the detailed distribution of the Spring Gentian in Teesdale, Yorkshire, more to the soil and underlying rock. Elkington also pointed out that the Irish (Burren) and British (Teesdale) populations of Spring Gentians are clearly distinguishable, and that the populations in the Alps and Pyrenees are significantly different from either.

A. R. 'Roy' Perry (1938–2014), a bryologist from the National Museum of Wales, visited the Burren in the 1960s. He reviewed the lichens of the area and his report is included in Webb and Scannell's 1983 *Flora*. Perry stated that the Burren was of exceptional interest for mosses and liverworts. Also for Webb and Scannell, Michael Mitchell (1934–) of the National University of Ireland Galway (NUIG) reviewed the lichens of the Burren. He subsequently wrote *Lichens of the Burren Hills and the Aran Islands* with Ph.D. student D. M. McCarthy (McCarthy & Mitchell, 1988). Michael Guiry (1949–) of the NUIG reviewed the marine algae for Webb and Scannell, commenting that there was nothing very unusual about the individual species found in the Burren but that the large number of species on a single shore was remarkable, as was the size range of many of the algae

Since the publication of the *Flora of Connemara and the Burren* in 1983, several plant species rare or new to the Burren, discussed below, have been found or their distribution redefined.

The status of Dog's Mercury (*Mercurialis perennis*) in Ireland has been a matter of conjecture for many years. While common in woodlands throughout Britain, it is rare in Ireland and most sites are close to estate woods, suggesting introduction. In his 2010 *New Flora*, Stace did not consider it a native. However, Webb and Scannell (1983) believed it probably is a native, with a slight doubt about its status in the Burren. They recorded it as occasional in two areas. A new location in the Burren was discovered in June 1981, on the eastern edge of a woodland dominated by Ash and Hazel, approximately 3 kilometres north-west of Ennis (Akeroyd & Parnell, 1981). Botanist Tom Curtis discovered another new site south of Corrofin in 1981, with the same associated species (Curtis, 1981).

Mudwort was refound by Tom Curtis and H. McGough in May 1985, growing in shallow solution hollows on coastal limestone – it had not been seen here since 1900 (Curtis & McGough, 1987). They also noted an odd habitat for Mountain Avens, growing on a flat fen peat, 25 cm deep, which was inundated by rising waters on the north-west shore of Muckanagh Lough. This location was most unusual, as Mountain Avens is normally associated with free-draining sites and not linked with wetland species. The frequent deposition of shell sand here may have prevented acidification and produced better soil drainage (Curtis & McGough, 1984).

One of the most remarkable botanical events in recent years was the rediscovery of Arctic Sandwort (*Arenaria norvegica* ssp. *norvegica*) in 2008 after an absence of 47 years (Walker *et al.*, 2013). A colony of this sandwort was originally discovered by Jack Heslop-Harrison during a university field trip to the Burren in 1961, 'on the southern slope of Gleninagh Mountain, overlooking Caher Lower'

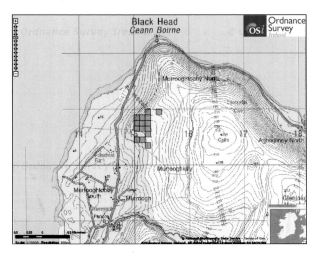

FIG 38. The distribution of Arctic Sandwort (*Arenaria norvegica* ssp. *norvegica*) in the Burren. Magenta squares represent the plant's presence within 100 m × 100 m grid cells. Reproduced with the permission of Kevin Walker.

(Heslop-Harrison *et al.*, 1961). The habitat was described as 'shallow crevices and solution hollows on an area of limestone pavement at *c.*800 feet'. A specimen was collected and its identification was indisputable; indeed, seeds collected from the plants were germinated. The discovery of this arctic–alpine species in the

FIG 39. The green road running from Murroogh to Black Head, where the Arctic Sandwort (*Arenaria norvegica* ssp. *norvegica*) was rediscovered in 2008. The few plants on the road survive considerable recreational traffic. May 2016. (Fiona Guinness)

FIG 40. Arctic Sandwort (*Arenaria norvegica* ssp. *norvegica*), originally found in the Burren in 1961, was thought to have become extinct until its rediscovery. (Fiona Guinness)

Burren marked its most southern distribution in the world. Despite subsequent searches, it was not relocated until the recent rediscovery. In their 1983 *Flora*, Webb and Scannell assumed that the plant was either extinct or that it still grew in a small quantity. They added that it would take about a dozen botanists a week's hard work to demonstrate conclusively that the plant was absent. Then, in 2008, British botanist Kevin Walker, from the BSBI, found two small colonies by chance, on gravelly soils in the middle of the green road from Murroogh to Black Head. Walker returned with 18 helpers in 2009 for more detailed studies, and together they located and mapped some 249 individual plants.

The discovery of Arctic Sandwort throws up intriguing questions about its pre- and post-glacial history. Did it survive *in situ* in areas free from ice? Or did it arrive by direct overland migration from northern Europe at the end, or towards the end, of the last ice age? Analysis of the genetics of the Irish, British and European Arctic Sandwort populations suggests that the Burren plant is a post-glacial arrival rather than a survivor, as it has limited genetic variability (Howard-Williams, 2013).

BURREN FAUNA

Recording of the Burren's fauna was much slower to get underway than the botanical work. It was the Lepidoptera that first attracted attention. The earliest catalogues of Irish Lepidoptera were published by the Reverend Joseph Greene (1852–1906) and the Reverend Arthur Ricky Hogan (1832–80) in 1854 and 1855, respectively, printed in the first two volumes of the *Natural History Review*. The lists were obviously not complete at that stage, but about 636 species were included. One of the first visiting lepidopterists from Britain was Edwin Birchall (1819–84), born in Leeds and a sometime resident of Dublin. He made field trips to Co. Wicklow and Co. Galway, and also visited Killarney, but apparently did not venture into the Burren. He was described as 'one of the ablest and most gifted Lepidopterists in his day, a man of European repute' (Anon., 1884). In his farewell address in 1884 to the officers and members of the Council of the Entomological Society of London, the President, Joseph William Dunning (1833–97), while lamenting the passing of Birchall, remarked: 'Is it too much to hope that the Entomology of Ireland may no longer be left to the casual investigation of a Yorkshire Lepidopterist and that the day is in hand when the Irish themselves will study the insects of Ireland and establish an Entomological Home Rule' (Dunning, 1884).

Killarney, Co. Kerry, Howth Head, Co. Dublin and Connemara, Co. Galway were the favoured locations for visits by Victorian lepidopterists. The Burren was still waiting to be discovered – although it would later become akin to Aladdin's

Cave, with many treasures and surprises. It has now been determined to hold more species of butterflies and moths than any other region in Ireland. But why is this? It is probably due to a combination of many factors, including the variety of habitats, the abundance and diversity of the flora, the fact that agriculture is less intensive and more pastoral here than in most other parts of Ireland, and that there is relatively little use of pesticides or herbicides.

In June 1922, the first surprising discovery was made, of the Pearl-bordered Fritillary (*Boloria euphrosyne*). The Cork naturalist and lepidopterist Robert Albert Phillips (1866–1945) was walking with his friend H. Fogarty in the limestone cragland at Clooncoose, near Kilfenora (Fig. 30), when they noticed 'large numbers of a pretty butterfly flitting about in the sunshine' (Phillips, 1923). Later, the one captured specimen was sent to Arthur Wilson Stelfox (1883–1972) at the Natural History Museum in Dublin, who identified it as the Pearl-bordered Fritillary (Phillips, 1923). Nearly 50 years earlier, in his 1866 'Catalogue of the Lepidoptera of Ireland', Birchall had confidentially predicted that the species would be found in Ireland. It is arguably Ireland's rarest and most endangered butterfly, one of the 34 resident and regular migrant species recorded. In a special survey in May and June 2011, the Pearl-bordered Fritillary was found in 44 sites, principally on

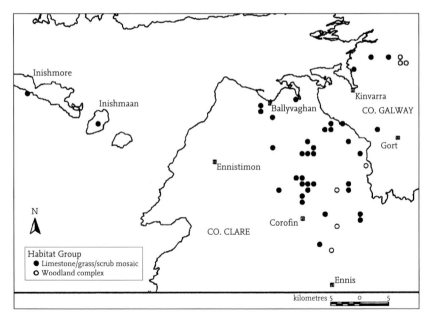

FIG 41. Distribution map of the Pearl-bordered Fritillary (*Boloria euphrosyne*) in Ireland, 2006–11, and a classification of the macrohabitats in which it occurs. From Regan and Meagher (2011).

FIG 42. Pearl-bordered Fritillary (*Boloria euphrosyne*). (Philip Strickland)

a mosaic of limestone pavement, calcareous grassland and scrub on the karst landscape of the Burren, south-east Co. Galway and the Aran Islands (Regan & Meagher, 2011). Its distribution in Britain has declined by 61 per cent since 1970, and similar declines have occurred in Belgium, Denmark, Germany, Lithuania, Luxembourg and the Netherlands.

The discovery of the Burren Green (*Calamia tridens occidentalis*) in August 1949 galvanised lepidopterists further, as the moth had never been seen or recorded in Ireland or Britain. It was found near Ballyvaghan by Captain William Stuart Wright from Co. Antrim, who was in the Burren on a botanical collecting trip. After catching the moth, a female, he placed it in an empty matchbox and sent it off to expert British lepidopterist Eric William Classey (1916–2008) to confirm its identification. In 1951, Wright published what must be one of the shortest and most modest notes on the discovery of a species new to Ireland or Britain. Was this moth a stray or migrant to the Burren, or was it a resident? In August 1950, an expedition of lepidopterists from Britain headed for the Burren in search of the answer. An amusing narrative of the expedition, by Hugh Robinson, is presented in a 1951 paper by Classey and Robinson. The expedition established that the moth was indeed a resident species of the Burren – ova and larvae were found. The additional moths collected were presented to Edward Alfred Cockayne (1880–1956), an eminent paediatrician, entomologist and former president the Royal Entomological Society of London, who discovered that the Burren Green differed from the nominate Continental form and was a separate geographical race or subspecies. The insect was subsequently named ssp. *occidentalis* Cockayne (Cockayne, 1954). The European race, ssp. *tridens*, is found in Scandinavia and central Europe.

The Burren Green find inspired more expeditions from Britain to the Burren in June 1951 and July 1952, as well as a series of visits to investigate the fauna over a 15-year period, collecting in nearly every month from April to September between 1951 and 1965 (Bradley 1952, 1953; Bradley & Pelham-Clinton, 1967). The latter study, by John Bradley (1920–2004) and Edward Charles 'Teddy' Pelham-Clinton (1920–88), included a comprehensive list of the Lepidoptera of the Burren recorded during their investigations along with those undertaken previously. Their paper remains the most comprehensive summary of the Lepidoptera of the Burren.

FIG 43. The Burren Green (*Calamia tridens* ssp. *occidentalis*), Cockayne, Burren, August. (Philip Strickland)

The Burren continued to conjure up surprises. The Irish Annulet (*Gnophos dumetata*) was discovered by Peter Forder and his wife in August 1991 (Forder, 1993). It is extremely rare, found only in a few 10 kilometre squares in the Burren and nowhere else in Ireland or Britain (Chapter 5). Both Philip Strickland and Jesmond M. Harding have recently published their findings on the moths and butterflies recorded in the Burren (Harding, 2008; Strickland, 2016).

Other principal invertebrate surveys in the Burren have included visits from British zoologists. The August 1959 investigation by Owain Westmacott Richards (1901–84), of the fauna of limestone pavement with grikes near Fermoyle, uncovered three species new to Ireland and Britain, two of which had Lusitanian distributions: the hemipteran *Rhopalopyx monticola*, found elsewhere only in the Pyrenees; the coleopteran *Apion dentirostre* (now a synonym of *Apion (Ceratapion) carduorum*), at the time found elsewhere only in southern Spain; and the dipteran *Sarcophaga soror*, found elsewhere in central Europe (Richards, 1961). Ian Lansbury studied the Hemiptera, Coleoptera, Diptera and other invertebrates in the Burren and on Inishmore in May 1959. He found 34 aquatic and semi-aquatic Hemiptera from the Burren but none of biogeographical interest (Lansbury, 1965). Michael Morris collected weevils and other insects from 10 sites within the Burren in June 1965. He also recorded *Apion (Ceratapion) carduorum*, and in addition found *Otiorhynchus arcticus*, known elsewhere only from Scotland and Scandinavia (Morris, 1967).

Concerning other invertebrates, the freshwater beetle *Ochthebius nilssoni* was first discovered at Lough Briskeen in south-east Co. Galway in August 2006, and later found in two marl lakes – Lough Cooloorta and Lough Gealáin – in the Burren (Chapter 5). Of less obvious biogeographical relevance to the Burren, but nevertheless startling, was the discovery of the fairy shrimp *Tanymastix stagnalis* (Young, 1976). Originally found in Rahasane Turlough in Co. Galway by Rod Young in 1974, some 19 kilometres north-east of the Burren, it was then located in 1976 in a pool south-east of Slievecarran by Jack N. R. Grainger (Grainger, 1976). Known from only six locations in Ireland, the aquatic invertebrate occurs in shallow pools that dry out in the summer.

REPTILES, MAMMALS AND BIRDS

While botanists and lepidopterists historically championed the ecology of the Burren, later visitors studying vertebrates were thinner on the ground. Nevertheless, a few unusual discoveries have been made, such as the recording of Slow-worms (*Anguis fragilis*) in 1977, new to Ireland. A local farmer, S. O'Kelly,

FIG 44. Slow-worm (*Anguis fragilis*), a recent addition to the Burren fauna that was almost certainly introduced in the early 1970s. (Fiona Guinness)

found a male Slow-worm near Killeenmacoog North, some 4 kilometres south-east of Slievecarran, and sent it to T. K. McCarthy at NUIG for identification. Local people had reported the occurrence of the animals during previous years (McCarthy, 1977). These reptiles were almost certainly introduced to the Burren in the early 1970s, and today they can be seen at the base of Slievecarran and in other locations in eastern Burren (B. Dunford, pers. comm., 2017). They are widely distributed throughout England, Wales and Scotland, including the Outer Hebrides, and through most of Europe and into western Asia.

A less successful introduction was the Green Lizard (*Lacerta viridis*) – one was observed by David Cabot in April 1962. Cabot was subsequently informed by John S. Jackson, Keeper of Natural History at the National Museum in Dublin, that eight males and seven females had been released by someone in the Burren during the summer of 1958 (Cabot, 1965). One was certainly still alive after four years, but none has been seen since 1962. It is assumed they all failed to survive. The species occurs in mainland Europe and several attempts have been made to introduce it to Britain. There is one viable colony there, on Boscombe Cliffs in Dorset, where individuals were discovered in 2002. Since then, the colony has thrived and spread (Mole, 2010). Praeger once described those who introduced alien species as 'forgers of nature's signature' (Viney, 2003).

With regard to visiting mammalian researchers, in 1979 R. N. Gallagher and James S. Fairley of NUIG carried out a population study of Wood Mice (*Apodemus*

FIG 45. Juvenile Wood Mouse (*Apodemus sylvaticus*). Caher River Valley, July. (Carl Wright)

FIG 46. Leveret of the Irish Hare (*Lepus timidus hibernicus*). The species is frequent in the Burren, particularly on the uplands. (Fiona Guinness)

FIG 47. Robin (*Erithacus rubecula*), one of the commonest breeding birds in the Burren and on all three Aran Islands. It is, however, scarce in exposed coastal areas and in grassland that lacks shrubs. (Fiona Guinness)

FIG 48. Male Yellowhammer (*Emberiza citronella*). The species used to be more widespread and frequent in the eastern part of the Burren, but today is declining. (John Fox)

sylvaticus) in the Burren at Carran, close to the field study station established there by NUIG (Gallagher & Fairley, 1979). The density of mice in the woodland habitat surpassed comparable estimates from all previous Irish studies.

More attention has been given to the bird populations of the Burren by visiting ornithologists. Preliminary studies were carried out by David Cabot in April 1962 as part of the BES Burren project (Cabot, 1999). The total area surveyed that year was approximately 120 hectares, consisting mostly of Hazel woodland, a mixture of low (up to 4 metres in height) and tall (greater than 4 metres) scrub. Twenty species of passerines were recorded, the commonest being Chaffinch (*Fringilla coelebs*; 18 per cent of all 290 territories recorded), followed by Robin (*Erithacus rubecula*; 14 per cent), Blackbird (*Turdus merula*; 11 per cent), Willow Warbler (*Phylloscopus trochilus*; 11 per cent), Wren (*Troglodytes troglodytes*; 9 per cent), Great Tit (*Parus major*; 6 per cent), Woodpigeon (*Columba palumbus*; 6 per cent), Dunnock (*Prunella modularis*; 6 per cent), Yellowhammer (*Emberiza citrinella*; 3 per cent) and Chiffchaff (*Phylloscopus collybita*; 3 per cent). Some 16 years later, Richard Moles completed more detailed surveys in April and July of breeding passerines in two areas of scrubland totalling 28.4 hectares (Moles, 1982). His findings confirmed the Chaffinch as the commonest species in tall scrub, followed by Willow Warbler, Robin and Wren. Tall scrub contained seven times more breeding bird territories than low scrub.

In 2002, Liam Lysaght published his *Atlas of Breeding Birds of the Burren and the Aran Islands*. With the help of 45 recorders, he produced the first Irish regional bird atlas, based on tetrads measuring 2 kilometres square. Fieldwork was spread over four years from 1993 to 1996. The results provide an excellent baseline of the breeding species at the time, against which future comparisons can be made.

OTHER STUDIES

Tracing the history of vegetation and land use in the Burren since the end of the last ice age has provided a remarkable picture of the evolution of the landscape and its vegetation. William A. Watts (1930–2010) of Trinity College, Dublin, led the way when he cored a late-glacial site on the Burren karst limestone – a small, richly calcareous fen at Gortlecka near Mullagh More (Watts, 1963). He also investigated the pollen stratigraphy at Lough Goller from sediments in a small lake in a drumlin field about 2 kilometres south of Lisdoonvarna. In recent years, research of the Burren's vegetation history and land use has been actively pursued by Michael O'Connell and his co-workers at NUIG (Jeličič & O'Connell,

FIG 49. The modern descendants of the wild horses of Kilcorney. (Fiona Guinness)

FIG 50. Cave entrance at Kilcorney. Who knows what might lie behind this mysterious-looking opening? (Fiona Guinness)

1992; Feeser & O'Connell, 2009, 2010; O'Connell, 2013; Molloy & O'Connell, 2014). Discoveries from these investigations are the subject of Chapter 4.

The indefatigable Frederick Foot provided the first professional descriptions of the Burren's geology in the early 1860s (Foot 1860, 1863). The Burren is the most important area for caves in Ireland, and a veritable speleologist's paradise. There are more than 300 caves and over 56 kilometres of them have been surveyed since the late 1940s, many by the University of Bristol Speleological Society (UBSS). The first recorded explorer into the Burren caves was Co. Clare man Charles Lucas (p. 18). In 1736 he examined the cave at Kilcorney, now known as the Cave of the Wild Horses (Lucas, 1740). Local legend has it that when the cave flooded, a fairy herd of wild horses came out to feed on the rich limestone pasture of the turlough. One of these animals was captured and put to stud with local horses, from which the famous strain of Clare horses, noted for their high spirits and fierceness, came forth.

Lucas's account of his Kilcorney visit is the earliest report of the exploration of any cave in Co. Clare, but a more extensive exploration of the cave had to wait until 1912, when C. A. Baker carried out his survey (Baker & Kentish, 1913). The cave is set in the enclosed depression and turlough of Kilcorney, some 11 kilometres north of Corrofin. It descends to a depth of 70 metres and floods to the level of the turlough, with waters bubbling out of the cave entrance into the adjoining field. The first section of the cave system – 518 metres long – is named Kilcorney 1 and otherwise known as the Cave of the Wild Horses. The total length of the whole cave system is 1,285 metres. Entering the extensive system by dropping down through a swallow hole requires some courage, a sound knowledge of the water regime and a good weather forecast.

Following in Lucas's footsteps were the intrepid Foot and fellow geologist George Henry Kinahan (1829–1908), both of whom worked for the Geological Survey of Ireland. Together, they examined many of the Burren caves during the middle of the nineteenth century, including Kilcorney (Kinahan, 1875). Catching the flavour of the cave environments, Kinahan wrote, 'The rock throughout the entire neighbourhood is drilled and bored with caverns, often of great extent, with numerous passages and windings, serving in many cases, as watercourses for the "buried" rivers which give rise to the sink-holes of the turloughs for which the district of the Burren is famous.' Foot himself visited almost every known cave in the Burren.

In the mid-1930s, members of the Yorkshire Ramblers Club became active in the Burren and explored the caves on the east side of Slieve Elva. In 1948, the first cavers from the UBSS arrived in the Burren. Thereafter, the society became extremely active in exploring and defining the extraordinary cave systems of the

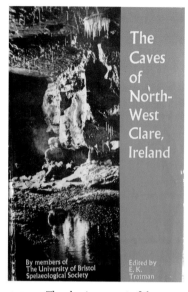

The
Caves
of
North-
West
Clare,
Ireland

By members of
The University of Bristol
Spelaeological Society

Edited by
E. K.
Tratman

FIG 51. The classic account of the exploration of the Burren caves (Tratman, 1969).

area. An account of their work up to the mid-1960s is detailed in *The Caves of North-west Clare, Ireland* (Tratman, 1969).

The most famous of all caves in the Burren today is Aillwee Cave, near Ballyvaghan, on the western flank of Aillwee Hill. It is one of two major cave systems discovered to date on the high (non-shale) Burren, and is also currently the only Burren cave that is open to the public (the Doolin Cave is also accessible but lies just south of our area) (Drew & Cohen, 1980). The cave was discovered by accident when a local farmer, Jack McGann, was up on the western flank of Aillwee Hill in 1940 rounding up sheep and his dog chased a rabbit into the entrance. McGann didn't mention the cave for nearly 30 years, when visiting speleologists then excavated and mapped it, finding it to be 210 metres long. Development work commenced in 1976 to open the cave to the public, and by 1995 more than a million visitors had passed through.

The cave is approximately 1.5 million years old and is about 250 metres under the surface. It consists of a tube-like tunnel that was eroded by waters from glacial melt and other sources. The water collected from a wide area, seeping at first through natural fissures and lines of weakness in the limestone, then

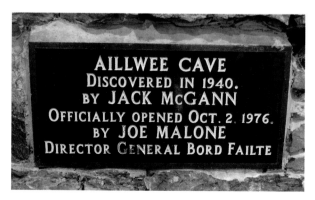

AILLWEE CAVE
DISCOVERED IN 1940.
BY JACK McGANN
OFFICIALLY OPENED OCT. 2. 1976.
BY JOE MALONE
DIRECTOR GENERAL BORD FAILTE

FIG 52. Plaque at the entrance to Aillwee Cave. (Fiona Guinness)

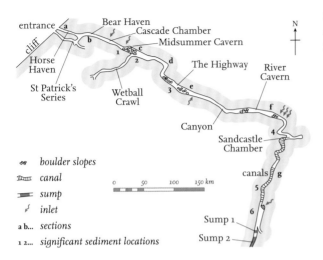

FIG 53. A map of Aillwee Cave. From Drew and Cohen (1980).

flowed as an underground river through the rock, wearing it away to create the tunnel. When ice melt was complete, water levels in the tunnel dropped substantially; today, there is still some running water but this is at a much lower level. Near the cave entrance, a tooth of a Brown Bear (*Ursus arctos*) was found (Fig. 54), and later the upper part of a jaw of a large male. Shallow pits with

FIG 54. Remains of a Brown Bear (*Ursus arctos*) in 'Bear Haven', Aillwee Cave. (Fiona Guinness)

FIG 55. Stalagmites – upward-growing mounds of calcium deposits, precipitated from dripping water – in the 'Mud Chamber', Aillwee Cave. These are approximately 1,000 years old. (David Cabot)

FIG 56. Straw stalactites hanging from the roof of Aillwee Cave. They grow rapidly – approximately 1 cm every 10 years – before eventually breaking off. (David Cabot)

probable claw marks are found in the so-called 'Bear Haven', which is thought to have been a hibernation site.

Evidence from the Polar Bear (*Ursus maritimus*) genome suggests that this species hybridised with the Brown Bear in Ireland at one time, presumably when polar ice spread southwards to cover most of the country. The research puts the incidence of this meeting in the last 20,000–50,000 years, which ties in with the presence of sea ice around Ireland around 22,000 BP (years before present). Since that time, all Polar Bears have carried some genes from Irish Brown Bears, possibly from the Co. Clare population. The latter became extinct in Ireland about 3,000 years ago (Ceiridwen *et al.*, 2011).

The total length of Aillwee Cave is just over 1 kilometre, although only 300 metres are open to the public (Drew & Cohen, 1980). Calcite from the interior of the cave has been dated back more than 350,000 years, determined by measuring the amount of radioactive uranium still present (Drew, 1986). The smaller stalagmites on the floor are more modern features, approximately 1,000 years old. Their staining is due to impurities of iron compounds.

Other Burren caves were searched early on for prehistoric animal bones. The caves at Edenvale, Newhall and Barntick near Ennis, Co. Clare, were explored towards the end of the nineteenth century by an enthusiastic team led by Richard John Ussher (1841–1913), the first renowned Irish ornithologist. The Clare cave investigations followed the successful work on the caves at Kesh, Co. Sligo, which were organised by a committee appointed by the Royal Irish Academy. Although the Edenvale and Newhall caves are just outside the limits of our area, they throw light on the fauna that must have been present in the Burren. Ussher sent more than 50,000 bones from the Edenvale cave, and some 20,000 from the Newhall and Barntick caves, to the National Museum in Dublin. Most of these have been subsequently identified and radiocarbon dated. In a 2017 review paper by Nigel Monaghan, the age of many of the mammal bones recovered from the Clare

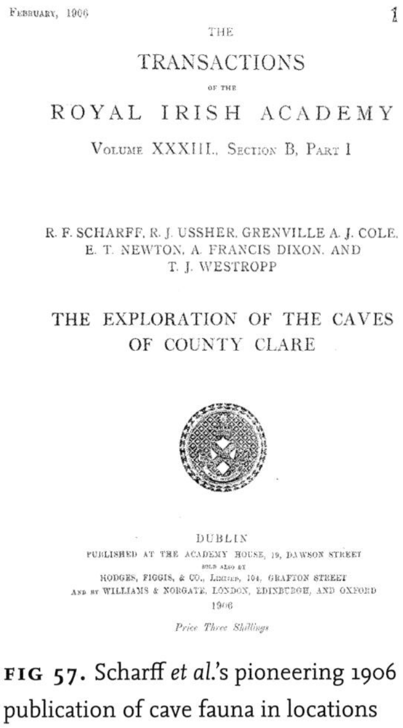

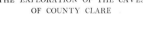

FEBRUARY, 1906
THE
TRANSACTIONS
OF THE
ROYAL IRISH ACADEMY
VOLUME XXXIII., SECTION B, PART 1

R. F. SCHARFF, R. J. USSHER, GRENVILLE A. J. COLE,
E. T. NEWTON, A. FRANCIS DIXON, AND
T. J. WESTROPP

THE EXPLORATION OF THE CAVES
OF COUNTY CLARE

DUBLIN
PUBLISHED AT THE ACADEMY HOUSE, 19, DAWSON STREET
SOLD ALSO BY
HODGES, FIGGIS, & CO., LIMITED, 104, GRAFTON STREET
AND BY WILLIAMS & NORGATE, LONDON, EDINBURGH, AND OXFORD
1906
Price Three Shillings

FIG 57. Scharff *et al.*'s pioneering 1906 publication of cave fauna in locations just outside the Burren.

caves are listed, giving a picture of the interglacial and Holocene fauna. One remarkable discovery was made recently from a bone that was collected in the Edenvale Cave in 1903. This Brown Bear patella had languished in the National Museum for 103 years, but when it was examined it revealed that it had been butchered by humans. A radiocarbon date of 12,500 BP was established, thus indicating the presence of humans in Ireland 2,500 years earlier than previously thought. The discovery was made by Marion Dowd from the Institute of Technology, Co. Sligo, and Ruth Carden of the National Museum of Ireland (Dowd, 2016).

This review has attempted to include the principal scientific investigations into the natural history and speleology of the Burren, focusing on the early explorers. Work continues today, with many discoveries being made each year. New scientific methods such as genetic analysis are revealing answers to some of the unresolved questions on the origins and history of the organisms of the Burren, their unusual ecology and, most importantly for the area's future, the impact of, and interrelationships between, farming, changing land use and tourism.

CHAPTER 3

Shaping the Landscape

T HE BURREN IS ONE OF THE MOST distinctive landscapes in Ireland. It is also one of the best examples in the world of a glacio-karst landscape, with all the classic features easily seen in a relatively small area. These include limestone pavement with clints, grikes, kamenitzas, karren and runnels; dolines; glacial erratics; dry valleys; turloughs; swallow holes; springs; and caves.

The word karst derives from the Karst (Kras in Slovene), a limestone plateau that lies between Slovenia and Italy. It is about 60 kilometres long and up to 20 kilometres wide, and covers an area of about 440 square kilometres (the Burren is 465 square kilometres in area). The Slovene word *kras* means 'bleak, waterless place', and the region was first described in 1689 by Janez Vajkard Valvasor (1641–93), a pioneer of the study of karst features. He introduced the word karst to European scholars, describing the phenomenon of underground rivers in his account of Lake Cerknica (Valvasor, 1689). He was elected as an early Fellow of the Royal Society for this work.

FIG 58. Janez Vajkard Valvasor (1641–93), a pioneer of the study of karst phenomena in Slovenia.

GLACIO-KARST LANDSCAPE

The distinctive karst features of the Burren were formed by the erosion of the area's limestone by water to produce a 'dissolution dominated landscape' (McNamara, 2009). As rain falls, it absorbs atmospheric carbon dioxide to form a relatively weak acid, carbonic acid, which then dissolves the main constituent of limestone, calcium carbonate. This produces calcium bicarbonate, which is soluble, so it trickles away with the water. The dissolution continues underground to enlarge cracks in the rocks or, where there is a considerable volume of water, to hollow out caves. In certain conditions the bicarbonate loses its carbon dioxide, reverting to carbonate and building dripstone features in the caves such as stalactites and stalagmites. So, rain, which has a certain regularity on the west coast of Ireland, has been acting on the Burren's limestone for a very long time – since the Tertiary period, in fact, 66–2.6 mya (million years ago), before the Pleistocene.

What is clear nowadays is that the Burren was glaciated, smoothed by passing ice sheets that removed most of the earlier relief and left the rounded hills we know today. The Pleistocene epoch, during which multiple glaciation events occurred, commenced some 2.6 mya and concluded at 11,700 years ago (Cohen & Gibbard, 2011). Karst landscape itself is not a rare landform worldwide, for it occupies approximately 11 per cent of the Earth's surface, including huge areas in China (more than 600,000 square kilometres) and in Australia (500,000 square kilometres) (UNESCO, 2017). Glaciated karst is decidedly rare, however, and the Burren is one of the best examples in the world. Smaller areas occur around the Alps and in other mountain ranges where glaciation has been a feature in the past. Slovenia has extensive areas of karst, comprising about half of the country, and its government-funded Karst Research Institute is the cradle of the discipline of karstology.

RICH BIODIVERSITY

A frequent characteristic of karst areas is that they are 'hotspots' for biodiversity, harbouring higher numbers and densities of species, subspecies and varieties of flora and fauna than would be expected. This is well exemplified by the Burren's remarkable flora and rich invertebrate fauna, but is just as true in south-west China or Slovenia.

The reasons for the biological richness are complex and are related to the thin soils, which are washed underground as the karstic drainage develops, and to the multiple microclimates that are found in the dissected terrain. There is also an abundance of calcium in the environment, which produces an unusual soil with a high pH and a low availability of phosphate. The karst landscape also precludes most forms of agriculture apart from grazing. Ancient karstic landscapes that have not been glaciated often have a very rich fauna in the shallow surface layers of the groundwater and in the caves beneath, but the Burren seems to have lost this, if it ever existed. At a latitude of approximately 53°N, the area benefits from a so-called 'storage-heater effect', absorbing heat during sunny weather and radiating it gradually, thereby prolonging plant growth into the winter months and allowing southern species to survive.

GEOLOGICAL HISTORY OF THE BURREN

The geological history of the island we know as Ireland today, and the Burren in particular, is long and complex. Limestone and shale were laid down as sediments in the sea during the Carboniferous Visean stage, 347–331 mya, when Ireland was part of the supercontinent of Laurasia. There were landmasses to the north and south, with a marine basin in between. Most of what would become central Ireland was covered by shallow tropical waters. There was more carbon dioxide in the atmosphere, derived probably from volcanic eruptions, and the climate seems to have been much warmer than today – the average global temperature was 20 °C, compared to 12 °C in modern times. There was also more oxygen in the atmosphere, with levels at 35 per cent by the end of the Carboniferous – the highest ever recorded – compared with 20 per cent today.

Formation of Carboniferous limestone
A vast array of corals, brachiopods, crinoids and gastropods swarmed in the shallow tropical Visean seas, extracting calcium carbonate from the water for their shells and skeletons. Some of the animals lived on the sea floor, such as corals and shellfish, while planktonic ones – commonly Foraminifera – fell to the bottom when they died, like relentless snow. Their shells and skeletons contributed partly to the limestone we see today, with the remainder consisting of calcareous mud or micrite derived from recrystallisation of other remains. As the deposition of limestone was progressing, there were fluctuations in water levels in the seas due to tectonic movements and the onset and decline of an ancient ice age at the poles. Sometimes, the surface was raised relative to the sea and soil developed to form the mudstone bands we see in the rocks today. Then everything would sink (or the sea would rise) and the oceanic communities would re-establish themselves for another few million years.

Many of the marine creatures – shellfish, corals, sea lilies and others – became fossilised, and their imprints can be seen all over the Burren today. The main deposition of limestone continued for about 20 million years towards the end of the early Carboniferous period. This mass of sediment became the 800 metres of solid limestone that is present today, gradually transformed into rock by its own weight and by the weight of succeeding levels of sediment that have since been removed. The resulting geological record seems full of abrupt changes from one environment to another, but the time span involved is so long that there is no reason to suppose that these changes were any more dramatic than those occurring today in different parts of the world.

Through the sequence of grey Burren limestone are thin deposits of silica. These are blackish chert, usually 5–20 cm thick, and are derived from skeletons of microscopic diatoms and radiolarians, and possibly from sponge spicules. How such layers formed is debatable, but it seems connected with the density of the material. Silica is denser than limestone and so may have settled at the base of the various strata as they formed. The resulting rock has layers of nodules and concretionary masses. Chert is hard, like flint, and when broken forms conchoidal fractures with sharp edges, ideal for use as cutting/scraping tools or in weaponry (arrowheads). These qualities were well appreciated by the neolithic settlers in the Burren during the Bronze Age. Archaeological excavations carried out at the megalithic Poulnabrone portal tomb in 1986–88 by Ann Lynch and others from the National Monuments Service revealed for the first time evidence of chert tools in the form of three arrowheads and three scrapers (Lynch, 2014). The chamber, dated by radioactive carbon techniques, was used by neolithic people over a 600-year period from 3800 BCE to 3200 BCE.

Later in the Carboniferous period (326–313 mya), the seabed sank and the sea was too deep for the previous marine ecosystem. Fine sediment, derived from a continent at some distance, was dispersed widely in the water and settled slowly to form the Namurian shales, which still cover the limestone to the south of the Burren. Life in the deep waters at this time was meagre – little light reached the sea floor, oxygen levels were low, and organic matter and sulphides accumulated. Cephalopods (squids, octopus and nautilus) were the principal form of life, occupying a zone high in the water column – they comprise most of the fossils found in the shales. Pyrite, or fool's gold (an iron sulphide), also developed chemically in the shales.

There was then a period when a landmass was closer and coarser riverborne sediments were laid down in fluctuating sea-water levels. Muds and sands dominated about 318 mya, forming the horizontal layers of sandstone, siltstone and mudstone seen today in the awe-inspiring 244 metre-high Cliffs of Moher.

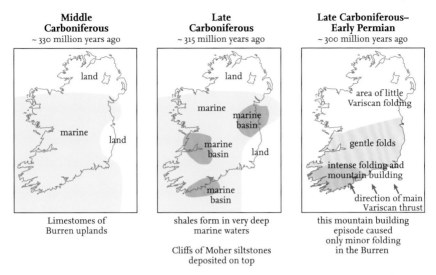

Middle Carboniferous	Late Carboniferous	Late Carboniferous– Early Permian
~ 330 million years ago	~ 315 million years ago	~ 300 million years ago

Limestomes of Burren uplands

shales form in very deep marine waters

Cliffs of Moher siltstones deposited on top

this mountain building episode caused only minor folding in the Burren

FIG 59. Maps showing the ancient environments in which the rocks of the Burren (and elsewhere in Ireland) were deposited during the Carboniferous and Early Permian periods, from 330 million years ago to 300 million years ago. (From Burren Geopark)

There was a complex deposition of these materials, changing with the flows in an estuarine environment and resulting in beds that are rarely more than a metre thick. During the mudstone phases, there were calm sea conditions and unknown creatures – possibly a form of gigantic woodlouse, snail or worm – moved through the soft sediment, leaving worm-like impressions and trace fossils. These fascinating tracks cross and recross the sediments in a dense arrangement,

FIG 60. Tracks left by tunnelling animals in sandstone slabs removed from the Cliffs of Moher and now set on a pathway in the garden at the Burren Perfumery. (David Cabot)

and are seen in buildings all over Ireland where the mudstone has been used, usually as flagstones on floors but, in the Liscannor area – where the stone comes from – occasionally as slates.

Other sand outwash materials were transformed into sandstones and shales, seen as outcrops on Slieve Elva and areas south of Lisdoonvarna. Occasional shallow phases are shown by ripple marks on the rock surface, but there was also subsidence and reinvasion by marine animals. One such deposit is left as a

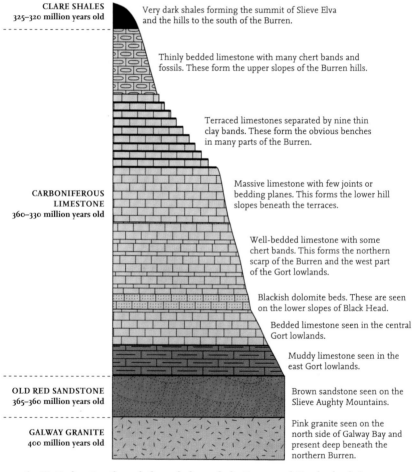

CLARE SHALES
325–320 million years old

Very dark shales forming the summit of Slieve Elva and the hills to the south of the Burren.

Thinly bedded limestone with many chert bands and fossils. These form the upper slopes of the Burren hills.

Terraced limestones separated by nine thin clay bands. These form the obvious benches in many parts of the Burren.

CARBONIFEROUS LIMESTONE
360–330 million years old

Massive limestone with few joints or bedding planes. This forms the lower hill slopes beneath the terraces.

Well-bedded limestone with some chert bands. This forms the northern scarp of the Burren and the west part of the Gort lowlands.

Blackish dolomite beds. These are seen on the lower slopes of Black Head.

Bedded limestone seen in the central Gort lowlands.

Muddy limestone seen in the east Gort lowlands.

OLD RED SANDSTONE
365–360 million years old

Brown sandstone seen on the Slieve Aughty Mountains.

GALWAY GRANITE
400 million years old

Pink granite seen on the north side of Galway Bay and present deep beneath the northern Burren.

FIG 61. Vertical section through the rocks beneath the Burren and Gort lowlands (not to scale). From Simms (2006).

stratum of phosphate-rich material, made up of fish bones and teeth mixed with the remains of invertebrates. This layer reached a thickness of 2 metres near Doolin and Noughaval, where it was mined during the Second World War. Some 105,000 tonnes were extracted, for use as a constituent of fertiliser.

By the end of the Carboniferous period, 300 mya, the rocks of the Burren had been formed. The massive deposit consisted of relatively pure limestone, capped by later layers of shale, sandstone and other sedimentary rocks that were perhaps 2.5 kilometres thick in total. This cap protected the limestone from erosion. The same rock type forms the bedrock of much of central Ireland and is the most extensive area of carboniferous limestone in Europe. Here, however, the protective cover was eroded earlier than in the Burren, and the limestone weathered into a lowland, where it is now concealed by glacial till, peatlands, lakes and rivers. We can thank chance for the survival of the rocky skeleton of the Burren, and the fact that the uplands have been revealed relatively recently in geological time and have not yet been worn down and dissolved away.

The Burren hills and their terraces

The flatness of the limestone beds in the Burren is one of the area's most distinctive features and creates the striking appearance of its limestone pavements. The rock escaped the effects of the mountain-building forces that created the Alps and the mountains of Co. Cork and Co. Kerry at the end of the Carboniferous. All that happened was some gradual cracking, jointing and bending, the last giving us the classic syncline seen in Mullagh More (Fig. 62).

Many of the Burren's flat-topped hills have characteristic terracing, which introduces a dramatic feature into the landscape (Fig. 64). The most convincing explanation for these focuses on the presence of thin bands of mudstones, 15–50 cm deep, that run through the limestone. Being softer than the limestone, these layers are progressively dissolved by water, which eats into the hillside. The overlying deposits or blocks of limestone become unsupported, eventually collapsing under their own weight. The fallen and broken limestone then accumulates as debris at the base of each terrace. The sequential weathering of the mudstone layers, eating into the hillside, gives rise to the tiers, with the oldest (exposed the longest) at the top (Fig. 65).

Underground formations and tectonic movements in the Burren occurred at a depth where the surroundings were warm. Fractures and joints that formed in the limestone were then invaded by mineral-rich fluids. Forcing their way up through the crevasses, these solidified as veins of whitish calcite – a carbonate mineral and the most stable form of calcium carbonate. The veins are clearly visible as lines in the limestone shore and cliffs south of Black Head, where

FIG 62. Mullagh More in the distance, displaying its syncline. The surrounding rocks are more or less bedded horizontally. (David Cabot)

FIG 63. Horizontal bands of mudstones in terraced cliffs of limestone near Poulsallagh. (David Cabot)

FIG 64. Terracing in the south-eastern Burren being engulfed by spreading woodland. (Fiona Guinness)

the calcite is coloured by iron traces (Fig. 66). Other minerals dissolved in the superheated fluids – including fluorite, galena and pyrite – also found their way into the limestone fissures and fractures. Deposits of lead and silver have been discovered on the west side of Slievecarran and north of Fisherstreet at Doolin. Fluorite occurs throughout the Burren and was worked from an open-cast site in the parish of New Quay in the 1960s.

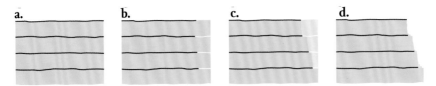

FIG 65. The formation process of terracing on the Burren hills. a. Layers of limestone separated by thin horizontal bands of soft mudstone. b. Water attacks the mudstone and washes it away. c. The limestone is now left unsupported and is unstable. d. The limestone collapses, forming a terrace. (From Burren Geopark)

FIG 66. Calcite veins stained with iron on the shoreline rocks south of Black Head.
(David Cabot)

For the last 50–60 million years, the Burren probably remained above sea-level, having been uplifted gradually over the preceding 200 million years. Once exposed, the landscape then became subjected to weathering and erosion, and the protective rock overlying the limestone was slowly removed. The limestone on the eastern side (now the Gort lowlands) was exposed about 25 mya and slowly rotted, perhaps into the exotic shapes that are seen today in unglaciated countries. The hills of the Burren were still mostly protected, although the north-facing valleys at Ballyvaghan and Turlough probably formed from ordinary (surface) run-off. There was also some deep erosion in the limestone, forming caves – the Doolin Cave seems to be of this age. This feature occurs underneath a dry valley that was abandoned by its stream when the water found a patch of limestone in its bed and eroded a passageway through it.

The major impact on the hills of the Burren, however, was the arrival of ice sheets and glaciers during the Pleistocene epoch, commencing some 2.6 mya. The Pleistocene included many cycles of cold glacial and warmer interglacial phases when the weather was probably warmer than today. The first major period of ice recorded in terrestrial sediments (the Munsterian advance) occurred 374,000–130,000 years ago, after which the climate warmed to something approaching today's levels. The flora included many familiar species, and one can assume that the limestone began to erode under the influence of rain seeping through the soil and vegetation. The most recent glacial period in Ireland was called the Midlandian, the last phase of which commenced around 32,000 years ago, when the island was largely ice-free. Flowline evidence suggests that ice from Scotland advanced south-west across parts of Ireland, probably coalescing with ice from local mountains. This formed the nucleus of the expanding ice sheet, which spread further as the weight of snow accumulated on it. The ice sheet extended off the west coast of Ireland, out to the edge of the continental shelf, but there is some conjecture as to whether it covered all the mountains in Ireland.

The last glacial maximum (LGM) occurred between 26,500 and 19,000 years ago, and there are conflicting opinions as to whether the ice covered everything or could have left exposed nunataks, areas of ice-free high ground. There is some evidence, for example, that the summit of Truskmore (647 metres) in Co. Sligo was exposed (Coxon, 1988). However, the current consensus is that during the LGM the Irish ice sheet not only covered all of Ireland, but also extended to the Atlantic shelf edge in the west and far out onto the Celtic Sea shelf in the south, and ran together with ice from Scotland, north-west England and Wales.

THE PRE-MIDLANDIAN INTERSTADIAL

Prior to the onset of the final, Midlandian glacial phase 130,000–114,000 years ago, Ireland enjoyed an ice-free but cool period without many trees. At this time, Woolly Mammoth (*Mammuthus primigenius*), Spotted Hyena (*Crocuta crocuta*), Giant Deer (*Megaloceros giganteus*), Wild Horse (*Equus ferus*), Brown Bear (*Ursus arctos*), Reindeer (*Rangifer tarandus*) and Red Deer (*Cervus elaphus*) were abundant, along with smaller species such as Arctic Fox (*Vulpes lagopus*), Stoat (*Mustela erminea*), Collared Lemming (*Dicrostonyx torquatus*), Norwegian Lemming (*Lemmus lemmus*) and Mountain Hare (*Lepus timidus*). The landscape was generally covered by grassland and shrubs, ideal for all the herbivores. Although they died out in the Midlandian, many of these mammals migrated back to Ireland once the glacial ice melted, but all subsequently became extinct, apart from the Stoat and Mountain Hare, and possibly the Red Deer. Much of the interglacial flora suffered a similar fate, although a few hardy arctic–alpine species did survive.

Among the flora present during this pre-Midlandian period was Fringed Sandwort (*Arenaria ciliata*) and Alpine Saxifrage (*Saxifraga nivalis*), along with several other related species such as Yellow Saxifrage (*S. azoides*) and

Purple Saxifrage (*S. oppositifolia*). Fringed Sandwort flourished on the steep Carboniferous limestone of Benbulbin, Co. Sligo, where it had almost certainly escaped earlier extinction, perhaps surviving in an ice-free zone or on a nunatak. Fossil evidence from the foot of Benbulbin revealed the species' presence there approximately 35,000 years ago, just prior to the onset of the Midlandian glaciation phase. More recent work shows that the Benbulbin plant is genetically distinct from other populations on mountains from the Pyrénées and Alps to Svalbard in the Arctic Circle. This implies that it is not a recent arrival in Ireland but has continued to evolve here, collecting random mutations over a long period, which has lead to 'private' sections of the genome (Howard-Williams, 2013). The time involved is at least 150,000 years, and possibly up to a million years, and it therefore could have survived many of the cold phases of the ice age. The cliffs of Benbulbin, at approximately 526 metres above sea-level, may have risen above the glaciers for a time, acting as a nunatak. However, in some places such as the Irish Midlands the ice was 1–1.2 kilometres deep, making permanent occupation of Benbulbin unlikely. It may have been the case that Fringed Sandwort was widespread across the lowlands and made periodic returns to the mountain when ice (or tree cover) intervened.

The LGM in Ireland during the late Midlandian was some 28,000–27,000 BP, and then in about 22,000–20,000 BP there was a rapid decline in the ice cover. By 16,000 BP, most of the landscape was once again ice-free.

ICE SHEET MOVEMENT AND GLACIAL DEPOSITION

In the Burren, the Midlandian ice sheet arrived initially from Co. Galway to the north. It rode up onto the hills, planing the surface and stripping away any loose material. Ice moves only by the action of gravity; once snow accumulates to a depth of about 50 metres, its weight causes the compacted ice at its base to flow. Ice movements are slow in the extreme, and the waxing and waning of phases takes thousands of years. Major advances and retreats seem to have occurred every 41,000 years or so in the period before 0.9 mya (Coxon & Waldren, 1995). Each time, the ice would have carried away more of the overlying shales, revealing a greater expanse of limestone as time went on and leading to the formation of the limestone pavement. The ice would have carried a burden of soil and rock frozen into its base and left erratics such as we see today. The granite erratics found on the northern Burren pavements and particularly on the Aran Islands are derived from Connemara. There seem to have been six separate advances and retreats during the Midlandian period, 100,000–20,000 years ago.

Each new ice advance usually erases all traces of preceding ones, but in the last one we know that the ice moved in from the north-east, further smoothing the surface and again bringing glacial till and erratics. This is shown by the few glacial striae that have survived modern erosion. The striae are scratches or gouges cut into bedrock by stones frozen into the base of the moving ice. Some occur close to sea-level at New Quay where the overlying soil has been removed in modern times. Others have been found in Glennamanagh, 152 metres above sea-level, and at Gortlecka.

FIG 67. Glacial striae in a north-east to south-west orientation on the cliff top at the Flaggy Shore, near New Quay. (David Cabot)

Moving ice also plucks limestone boulders from the lee of hills as it passes – many of the erratics at Poulsallagh are of quite local origin. In places, the glacial till was deposited in drumlins, elongated ridges of material that occur, for example, north of Fanore in the lee of the hills. Their long axes are also evidence of the direction of the ice movement.

It is not certain how extensive and deep this last ice advance was. Limestone erratics have been found as high as 300 metres above sea-level on the northern slopes of Slieve Elva and up to 240 metres altitude on the southern slopes, showing that the ice sheets must have been at least 300 metres deep at some stage (Farrington, 1965). Higher local summits may not have been covered at this time. In the words of the soil scientist T. F. Finch (1966):

The picture emerges of a south-westerly moving ice sheet which did not surmount the two mountains mentioned but left them as periglacial areas with the attendant frost heaving. The ice carried limestone and left it in its tracks and therefore the moraine limits are obvious on the shale and grit of the mountains. Slieve Elva and Knockauns Mountain were left standing as nunataks near the margin of the most recent glaciation. The edge of the ice sheet was indeed no farther away than Doolin, five miles to the south.

If Finch's hypothesis is correct, it is possible that several of the arctic and montane species survived the last glacial period in refugia on the summits of these two hills or to the south. It would also indicate that erratics were carried on the surface of the ice, which seems unlikely, rather than frozen into lower sections.

When seen today, glacial till seems to have been deposited as irregular mounds and ridges in the Burren. Sometimes it is found in patches on the sides of hills, and at other times it is spread more thinly. However, the frequency of glacial erratics is a guide as to what actually happened. An ice sheet that could transport these rocks must also have brought huge quantities of finer material that was deposited everywhere when the ice finally melted. If the material ended up on a limestone surface that had already been karstified to some extent during the interglacial, some of it would inevitably be washed away into the cracks, grikes and caves. As the grikes were widened and deepened by rainwater seeping down through the soil, this process would have increased. Perhaps the odd patches of glacial till that now remain are the last vestiges of much larger amounts. The runnels on the steep slopes of drift or glacial sediment on the east side of the Carran depression suggest that such erosion is taking place even now (Fig. 68).

FIG 68. Erosion runnels on steep slopes of glacial sediment, east of the Carran depression, or polje. The runnels are evidence of rainwash erosion. (David Cabot)

FIG 69. Limestone erratics near Black Head. The mountains of Connemara can be seen in the distance across Galway Bay. (Fiona Guinness)

EROSION OF THE LIMESTONE

A recurring question for many people visiting the Burren is: at what rate is the limestone being dissolved by the mildly acid rain? There have been several attempts at estimating the rate of erosion from the amount of material dissolved in outflowing rivers as well as from *in situ* measurement. Jean Corbel (1957) estimated a rate of about 10 mm in 100 years while Paul Williams obtained a figure of 5 mm from the carbonate carried by the Fergus river. However, these included underground dissolution from grikes, soil and caves as well as from the surface. More accurate assessments for the surface of temperate limestones vary from 1.5–5.0 mm per 100 years (Williams, 1966; Sweeting, 1966). In the particular case of the solution hollows, Philip Doddy and Cilian Roden (in press) have recently found a mean value of 7.1 mm per 100 years because cyanobacteria hasten the process through respiration. The mean of the 15 highest values in this study was actually 19.4 mm, something that would be visible over a lifetime.

Some glacial erratics are poised on plinths, set on the limestone pavement (Fig. 70). Here, the erratic has protected the limestone immediately underneath from erosion by the rain. The gap between the base of the boulder and the

FIG 70. A glacial erratic poised on a plinth or pedestal of uneroded limestone. Note the circular depression caused by water dripping off the erratic. (David Cabot)

limestone pavement therefore provides a rough indication of the rate of erosion of the pavement since the final melting of the last glaciers some 16,000 years ago.

A further question concerns the erosion of the glacial drift since the end of the last glaciation, before and after the soil was vegetated. A large amount of the material seems to have been removed, but there is no evidence of it being deposited in the lower parts of the region's drainage system or in the lakes. At the onset of melting, the sea-level was generally 120 metres below the current level, so it is unlikely that the first washout will be recorded (Rohling *et al.*, 2009). Moreover, the Burren lakes – Bunny, Muckanagh, Ballyeighter and Cullaun – are probably of recent origin, fed mostly by underground, not surface, waters, so it is unlikely that they would have been the recipients of the clay and silt (Farrington, 1965).

It seems most likely that the limestone pavement we see today was covered by a layer of soil and that this was eroded away from the very beginning. A pause would have occurred as trees protected the soil with their roots, but the process would have started again with each clearance or tillage of the soil by farmers. By the late Bronze Age, around 1600 BCE, much of the soil had been removed but a shallow layer persisted to support open woodland and clearings. Eventually, the only soils that remained were in the valleys where the gradient

FIG 71. A low mound of glacial sediment can be seen in the background here, with one erratic boulder in the foreground. (Fiona Guinness)

FIG 72. A commercial vegetable farm on soil derived from glacial sediment. Near Shanvally, May. (Fiona Guinness)

to the sea is lower and more material blocks up any underground drainage channels. Today, these areas support rich pastures for cattle, some sheep farming, and crop growing.

SURFACE FEATURES

As mentioned earlier, the limestones in the Burren are approximately 800 metres thick. About 300–400 metres of the rock emerge above ground in the northern parts of the area, while in the southern sections of the Burren, the summits reach only 100 metres. The hills are horizontally stratified, many with flat tops. A small amount of folding occurred at the southern end in response to mountain building further south, as seen in the syncline of Mullagh More.

The bare limestone pavement is made up of smooth grey blocks, or clints, many displaying surface solution features (karren): ridges, fluting and hollows (kamenitzas). Vertical linear fissures, or grikes, cut deep into the pavement on the lines of pre-existing cracks, in some locations creating either a modular or

FIG 73. Fossil corals. Branches of the coral have been cut in cross section on top of the rock, while a few have been cut longitudinally on the side. Flaggy Shore, near Kinvarra. (David Cabot)

a crazy patchwork pattern of rocks. Fossils of corals, gastropods, crinoids and brachiopods pepper some of the limestone rocks, and are especially visible in the pavements. Elsewhere, the pavement is broken up and shattered, occurring as a fragmented jumble of sharp, jagged rocks that are balanced on each other and difficult to cross.

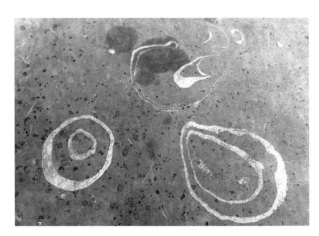

FIG 74. Fossil brachiopods, cut horizontally and now found on a limestone doorstep at Deelin More House. (Fiona Guinness)

There are other, larger, landscape features typical of limestone country. Many dry valleys occur, some with rocky walls, and there are also small circular depressions or dolines (from the Russian *dolina*, meaning 'valley' or 'plain') scattered across the landscape. These are often funnel-shaped in profile, created mostly by underground erosion eating away the supporting rock, followed by collapse. Poulavallan at the head of the Glen of Clab is a classic example, where cliff edges remain after the collapse of a cave in the centre.

Poljes (from the Serbo-Croat word meaning 'valley') are larger depressions, usually overlying caves and underground water systems. The largest polje in the Burren is at Carran, and is some 3.2 kilometres long, 1.6 kilometres wide and 61 metres deep. Such depressions are formed over many millennia, by gradual water erosion and collapse. Sometimes dolines combine to produce a polje. Most poljes contain turloughs or temporary lakes, because their collapse has brought their floors to a level near the water table. When this rises in winter, water flows out into the depression through one or several swallow holes. As the water table falls – generally in spring – the turlough then empties. Other turloughs occur in the lowlands, but even here there may be collapse features near the swallow holes.

FIG 75. The classic large doline at Poulavallan, Glen of Clab. It has been estimated that there are more than 1,500 dolines in the Burren that have an area greater than 100 square metres. (Fiona Guinness)

FIG 76. Glacial till or sediment on the west side of Slievenagapple. The ruins of the medieval Drumcreehy Church and burial ground are seen in the foreground, while the Blake-Forster Monument of 1912 is on the right. (Fiona Guinness)

The Burren is probably one of the best locations in Ireland and Britain for its labyrinth of caves, carved out over many thousands of years by underground streams. Many are 'fossil' caves because they have been abandoned by the stream that made them in favour of a new route deeper in the rock. Others are active and some pose a daunting challenge to intrepid speleologists. There are many entrances to this underworld, some hidden by vegetation, and others open and clearly seen at the base of cliff faces.

RECENT IMPACTS

There is little doubt that mankind has been instrumental in causing at least some, and perhaps most, of the soil erosion that has created the modern Burren. The arrival of coastal settlers at Fanore has been dated to about 8000 BCE, but the first people in the interior to have left traces were active at about 4000 BCE. Excavations carried out at the megalithic Poulnabrone portal tomb by Ann Lynch and others (see earlier) dated it as far back as 3800 BCE. The sophistication of the structure implies that a considerable level of organisation existed among the local people, and its position on a prominent rise in what would have been a relatively treeless area suggests that it dominated the surroundings. There was certainly enough human activity to contribute to soil erosion locally.

In summary, the limestone rocks of the Burren began to form more than 300 mya. Sediment accumulated on the seabed in layers from shelled animals in a tropical sea and then became covered by a huge thickness of sand and mud. The weight and pressure of the overlying strata turned the sediment, shells and skeletons into fossil-filled limestone and allowed minerals to infiltrate the cracks.

FIG 77. Poulnabrone portal dolmen, a stark, inspiring reminder of early Burren farmers. (David Cabot)

FIG 78. A bank of glacial till or sediment in the south-east Burren. (Fiona Guinness)

At that time the rocks were located close to the Equator, but were subsequently pushed northwards to their present position by continental drift. They were also raised up towards the surface by unusually gentle crustal forces, keeping their original flatness. In this position, the overlying sediment was eroded from about 25 mya, revealing the buried limestone in the central plain of Ireland. This process worked progressively westwards to Gort, reaching the Burren hills just as the ice age began. A few small exposures of limestone had appeared by that time and rainfall had dissolved cracks and passageways into it, forming caves. The main exposure occurred through the action of moving ice as the ice sheet waxed and waned, leaving soil and rocks in its wake. In the warmer periods, the dissolution of the rock continued apace, and vegetation spread from the south to cover the soil, retreating again with the onset of cold.

The last ice advance occurred about 28,000 years ago and seems to have been less severe than preceding ones. It continued the process of physical erosion, flattening the profile of the hills by Galway Bay and carrying rocks and soil from further north. Some plants may have clung on in nunataks that remained above the ice sheet, but most migrated again to the south or westwards as sea-levels dropped. The final melting of the ice allowed plant life and trees to spread once more from the south and the erosion of the limestone to recommence, forming the surface patterns of hollows, cracks and pavement we know today. There is now practically no surface run-off and all the rain sinks underground, where it continues its erosive work. It has taken most of the soil with it through the thousands of years, helped by man's clearance of the trees. This has allowed the great expansion of flowering plants.

CHAPTER 4

Vegetation History and the Impact of Man

To understand the Burren today, it is essential to go back in time to examine what vegetation grew there in the past and, if possible, how it was influenced by man. There are several locations in the area where research on this topic has been carried out: Gortlecka in the south-east corner, Inisheer in the Aran Islands, and the uplands in the north and east Burren. The analysis of pollen, seed and plant fragments from sediments in these areas goes some way to revealing the chronology of the vegetation from the late-glacial and early Holocene period onwards. Different species produce different amounts of pollen, so the figures have to be corrected to determine the actual vegetation growing at any one point. This information can then be combined with research from further afield to create a possible local vegetation history since the last ice age.

GORTLECKA

The earliest Burren vegetation records are from a lake in Gortlecka, where cores were taken in the 1960s by Bill Watts (Watts, 1963). These indicate that at the start of the Holocene, after the final retreat of the ice (approximately 11,750 BP), a pioneer vegetation spread into the Gortlecka area. Juniper (*Juniperus communis*) was the earliest woody species to appear, followed soon after by Aspen (*Populus tremula*), birch (*Betula* spp.) and willow (*Salix* spp.) thickets. The Guelder-rose (*Viburnum opulus*) was also present at this time as a shrub. Then came Scots Pine (*Pinus sylvestris*) – one of its earliest recorded occurrences in Ireland – which colonised simultaneously with Hazel (*Corylus avellana*). The last trees in this succession were oaks (*Quercus* spp.) and Wych Elm (*Ulmus glabra*), which followed after an interval of perhaps 200–300 years. This sequence is characteristic of many other areas in Ireland, with the exception of those with acid soils, where elms (*Ulmus* spp.) did not venture. The occurrence of Scots Pine in a limestone area may seem strange, as it is normally associated with acid conditions. However, there was probably much more soil on the limestone at the time than there is today and, arriving earlier than other tall canopy trees, the species was able to establish itself and maintain a strong population.

The Gortlecka sediment cores contain a high percentage of oak for the entire Holocene, and this also seems remarkable, considering how scarce the species is in the Burren today. Cores also show a large decline in elm pollen about 5,100 BP, in common with the rest of Ireland. The reason for this is now thought to be disease, which is also the cause of the decline in elms in the twentieth century. Thereafter, Yew (*Taxus baccata*) and Ash (*Fraxinus excelsior*) become important in the landscape, filling the gaps created. The relatively high frequency of Yew pollen in the cores suggests that there were considerable stands throughout the Burren during the mid-postglacial period. The species' presence in the Burren peaked at about 5,550 BP, coinciding with the first major occupation of the area by man. It appears then to have dwindled and generally disappeared as woodland about 2,000 BP. It has been argued that if grazing pressures were relaxed today, especially from goats, Yew would be an important constituent of the vegetation of the Burren. It is even conceivable that Yew woodland is the potential natural modern vegetation of much of the Burren because no other species can establish itself in its shade (Watts, 1984).

At Gortlecka, the earliest record of Scots Pine dates from about 10,550 BP, and the species seems to have disappeared at about 1170 ±150 BP (850 CE). This seemed a late date to the researchers, as elsewhere in Ireland the tree disappeared from the pollen profile much earlier, usually at about 4,050 BP. The Gortlecka core suggested that the species persisted much later in the Burren, until early medieval times, and raised the possibility of its persistence to the present day. As the Scots Pine and Yew woodlands dwindled under the influence of man, grasses and other herbaceous species multiplied. Mugworts (*Artemisia* spp.), meadow-rues (*Thalictrum* spp.), chickweeds (Caryophyllaceae) and meadowsweets or dropworts (*Filipendula* spp.) all increased as open ground became available, reflecting woodland clearance and disturbance for tillage in the early first century CE.

This vegetational history contrasts with what is seen today. The area surrounding the lake is karst limestone with large stretches of bare rock. The modern vegetation is sparse, with distinctive Burren species such as Mountain Avens (*Dryas octopetala*) and Spring Gentian (*Gentiana verna*), as well as Shrubby

Cinquefoil (*Potentilla fruticosa*). Nearby are woody species, including small stands of Hazel and, below Mullagh More, a relatively recent Ash wood with some Wych Elm. Downy Birch (*Betula pubescens*), Holly (*Ilex aquifolium*) and Rowan (*Sorbus aucuparia*) also occur in the wood, while the margins contain Hawthorn (*Crataegus monogyna*) and Blackthorn (*Prunus spinosa*), both relatively resistant to grazing by feral goats. Yew is always eaten by the goats and is mostly confined to the inaccessible cliff faces of Mullagh More. Patches of Bracken (*Pteridium aquilinum*) and Heather (*Calluna vulgaris*) occur outside the woodland, and there is much Mountain Avens and Hoary Rock-rose (*Helianthemum oelandicum*) on the higher rocky shelves and cliffs (Watts, 1984).

INISHEER

A similar pattern of vegetation development was found to have occurred on Inisheer, one of the Aran Islands. Detailed pollen records from late-glacial and postglacial sediments from Loch Mór were examined by Michael O'Connell and others at the NUIG (Molloy & O'Connell, 2014). No vegetation is likely to have persisted through the last glaciation as the islands were overrun by ice, which extended many kilometres further out from the present coast. There was a phase of melting when the climate warmed approximately 15,000 years ago, and the islands were recolonised by some trees. Then, around 13,000 years ago, there was a severe dip in temperature (called the Younger Dryas period or the Nahanagan stadial), which may have been caused by an interruption of the North Atlantic Drift oceanic current. This phase lasted about 1,000 years and is marked by a return of tundra vegetation, including Mountain Avens.

Recolonisation recommenced about 11,700 years ago and, as at Gortlecka, is indicated by Juniper pollen, followed by Downy Birch and then Hazel. After this, a tall canopy woodland developed, of Scots Pine, along with some oaks and Wych Elm. Scots Pine first appears in the Loch Mór pollen cores at about 10,800 BP and the species seems to have survived until the early part of the first millennium CE, supporting the chronology obtained from Gortlecka. However, analysis of Scots Pine charcoal found in a kitchen midden on Inisheer suggests a later survival of the species to the fourteenth century.

Guelder-rose was present on Inisheer early on, as at Gortlecka, and was joined by Buckthorn (*Rhamnus cathartica*), Rowan or whitebeams (*Sorbus* spp.) and Common Ivy (*Hedera helix*) after a few hundred years. It was another 400 years before Holly appeared, which is a surprise in view of its importance in contemporary Irish woods. These shrubs all occur somewhere on the Aran

Islands today, although it seems that Buckthorn disappeared from Inisheer in medieval times.

The pollen rain record suggests that the early woodlands were of an open nature, providing shelter for various woody and herbaceous species. One of the surprises at Loch Mór is Ribwort Plantain (*Plantago lanceolata*), whose pollen is well represented up to and after the decline in elm trees *c*.5,100 BP, most probably in the absence of humans. It seems that local habitat conditions favoured this species, which is usually associated with cultivation and grazing animals. However, there is no accompanying increase in grass pollen at the site, suggesting that people were not involved with the species' colonisation of the island. The presence of Sea Plantain (*P. maritima*) pollen implies that there could have been open rocky areas similar to those seen today over the whole Burren area. This then prompts the question as to when the surface soil cover disappeared (see below).

It is curious that Mountain Avens did not survive this forested period on Inisheer as it seems to have done on the mainland. Admittedly, the species is a small producer of pollen, being pollinated by insects rather than wind, so it may have lingered on somewhere. However, the evidence indicates that it disappeared from the Aran Islands in the late-glacial period, before the expansion of the tree cover. Another arctic–alpine species, Purple Saxifrage (*Saxifraga oppositifolia*), persisted longer; there are two Holocene records for the plant when woodland was present. It still occurs in Connemara but is absent from the Burren. In addition, a rock-rose (*Helianthemum*) species survived from the late glacial through tree colonisation up to about 7,650 BP. It is quite likely to have been the Hoary Rock-rose since this is so common in the Burren today, but it could also have been the Common Rock-rose (*H. nummularium*), which now occurs only on coastal limestone in Co. Donegal. Both of these species have high light requirements and must have been present in reasonable numbers to have been recorded in the pollen profile.

FIG 79. Lough Gealáin and the small lake from which sediment cores have been taken. (Steve O'Reilly)

UPLANDS IN THE NORTH AND EAST BURREN

Earlier work by the NUIG team followed the development of vegetation on three hills in the mainland Burren and covers a more recent period, from the Middle Bronze Age (around 1500 BCE) to the 1950s CE (Feeser & O'Connell, 2009). The research sites were on the uplands of the north coast (Cappanawalla), on a hill near Carran to the south-east (Gortaclare), and at Slieve Rua near Mullagh More. The hills rise to about 300 metres above sea-level, and are exposed and bleak in the extreme today, with windshorn vegetation nibbled by cattle and no tall woody plants. Instead, they support Burren vegetation typical of the modern era, on a mixture of limestone pavement, grikes and patches of shallow soil. Common plant species are Blue Moor-grass (*Sesleria caerulea*), Mountain Avens and Bloody Crane's-bill (*Geranium sanguineum*). However, there is more peaty soil on the hills than might be expected, especially on Cappanawalla, as organic material builds up in hollows in the limestone that are quite wet in winter. This allows Heather, Bearberry (*Arctostaphylos uva-ursi*), Crowberry (*Empetrum nigrum*) and other heath species to grow, and these seem to represent the current climax cover, adapted to the soil, climate and exposure (Ivimey-Cook & Proctor, 1966).

The fact that peat had accumulated was critical to this research, as it preserves pollen from the vicinity. Short sections were taken down to the rock and analysed, layer by layer, to elucidate the vegetation history. Pollen analysis and radiocarbon dating were combined with an examination of coprophilous fungal spores (those associated with the dung of particular domestic animals). The grikes were also investigated for signs of soil erosion, if and when it occurred.

It appears that the landscape on these mainland hills, as on Inisheer, was formerly covered by an open woodland of Scots Pine and Hazel. This was prevalent during the Middle to Late Bronze Age, from about 1500 BCE (when the pollen record begins) to 500 BCE (Feeser & O'Connell, 2009). It seems that there was more soil on the surface then than today, and that this supported a heathy vegetation, quite possibly with Mountain Avens, Bearberry, Crowberry and Bloody Crane's-bill. Such a community occurs on limestone soils today in Norway and Sweden. Many of the cores taken were found to contain charcoal (which is useful for radiocarbon dating), showing that burning vegetation as a form of management was important in the uplands at this time. Pines are especially flammable and would guide farmers to potential places for future stock rearing. The pollen records also indicate a species-rich grassland community, with daisies (Asteraceae), chickweeds and, again, much Ribwort Plantain – a species that still occurs in every 10 kilometre square throughout Britain and Ireland. As Scots Pine

FIG 80. The top of Cappanawalla today, the former site of pine forest. (Roger Goodwillie)

declined, there was a corresponding rise in Sea Plantain, which remains common in the Burren currently. From this, we may presume that Mountain Avens also expanded its range, as it exploits the same habitat as Sea Plantain. Many of the current Burren species would additionally have benefited from the decline in Scots Pine, including Burnet Rose (*Rosa spinosissima*) and Carline Thistle (*Carlina vulgaris*).

The dates that Scots Pine disappeared vary from one site to another. At Cappanawalla, the pollen trace peters out at about 400 BCE. While the species was declining, Hazel was still present in the uplands and some pollen of oaks, Wych Elm and Downy Birch has been identified in the cores. In the eastern Burren, by contrast, Scots Pine survived much later, until about 1400 CE at least (Gortaclare), and probably longer. Grasslands increased as the woodland declined, while heathland species also expanded, suggesting soil acidification. Cereal-type pollen appears in the record at all locations but was probably derived from the valleys, where the soil was thicker and there were fewer rocks.

A site in the Caher River valley (Lislarheenmore) was also analysed. Scots Pine disappeared here in 1000 BCE, leaving Hazel, oaks and Alder (*Alnus glutinosa*) as the main tree species. The record suggests an open landscape, still with some Hazel but also with Ash, which does not occur on the hill sites. Grazing was probably important in addition to tillage, and is indicated by the presence of members of the buttercup (Ranunculaceae) and legume (Fabaceae) families. Heather pollen first appeared on these lowlands in about 1000 CE; it is not found in the earlier part of the core, which starts at 1400 BCE.

In 1100–1700 CE, Hazel declined to insignificance and grassland pollen continued to expand. This period was characterised by several different grass species and a variety of broadleaved herbs, including some characteristic of a heathy community. Ribwort Plantain, Daisy (*Bellis perennis*) and Groundsel (*Senecio vulgaris*) were identified, along with Goldenrod (*Solidago virgaurea*) and Mountain Everlasting (*Antennaria dioica*). The high frequency of coprophilous fungal spores indicates upland grazing by cattle, which would have continued to suppress Hazel and may have occurred year-round, so the scrub had no chance of growing back. The heathy communities on the uplands probably had a similar structure to those seen today.

The expansion of agriculture coincides with the laying out of large field systems in the Burren uplands between 1650 and 1750, and the arrival of large estates and landlords. There was likely a shift from subsistence farming to a more organised system. Sheep farming became popular, and a jump in Tormentil (*Potentilla erecta*) pollen in the record may indicate closer grazing.

In the latter part of the eighteenth century (1700–1850), the population of Ireland started to rise rapidly thanks to cultivation of the potato. This was a wonder food that could yield enough for a family from spade culture alone and would grow in areas unsuitable for other crops. Remnant potato beds (cultivation ridges) are seen locally throughout the area, from the deep soils of the Poulavallan depression, at the head of the Glen of Clab, to the shoulders of Slieve Roe in the southeast. The population expansion prior to the start of the Great Famine in 1845 brought about increased pressure on trees and woody species, principally as sources of firewood since potatoes have to be cooked. There was little if any Hazel left to harvest by this stage, and other woody species such as Heather, Mountain Avens, and even Bracken roots and stems were all sought after. Grassland therefore expanded further. During the most severe period of the catastrophic Great Famine, between 1845 and 1852, more than 2 million people either died through starvation and disease, or emigrated from Ireland.

In the 'modern' period (1850–1950), Hazel experienced a comeback as the human population declined and farming pressures relaxed. Coprophilous fungal spores also declined in the pollen record, reflecting less intensive grazing and perhaps the seasonal use of the uplands.

On the Gortaclare uplands, the core taken dates back only as far as 100 CE, but the information on the vegetation is comparable to the other sites. In the period 100–1650 CE, there was widespread Hazel scrub. Birch, oaks and Alder were also being present on the uplands, while Scots Pine existed in only small numbers, before the species became locally extinct in about 1450, much later than at Cappanawalla. A heath vegetation then developed, with much Heather and Devil's-bit Scabious (*Succisa pratensis*). Ribwort Plantain again figures strongly as it is a major pollen producer. An increase in the number of coprophilous fungal spores in the period 1400–1650 indicates an increase in local farming activity. Burning of the vegetation was frequent at this time, particularly towards the end of the period, presumably when the population was increasing. Burning was seen as an important management technique, as it produces new growth for grazing animals. It is possible that some cereals were grown on the uplands or nearby, as the presence of weed pollen (Mugwort, *Artemisia vulgaris*) suggests cultivation. It has been argued that the increase in Mugwort pollen in different cores around Ireland from about 600 CE onwards is a reflection of the introduction of the mouldboard plough (Mitchell, 1976). However, the species is also considered a medicinal herb in many cultures and may have been tolerated once it appeared in tilled land. It would also have had some effect as a fallow crop, since it is deep-rooted and produces extensive rhizomes. Today, Mugwort is rather rare in the Burren (Webb & Scannell, 1983).

During the period 1650–1850, the Gortaclare landscape became open and virtually devoid of trees and shrubs. The rise in coprophilous fungal spores indicates that stock raising intensified, and at the same time Heather populations reduced under dual pressures from grazing and fuel. Towards the end of the seventeenth century, an increase in Bracken spores suggests further clearance of Hazel and the opening up of more grassland.

Archaeological remains on the Gortaclare and Cappanawalla uplands point to a human presence from the Bronze Age onwards, while the remains of stone and slab walls indicate that farming also took place at that time (Plunkett Dillon, 1985). The present-day open character of the landscape generally dates from the late seventeenth and eighteenth centuries, when the local human population was expanding rapidly, with consequent increased grazing pressure from domestic animals. Any remaining Hazel scrub was soon nibbled down. At this time, single stone walls were constructed, enclosing large areas of Burren uplands, and focusing and intensifying grazing pressure into these areas.

Today, a major factor in the maintenance of these upland vegetation communities is the tradition of winterage, the grazing of cattle (formerly sheep) during the winter and early spring (Dunford & Feehan, 2001). This custom is considered essential if the upland plant communities are to flourish and survive (Chapter 12).

ROCKFOREST

What could be called a Scots Pine woodland occurs today at Rockforest inside the Burren National Park, some 10 kilometres north-east of Corrofin. It would be more accurate to call it 'scrub with trees', as it is unlike any other planted or natural pinewood. The woodland is on low-lying limestone, 20 metres above sea-level, and covers 36 hectares. It consists of scattered Scots Pine trees, mainly stunted and gnarled, growing among Hazel and other scrub on limestone pavement. The ground flora is made up of many characteristic Burren species, such as Blue Moor-grass, Burnet Rose, Wood Sage (*Teucrium scorodonia*), Wild Thyme (*Thymus polytrichus*) and Broad-leaved Helleborine (*Epipactis helleborine*). It also includes much Heather and Bell Heather (*Erica cinerea*). It is quite far from the road, so was not visited for many years, and initial investigations into the history of the area suggest it could represent a relict Scots Pine wood.

FIG 81. Scots Pine (*Pinus sylvestris*) woodland with Hazel (*Corylus avellana*) understorey, Rockforest, May. This is possibly what the early neolithic woodland of the Burren looked like. (Fiona Guinness)

In June 2006, Jenni Roche took sediment cores from the deep basin of the nearby Rockforest Lough as part of her Ph.D. studies at Trinity College, Dublin (Roche *et al.*, 2010). They included a pine leaf that could be dated to 840 CE, amid other remains indicating that Scots Pine and Hazel were the dominant species at the time. There were also subsidiary amounts of birch pollen and, to a lesser extent, pollen from oaks and elms. The pollen rain suggested that there had been a mosaic of habitats, including some open grassland (indicated by plantain pollen) and some heathland. In addition, it illustrated a continuous presence of the woodland from 50 BCE to the present day. There was no sign of a decline in pine pollen despite considerable human activity in the area, including partial woodland clearance and mixed farming. On this basis, it seems that Scots Pine persisted at Rockforest in a small refuge or microrefugium, thus refuting the widely held hypothesis that the species had become extinct in Ireland.

To test the hypothesis further, and in particular to establish whether the Scots Pine pollen derived from plantations dating from the early eighteenth century, Alwynne McGeever and Fraser Mitchell carried out high-resolution pollen analysis on a sediment core taken from Aughrim Swamp, close to Rockforest Lough. These new investigations took place some years after Roche's original work (McGeever & Mitchell, 2016). The core's chronology was determined through radiocarbon dating and age-depth modelling. Again, results revealed a continuous record of Scots Pine spanning the last two millennia. But unlike the

FIG 82. Scots Pine (*Pinus sylvestris*) wood with Hazel (*Corylus avellana*). Many of the Scots Pine trees are small and gnarled. Rockforest, May 2017. (Fiona Guinness)

earlier results from Rockforest Lough, there was a marked decline in the pollen at about 460 CE, followed by a quick recovery that extends to the present day. There was a stable accumulation of sediment in the swamp, with no major erosion events, as shown by tests to determine the amount of organic matter in the core's horizons. A refinement in this research was the use of moss cushions (or polsters) as natural pollen traps, whereby the actual pollen rain could be measured rather than percentage values from a chemically treated core. These showed that the mosses beside the swamp today do indeed collect a representative sample of current pollen from the Rockforest woodland, and therefore that the cores could be relied upon to describe the vegetation.

These later results reinforced the theory that Scots Pine did not become extinct during its overall decline in Ireland over the last 1,550 years, but survived at Rockforest, independent of the eighteenth-century plantations. Based on the results, the species can be upgraded from extinct to the status of a native tree. The question of why it survived at Rockforest is hard to answer. Its final disappearance elsewhere in Ireland seems due to human exploitation, but prior to that, it may have been reduced naturally by the spread of blanket bog and the expansion of Alder. At Rockforest, the trees were part of a private estate and protected from being cut for timber or firewood. Even before this, there would have been little point in clearing the wood from the pavement as doing so would not have produced any useful agricultural or grazing land. The flat pavements

FIG 83. Scots Pine
(*Pinus sylvestris*) flowers.
Rockforest, early May.
(Fiona Guinness)

south and east of Mullagh More are some of the more forbidding expanses of stone in the whole area.

While the pollen analysis at Aughrim Swamp demonstrated that the Rockforest woodland was probably a microrefugium for the survival of Scots Pine in Ireland, no complementary faunal studies have yet been published to provide additional supportive evidence. A review in the 1980s by Martin Speight of the three groups of invertebrates dependent on decaying wood concluded that there was little reason to indicate that indigenous Scots Pine did exist before the eighteenth-century reintroductions (Speight, 1985). In Britain, it is possible to differentiate between indigenous and non-native Scots Pine woodlands, including commercial plantations, on the basis of their insect fauna. There is a discrete group of beetles that is more or less characteristic of the native Caledonian Scots Pine forest, although almost all members of this group seem absent from Ireland. The present insect fauna of conifer woodlands in Ireland consists largely of foliage-feeding species and some of their associated predators, which arrived with the forestry plantations from Europe. Speight does, however, note four species that are restricted to natural Scottish pine forests that do occur in Ireland. It will therefore be interesting to examine the Rockforest woodland for some of these insect indicators, although the small size of the stand may have militated against their long-term survival.

In addition, genetic studies are being carried out on the Scots Pine in Rockforest wood to assess the level of diversity. The results of these studies are eagerly awaited.

SOIL COVER

A question that has exercised botanists and other visitors to the Burren concerns the presence of naked limestone pavement, whether smooth or shattered: was it ever covered by soil and, if so, when did it become exposed? The absence of layered mineral residues in the Gortlecka and Rockforest lake sediment cores suggests that exposed limestone pavement in these areas is not a recent phenomenon, but dates back many thousands of years. However, this may be a reflection of the dependence of these lakes on groundwater, which is largely filtered, rather than on direct run-off. Some of the cave systems certainly contain soil.

There seems no reason why an ice sheet that transported multiple erratics to the Burren did not leave a skin of till (which then would have developed into soil) on the rock over which it passed. A gradual loss of this soil is illustrated by the material (and charcoal) washed into the grikes of the uplands. It seems that this process continued for many centuries. At Gortaclare, for example, the material was washed in close to 0 CE, while at Slieve Rua, three ages were determined, at 1050 BCE, 450 CE and 1150 CE (Feeser & O'Connell, 2009). Other research carried out in caves on the rate of calcite deposition has suggested that the soil cover disappeared substantially between 1000 BCE and 2000 BCE, as indicated by a reduction in the rate of accumulation of the calcite (Drew, 1982). The soil contribution of calcium ions ended, with only that from the intact rock continuing. This may, in fact, have been the end of a continuous process, as farmers have been exerting some level of soil disturbance in the area since at least 4000 BCE. It is hard to believe, for example, that Poulnabrone dolmen – dated to 3800 BCE – was constructed in deep forest.

The survival of light-demanding arctic–alpine or montane species from the late Holocene to the present day is an intriguing issue. At Cappanawalla, the pollen record suggests that there was an understorey of Bearberry–Mountain Avens heath, at least during that time (Feeser & O'Connell, 2009). Bearberry does not grow on limestone rock directly, so it must always have been on these upland soils. Mountain Avens, by contrast, will grow on open rock, so it is likely to have occurred in more sites, including those such as mountain cliffs and the exposed coast where trees could not grow. The survival of the flora through the forested period therefore requires a more complex explanation, one in which each species must be looked at separately. How, for example, did the sun-demanding Spring Gentian or Irish Saxifrage (*Saxifraga rosacea*) maintain themselves through this time? It is unlikely that the whole species-rich community came through the period together.

At Gortlecka, there is little evidence of pollen or other remains of the noteworthy Burren species that provide the botanical interest of the area today, apart from some leaf fragments of Shrubby Cinquefoil in late Holocene sediments but not in any earlier period (Watts, 1984). However, in some upland sites examined by Feeser and O'Connell (2009) remnants of Mountain Avens were discovered, dating from the late-glacial and early-postglacial periods, then disappearing soon afterwards. Leaf remains of Shrubby Cinquefoil were also found, but only in the upper parts of sediment cores, again dating to the early postglacial period. No evidence of their presence in earlier postglacial sediments was found, which one might have expected for the cold-tolerant species. Pollen grains of a rock-rose species, suspected to be Hoary Rock-rose, were present in small amounts throughout the postglacial period. As elsewhere, there was no sign of Spring Gentian or species belonging to the Mediterranean group. These species do not produce much pollen, and their seeds are small and easily missed, so it is possible that they may appear in some future coring work. However, their absence could suggest a late arrival as constituents of the Burren flora (Chapter 5). Macrofossils of the Holly-leaved Naiad (*Najas marina*) were found at Gortlecka, dating from the beginning of the Holocene to about 3,800 BP. Now extinct in Ireland, the species is found today only in one site in Britain, in the Norfolk Broads (Watts, 1984). Elsewhere, it occurs throughout temperate and tropical areas, in brackish or highly alkaline waters.

A little Spring Gentian pollen was found in later core studies: one grain was noted in the core from Lislaheenmore and two at Cappanawalla at a horizon dated to 400 BCE (Feeser & O'Connell, 2009). Mountain Avens pollen first appears at approximately 1100 CE, but the openness of the woodland (implied by the quantity of plantain pollen present) suggests that it could have grown among the trees much earlier, as well as on cliffs. It was then well set to flourish in the medieval period and through to the present day. As at Gortlecka, no remains of the Dense-flowered Orchid (*Neotinea maculata*) were found, nor any of the other Mediterranean group of species (Feeser & O'Connell, 2009). The pollen record provides tantalisingly meagre evidence of the history of many of the Burren specialities.

In conclusion, it seems that for 10,000 years since the first immigration of trees to the Burren, most of the region was covered by woodland, at first dense and, later, more open. Human impact gradually modified the habitat and created the bare limestone pavements, pasture and scrub that we know today. Soil loss was probably continuous since settlement, occurring wherever there was significant woodland clearance or tillage activity. The fissured structure of the rock assisted greatly in this process. The density of neolithic monuments and later settlements in the Burren is one of the highest in Ireland, indicating a considerable human population that lived largely by stock farming. Pressures on the landscape and remaining trees increased in line with population growth, prior to the Great Famine period of 1847–52. There was a scarcity of firewood and few, if any, trees survived outside guarded estates. Since the 1850s, scrub has been slowly spreading. The rate of spread of this increased at the end of the twentieth century as farming was influenced by outside markets and European Union (EU) subsidies.

The presence of a few of the more famous Burren plants is recorded occasionally in the pollen profile of core samples, but the past remains a closed book for most species. All that we can do is make an educated guess as to the likely scenarios for each species from our knowledge of their current distribution and ecology. Some possibilities are examined in the following chapters.

CHAPTER 5

Uniqueness of the Burren

O F ALL REGIONS IN IRELAND and Britain with high ecological value, the
Burren must take top prize for being the most confusing place for
a naturalist interested in biogeography or ecology. Nowhere else is
there such an eclectic mix of species and nowhere else are they so abundant – it

FIG 84. Profusion of flowers – Mountain Avens (*Dryas octopetala*), Bird's-foot Trefoil (*Lotus corniculatus*), Early-purple Orchids (*Orchis mascula*) and some Heather (*Calluna vulgaris*). Black Head, May. (Fiona Guinness)

is almost as if the whole flora of Ireland wanted to crowd into the region. Some 635 native plant species, or 70.5 per cent of Ireland's total of 900, have been found here, even though the Burren represents less than 0.5 per cent of the island's area (Dunford, 2002).

The well-known jumble of species, from the arctic and alpine regions, and from Atlantic coasts and the Mediterranean, has challenged botanists seeking explanations for their presence. In addition, various invertebrates living in the Burren seem to repeat the distribution patterns exhibited by the plants. Plants from different regions often thrive close to one another, and some of the high-altitude alpine and montane species descend to sea-level, while plants normally found in acid surroundings grow on the strongly alkaline rocks. Even today, it is hard to provide satisfactory explanations for all the questions that arise. The wealth and make-up of the plants was questioned by David Webb and Maura Scannell nearly 40 years ago in their *Flora of Connemara and the Burren* (1983). Both of them knew the plants of the Burren better than most other botanists at the time, and they asked: 'Why they should flourish so much more exuberantly in the Burren than elsewhere in Ireland is not at all clear.'

How plants got where they are has exercised the minds of botanists and naturalists for centuries. Carl Linnaeus (1707–78), the great Swedish systematist and architect of binomial nomenclature, was one of the first to ponder this question, and in doing so laid the foundations for the science of plant geography in the late eighteenth century. However, Alexander von Humboldt (1769–1859) is considered to be the father of phytogeography. *Essai sur la Géographie des Plantes* (1805), which the Prussian naturalist wrote with Aimé Bonpland (1773–1858), who accompanied Humboldt for five years during his travels through Latin America, teased out the issues of biogeography in some detail (Humboldt & Bonpland, 1805). In 1955, James Matthews (1889–1978) made an early attempt to establish different categories of the European range for the British and Irish flora. Thirty years later, however, Webb (1983) found it impossible to accept this in detail because of increased understanding of taxonomy and more comprehensive distribution information.

PHYTOGEOGRAPHICAL ISSUES

To illustrate the interesting features of the flora, one can examine a selection of the unusual plants that are associated with the Burren. First, there are four species that occur nowhere else in Ireland, being restricted to the Burren and the Aran Islands:

1. Hoary Rock-rose (*Helianthemum oelandicum* ssp. *piloselloides*). Local but abundant in the Burren, especially around Poulsallagh. Widespread but local on limestone pavement on Inishmore, and at one station in Inishmaan. It has been estimated that Ireland holds approximately 25 per cent of the total European population of this species (Wyse Jackson *et al.*, 2016).

2. The sea-lavender *Limonium recurvum* ssp. *pseudotranswallianum*. Local on maritime rocks from Poulsallagh to Black Head. Also on cliffs on the south-west side of Inishmore.

FIG 85. Above: Hoary Rock-rose (*Helianthemum oelandicum* ssp. *piloselloides*), May. (Fiona Guinness). Left: the sea-lavender *Limonium recurvum* ssp. *pseudotranswallianum*. Poulsallagh, May. (David Cabot)

FIG 86. Purple-Milk vetch (*Astragalus danicus*), June. (Ivar Leidus)

FIG 87. Arctic Sandwort (*Arenaria norvegica* ssp. *norvegica*). At Gleninagh on the green road from Fanore to Ballyvaghan, May. (Fiona Guinness)

3. Purple-Milk vetch (*Astragalus danicus*). Confined to south-east Inishmore and north-east Inishmaan, where it is very local, mostly on stabilised sand dunes, but fairly frequent over small areas.
4. Arctic Sandwort (*Arenaria norvegica* ssp. *norvegica*). Confined in the Burren to an area near Black Head, on the west side of Gleninagh Mountain.

Another species associated with the Burren and Aran Islands is Pyramidal Bugle (*Ajuga pyramidalis*). It is uncommon in the Burren, where it is found in the Poulsallagh area, but also occurs on Inishmore and Inishmaan, where it is more frequent. In addition, it has been recorded at outlying stations in Connemara and Co. Donegal and on Rathlin Island in Co. Antrim.

Within the large area of the Burren, these species are disproportionately located in the north-western coastal area and on the Aran Islands. In particular, the coastal zone extending for 8 kilometres from Black Head to Poulsallagh, and including the western flanks of Gleninagh Mountain, seems to be the core area.

Many commentators focus on the juxtaposition of species from different geographical regions growing side by side in the Burren. For example, Dense-flowered Orchid (*Neotinea maculata*) from the Mediterranean area occurs with Mountain Avens (*Dryas octopetala*) from the Arctic and Spring Gentian (*Gentiana verna*) from the Alps. While these three plants have fairly well-defined geographical 'homelands', the designation of other iconic Burren species poses problems, because botanists do not always share the same ideas on geographical centres.

Next, there is a group of 21 species that have their headquarters in the Burren and are more numerous here than elsewhere in Ireland. The first eight listed have larger populations in the Burren than are found anywhere in Ireland or Britain, according to Webb (1983). They are localised, however, and a quarter are absent from the Aran Islands; otherwise only 3 per cent of the Burren flora is missing from the islands.

FIG 88. Pyramidal Bugle (*Ajuga pyramidalis*). Poulsallagh, May. (David Cabot)

ABOVE: **FIG 89.** Curious bedfellows living within 20 cm of each other: Spring Gentian (*Gentiana verna*), an alpine species, and Dense-flowered Orchid (*Neotinea maculata*), from the Mediterranean region. South of Fanore More, May. (Fiona Guinness)

1. Maidenhair Fern (*Adiantum capillus-veneris*). Occasional and locally frequent in the Burren and on the Aran Islands, especially in grikes and sheltered rock crevices. Commoner on the Aran Islands than in the Burren. Scattered in the rest of Ireland, mainly coastal.

2. Irish Eyebright (*Euphrasia salisburgensis*). Common in the Burren, and more abundant on the Aran Islands, on thin soil over limestone pavement. Also in Co. Limerick, Co. Galway and Co. Fermanagh.

FIG 90. Maidenhair Fern (*Adiantum capillus-veneris*) in the sheltered, moist habitat of a grike. Black Head, May. (Fiona Guinness)

FIG 91. Left: Irish Eyebright (*Euphrasia salisburgensis*). Fanore, May. (Fiona Guinness). Below: Spring Gentian (*Gentiana verna*). Black Head, May. (Fiona Guinness)

3. Spring Gentian. Common throughout the Burren but less frequent on the Aran Islands. On a wide range of habitats, from limestone pavement and pasture to sand dunes, and even on peaty outcrops on limestone pavement. Spreads through limestone areas of Co. Galway and in one area of Co. Fermanagh.

FIG 92. Dense-flowered Orchid (*Neotinea maculata*), displaying well-developed seed capsules at the end of the flowering season. Slievecarran, May. (Fiona Guinness)

FIG 93. Thyme Broomrape (*Orobanche alba*) lacks chlorophyll and is fully parasitic on Wild Thyme (*Thymus polytrichus*). (David Cabot)

4. Dense-flowered Orchid. Locally frequent in dry grasslands and calcareous drift over limestone in the Burren and on the Aran Islands. Rarely in Co. Donegal, Co. Fermanagh and Co. Antrim.

5. Thyme Broomrape (*Orobanche alba*). Very local in western parts of the Burren where Wild Thyme (*Thymus polytrichus*) occurs in grassy places. Only on Inishmore in the Aran Islands, and rare there. Fairly frequent in coastal Co. Donegal, Co. Antrim and the Hebrides.

6. Mountain Avens. Common throughout the Burren and abundant in places, especially in upland areas. Surprisingly absent from the Aran Islands, but found rarely in Northern Ireland and, more commonly, in north-west Scotland.

7. Shrubby Cinquefoil (*Potentilla fruticosa*). Local on rocky and grassy ground subject to occasional flooding, especially around upper limits of turloughs in south-eastern parts of the Burren. Absent from the Aran Islands. In Britain, native only in the Lake District and Pennines.

8. Fen Violet (*Viola persicifolia*). Locally abundant in the Burren, sometimes growing in startling profusion – up to tens of thousands of flowers were recorded by Webb and Scannell (1983) at some sites. Restricted to the grassy zone below winter high-water levels around turloughs. Absent from the Aran Islands.

FIG 94. Above: Mountain Avens (*Dryas octopetala*). Gleninagh, May. (Fiona Guinness). Left: Shrubby Cinquefoil (*Potentilla fruticosa*). Mullagh More, May. (Fiona Guinness)

Species that are more generally distributed are as follows:

9. Hard Shield-fern (*Polystichum aculeatum*). Frequent on limestone pavement in the Burren, very rare on the Aran Islands. Although widespread in Ireland, it is common only on the Burren limestones.

10. Brittle Bladder-fern (*Cystopteris fragilis*). Very frequent in western and central parts of the Burren, especially in grikes on limestone pavement and on cliffs and walls, notably in damp spots. Not on Aran Islands.

11. Field Mouse-ear (*Cerastium arvense*). Locally frequent on grassland overlying limestone pavement, also on sand dunes and shingle beaches. Frequent and locally abundant on the Aran Islands.

12. Bloody Crane's-bill (*Geranium sanguineum*). Common throughout the Burren and on the Aran Islands. On limestone pavement and rocky grassland.

13. Mossy Saxifrage (*Saxifraga hypnoides*). Frequent, generally away from the coast, especially in the central Burren on limestone pavement or rocky grassland, often in the shelter of collapsed rock buildings and stone walls. Sparingly on the Aran Islands. Elsewhere in Ireland, it is occasional in Northern Ireland.

14. Spring Sandwort (*Minuartia verna*). Frequent and common, especially abundant on limestone pavement and grassland near the coast between Black Head and Poulsallagh. Not found east of Boston but present on the Aran Islands.

15. Stone Bramble (*Rubus saxatilis*). Very frequent on limestone pavement and on the Aran Islands. Elsewhere in Ireland, rather northern.

16. Squinancywort (*Asperula cynanchica*). Frequent and locally abundant in the Burren, especially in short grassland areas and on sand dunes. Present on the three Aran Islands.

17. Limestone Bedstraw (*Galium sterneri*). Common in both the Burren and on the Aran Islands, on limestone pavement and on dry grassland.

18. Mudwort (*Limosella aquatica*). Rare, in small pools and around lake margins and turloughs in the Burren. Not on the Aran Islands, but occurs in Co. Cork and on limestone areas in the north-west.

19. Wild Madder (*Rubia peregrina*). Frequent in most of the Burren and on the Aran Islands. On limestone and in grikes, hedges and stone walls.

20. Dark-red Helleborine (*Epipactis atrorubens*). Frequent, especially above 150 metres, in heath and limestone pavement grikes. Not on the Aran Islands, despite an erroneous record.

21. Blue Moor-grass (*Sesleria caerulea*). The dominant grass in the Burren on limestone pavement, fixed dunes and rocky ground. Also on all the Aran Islands. Locally common in limestone areas of western Ireland.

Two further species may be added to the above list:

1. Irish Saxifrage (*Saxifraga rosacea* ssp. *rosacea*). Local but sometimes frequent in the coastal zone between Black Head and Doolin, but uncommon in the rest of the Burren. It is more abundant on the three Aran Islands, where it is found on limestone pavement and broken stone. It also occurs in MacGuillycuddy's Reeks, Co. Kerry, in isolated coastal locations from Co. Cork to Co. Donegal, and in the Galtee Mountains in Co. Tipperary and Co. Limerick. Recorded in north Wales, Iceland and the Faroes, and in some mountains in central Europe. Despite being of boreal–montane origin, it grows almost at sea-level in the Burren.

2. Turlough Dandelion (*Taraxacum palustre*). Found commonly in turloughs in the Burren and in a few marshes in the Irish Midlands. Scattered in Britain and elsewhere in north-west Europe. Before the full extent of its presence in the Burren was known, it was estimated that the world population was fewer than 1,000 plants (Dudman & Richards, 1997). Another dandelion, *T. webbii*, seems to be endemic to Ireland, and again is found in turloughs.

Arctic–alpine flora

Webb (1983) examined the flora of Ireland in its European context while defining various groups of plants that shared distinct geographical origins, or 'homelands'. In his classification of plants, an 'arctic–alpine' species must be fairly widespread in the Arctic and subarctic regions of Europe, and must reappear at high altitudes up to at least 2,500 metres in the Alps and often also in the Pyrenees. The plant must be scarce or absent in the intervening areas. Species occurring at low latitudes, south of about 54–55°N, are excluded, as are those growing below an altitude of 800 metres, except in the immediate proximity of high mountains. Sixteen species meet Webb's definition as genuine arctic–alpines, mostly confined to montane habitats. Of these, two are found in the Burren:

1. Mountain Avens. Common throughout the area, descending almost to sea-level, but puzzlingly absent from the Aran Islands.

2. Roseroot (*Sedum rosea*). At Black Head and on the Aran Islands, where it is abundant on limestone pavement towards the western end of Inishmore, and on some cliffs, descending almost to sea-level. Also on the Cliffs of Moher, approximately 5 kilometres south-west of the Burren.

A strong candidate for this group is Arctic Sandwort, rediscovered in the Burren in April 2008, growing south of Black Head. It had eluded botanists since it was originally recorded in that area in 1961. Charles Nelson (2016) described it as an arctic–montane plant, whereas Webb (1983) classified it as a member of the arctic–alpine group.

Then there is a group of species with alpine or northern distributions but not found in the Arctic. Spring Sandwort, Bearberry (*Arctostaphylos uva-ursi*) and Spring Gentian are all truly alpine species, although they are also widely distributed in the lowlands of central Europe. Irish Eyebright, common in the Burren and on the Aran Islands, is considered to be a fairly orthodox alpine species. It is absent from Britain and its nearest station is 1,200 kilometres distant, in the Vosges Mountains. Pyramidal Bugle is categorised by Webb (1983) and others (e.g. Biological Records Centre (BRC) & BSBI, 2017) as boreal–montane. In the Alps, it grows at altitudes up to 2,700 metres. The Large-flowered Butterwort (*Pinguicula grandiflora*) has a double geographical classification. Webb lists it as an alpine species but it is also a Lusitanian plant due to its presence in the mountains of Spain and southern France.

The occurrence of so many arctic–alpine, arctic, alpine and boreal–montane species within the relatively small area of the Burren and Aran Islands raises the possibility that they may have survived the last glacial period.

FIG 95. Roseroot (*Sedum rosea*), Cliffs of Moher, August. (David Cabot)

Mediterranean–Atlantic flora

Turning to those species with a Mediterranean–Atlantic distribution, Webb (1983) listed 25 as occurring in Ireland. Of these, seven are found in the Burren and/or on the Aran Islands. The Dense-flowered Orchid, the most celebrated member, is firmly centred in the Mediterranean region. The other six members of this Mediterranean–Atlantic group in the Burren are:

1. Tutsan (*Hypericum androsaemum*). Particularly common in limestone grikes in the Burren and in woodland. Occasional on the three Aran Islands.
2. Navelwort (*Umbilicus rupestris*). Infrequent in rocky areas in the Burren and on Inishmore.
3. Portland Spurge (*Euphorbia portlandica*). Local in the Burren, on sand dunes at Fanore. Also on the Aran Islands, where it is locally frequent on boulder beaches on Inishmaan.
4. Tree-mallow (*Malva arborea*). Considered native on maritime cliffs and rocks at the western end of Inishmore. Almost certainly introduced on other islands and in parts of the Burren.
5. Sea-purslane (*Atriplex portulacoides*). Rare on rocks and stony shores in the Burren, and also found on Inishmore.
6. Wild Madder. Locally frequent in the Burren and on the three Aran Islands.

As mentioned above, the Large-flowered Butterwort has a Lusitanian range and so is listed as a Mediterranean–Atlantic species as well as an alpine one. The Burren is its most northerly location in Europe. While noting that the Maidenhair Fern is a member of the Mediterranean–Atlantic element (the same classification given by Webb (1983)), the *Online Atlas of the British and Irish Flora* states that it also occurs in central and eastern Asia, and in North America (BRC & BSBI, 2017). It is a tropical and subtropical fern, and, as such, could be regarded as a southern species, a member of the Mediterranean group of plants.

THE BURREN'S PHYTOGEOGRAPHICAL RIDDLES

The information above gives rise to four principal questions:

1. Why do four plant species occur exclusively in the Burren and a further two nearly exclusively so?
2. Why are some species more numerous in the Burren than elsewhere in Ireland and Britain?

3. Why is there such a mixture of species?
4. What is the route by which the species arrived?

The Burren is not the only karst region to exhibit such floristic eccentricities. A parallel occurs in the Karst region of south-west Slovenia, in the Škocjan Caves Regional Park. Here, there is a remarkable, dramatic collapsed doline. It is about 165 metres deep and contains representatives of ice age Alpine flora, from the cold mountain areas of Austria. Crusted-leaved Saxifrage (*Saxifraga crustata*), Auricula (*Primula auricula*), Alpine Yellow Violet (*Viola biflora*) and Kernera (*Kernera saxatilis*) are examples of such species. Growing 40 metres above these are plants characteristic of the Mediterranean or sub-Mediterranean region: Wild Asparagus (*Asparagus prostratus*), Maidenhair Fern, Prickly Juniper (*Juniperus oxycedrus*) and Rustyback Fern (*Asplenium ceterach*). An echo of the mixture of Burren species from different geographical regions is also found in some invertebrates, whose origins include both Alpine and Mediterranean regions (Peric, 2012).

Exclusivity of four species
The Hoary Rock-rose is restricted to areas of Carboniferous limestone in Britain and Ireland, and in the Burren it grows on outcrops or shallow soils near cliff tops, often on south- or south-western-facing sites. The Burren's relatively warm winter temperatures (seldom below 5 °C in January and February), infertile soil conditions and lack of competitive tall vegetation seem to provide ideal conditions for the species.

Records of the sea-lavender *Limonium transwallianum* in Ireland are now regarded as belonging to a separate taxon, *Limonium recurvum* ssp. *pseudotranswallianum* (Stace, 2010). In Ireland, the plant is restricted to steep-facing lowland maritime limestone cliffs, as occur only in parts of the western section of the Burren.

Purple Milk-vetch is found in machair, limestone heath and sandy places by the sea, on Inishmaan and Inishmore in the Aran Islands. The population there was first discovered in 1834 and remains stable. The islands were connected to the Burren until relatively recently – approximately 8,000 BP, based on computer modelling of sea-levels (Edwards & Brooks, 2008). No explanation has yet been offered as to the species' absence in suitable Burren habitats, just a few kilometres east across the sea, nor indeed for its presence on the Aran Islands at all. In Britain, it is found mostly in eastern coastal areas of England, northwards to Scotland, with outlying stations on the Isle of Man and the Hebrides. Its

FIG 96. Habitat for the sea-lavender *Limonium recurvum* ssp. *pseudotranswallianum*, near Poulsallagh. (David Cabot)

one-station phenomenon in Ireland might suggest introduction, but if so, why does it occur on two islands?

Arctic Sandwort in the Burren seems likely to have been a post-glacial immigrant from a southern refugium that also supplied populations in Britain and Iceland. There is comparatively little genetic divergence between them (Howard-Williams, 2013). The species has held on in unshaded places where trees are stunted (see below).

Abundance of some species

The answer to the question why some species are more numerous in the Burren than elsewhere in Ireland and Britain probably lies in a combination of factors. The peculiar soil conditions of the Burren, the exposed and porous rock structure, the habitat diversity, the mild winter temperatures, the removal of woodland vegetation and subsequent soil erosion, and the presence of grazing animals, both wild and domestic, together or separately have all had an influence on the plant species found here.

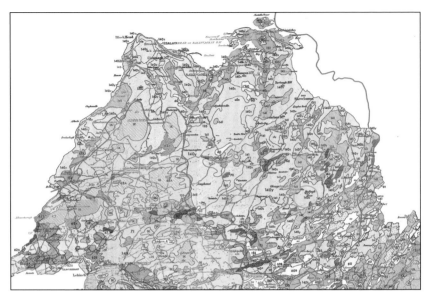

FIG 97. Soils of the Burren area (see key opposite). The Burren as defined by Webb (1962) is shown by the added red line. From Finch (1971).

Soil conditions and rock structure

Soil development in Carboniferous limestone country often produces shallow rendzinas, which are highly porous, free-draining and saturated with calcium carbonate. Levels of extractable calcium are approximately 24 times greater than in brown podzolic soils, magnesium approximately six times greater and ammonium approximately two times greater (Lee, 1998). Phosphate, although present, is largely unavailable to plants because of the high soil pH. Some of these factors promote broadleaved species at the expense of grasses and allow plants with higher light demands into the vegetation.

Other soil factors facilitate the occurrence of 'acid' or calcifuge species in the alkaline conditions of the Burren. In places, peat forms directly on the rock through the influence of climate, whereas elsewhere a previous, more acid soil may have been washed into the grikes. The survival of calcifuges is helped by the constant flow of rainwater through the soil, which leaches minerals away, but may also be a response to the lack of available phosphate.

Habitat diversity

Habitat diversity is relevant to plants on both the large and small scales. Large habitat features in the Burren, such as turloughs and soilless expanses of rock,

GREAT SOIL GROUP	SERIES etc.	ACRES Hectares	Per cent of total area
GREY BROWN PODZOLICS	59 Elton	47040 / 19037	6·04
	134 Kilfenora	4040 / 1635	0·52
	60 Patrickswell	17640 / 7139	2·27
	60A Patrickswell Lithic Phase	640 / 259	0·08
	60B Patrickswell Bouldery Phase	3760 / 1522	0·48
BROWN EARTHS	42 Baggotstown	520 / 210	0·07
	43 Ballincurra	1320 / 534	0·17
	45 Ballylanders	29480 / 11931	3·79
	49 Ballymackken	2440 / 987	0·31
	44 Derk	200 / 81	0·03
	51 Kilfergus	25080 / 10150	3·22
	135 Kinvarra	13880 / 5617	1·78
	135A Kinvarra Bouldery Phase	2960 / 1198	0·38
	136 Knocknaskeha	1160 / 470	0·14
	137 Tullig	7280 / 2946	0·94
	138 Waterpark	400 / 162	0·05

Numbers out of sequence refer to soils already mapped in other counties

GREAT SOIL GROUP	SERIES etc.	ACRES Hectares	Per cent of total area
BROWN PODZOLICS	54 Cooga	9800 / 3966	1·26
	56 Doonglara	2240 / 907	0·29
PODZOLS	55 Mountcollins	25680 / 10393	3·30
	77 Knockaceol	2000 / 809	0·26
	77b Knockaceol Bouldery Phase	1360 / 550	0·17
	78 Knockanimpaha	1800 / 729	0·23
	179 Knockanattin	760 / 308	0·10
	79 Knockastanna	1640 / 664	0·21
	79b Knockastanna Peaty Phase	2520 / 1020	0·32
	80 Seefin	280 / 113	0·04
RENDZINAS	140 Burren	200 / 81	0·03
	140x Burren* Rocky Phase	22040 / 8920	2·83
	140y Burren very** Rocky Phase	34040 / 13776	4·37
	140z Burren extremely*** Rocky Phase	18480 / 7479	2·37
	140A Burren Deeper Phase	360 / 146	0·05
	141 Kilcolgan	14360 / 5812	1·85
	141A Kilcolgan Bouldery Phase	7440 / 3011	0·96
LITHOSOL	83 Slievereagh	80 / 32	0·01

*25%,**50%,***75% rock

Key to the soils of the Burren area (see map opposite).

have clear influences on the vegetation, while on the small scale the intricacies of grikes and limestone pavement create abundant niches with differing microclimates, some exposed and some protected from grazing attention.

Climate
The mild winter temperatures of the Burren rival those of the peninsulas in Co. Cork and Co. Kerry, but their combination with limestone is unique in Ireland. The coastal Burren has a January mean above 6 °C, while the temperature in the Aran Islands averages above 6.5 °C. Humidity is high everywhere, as are wind speeds. Average annual rainfall on the Burren hills exceeds 150 cm, while in the lowlands annual rainfall is less than 100 cm. Snow in the Burren is unusual, and when it does occur it is usually of a short duration.

FIG 98. View from the top of Corkscrew Hill looking northwards after an April snowfall. (David Cabot)

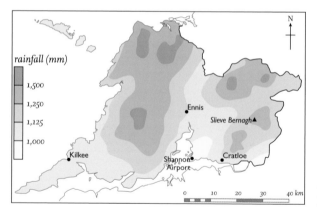

FIG 99. Rainfall distribution in Co. Clare on an average annual basis. From Finch (1971).

Links of the rarer Burren plant species with one or other of these factors is often suggested. Some of the more obvious ones are:

1. Dense-flowered Orchid must be favoured by the warm winters, although it also survives in Co. Roscommon, which has a mean winter temperature of 5 °C. The species additionally requires a low herbage level so as not to be swamped

by neighbouring plants. As in all orchids, its seeds rely on the presence of appropriate soil fungi to germinate and form a mycorrhizal association.

2. Maidenhair Fern prospers in the Burren thanks to the warm winters and also the sheltering grikes, which limit the negative effects of storms and salt.

3. Spring Gentian is intolerant of shading and dies out in scrub. It grows best in a free-draining soil but with plentiful water.

4. Irish Eyebright and Thyme Broomrape require the presence of Wild Thyme, which grows in unshaded calcareous conditions. Irish Eyebright is a hemiparasite on the species, while the Thyme Broomrape is a full parasite. The absence of tall, shading vegetation is therefore a critical factor for all three plants.

5. Fen Violet and Turlough Dandelion are found in turloughs, where they are submerged for part of the year. A lack of disturbance has favoured their persistence. In Britain, Fen Violet has disappeared from 19 of its former sites as a result of drainage, and today survives only in the Cambridgeshire fens.

6. Shrubby Cinquefoil grows on grassy or rocky ground that is subject to occasional flooding, especially around turloughs. It seems to withstand flooding of its rooting zone better than competing plants.

Mixture of species

There is no simple answer as to why the Burren holds such a mixture of species, except to say that they are there and able to grow and survive. Every terrestrial habitat in Ireland has had a succession of plant and animal species since the first tentative colonists started to grow as the ice melted. But in most other places, soil development, tree growth and agriculture have led to the simplification of communities and the huge decline of species we know too well. By contrast, the Burren has held onto most of the flora that has grown there in the last 12,000 years.

The primary reason for this must be the openness of the habitat – there are spaces for small species to grow, whether these have been caused by wind exposure on the west coast, a lack of soil on the cliffs and pavement, a limitation of grass growth by the low-phosphate conditions or a restriction of woody plants by endless grazing. Nowhere else in Ireland has a wild habitat been grazed so consistently and so intensively, without cultivation, for the whole of the postglacial period. The intensity of this use has had the unintended effect of hastening soil erosion on the porous rock skeleton, reversing the normal colonisation by bushes and trees, and holding the vegetation at an early stage of plant succession – the herb and dwarf shrub stage. There is no doubt that the Burren is an unintentional garden, a managed landscape where the grazing intensity has by chance developed to just the right level for the herbaceous flora to persist and spread. Left to itself, the Burren might now be a forested habitat, with odd cliffs breaking through a tree cover of oaks (*Quercus* spp.), Hazel (*Corylus avellana*) and Scots Pine (*Pinus sylvestris*).

Particular soil and climatic conditions add to the happy accidents of land use. The thinness of the Burren soils causes them to dry out regularly, restricting large perennials that are not adapted to them. In addition, the presence of insoluble chert and glacially derived loess allows plants that normally grow in acidic conditions to survive, while the constant rainfall leaches nutrients away that would otherwise be caught by deeper-rooting plants and in places promotes the accumulation of peat (Jeffrey, 2003). All these things limit grass growth and the sward that might otherwise be expected to develop under continuous grazing or the use of artificial fertilisers.

The route by which species arrived

The conclusion Webb drew from his 1983 study of the discontinuous distribution of 15 plants found only in Ireland and not in Britain (mostly the so-called Lusitanian species), three of which occur in the Burren and on the Aran Islands, was that 'the supposition of these discontinuously distributed flora surviving the last glaciation, or alternatively on now submerged lands off the West coast is less unlikely than any other alternative explanation'. Such *in situ* survival had been proposed earlier, in 1846, by English naturalist Edward Forbes (1815–54). More recent evidence for one of these Lusitanian species, St Dabeoc's Heath (*Daboecia cantabrica*), based on DNA analysis points to a postglacial arrival (see below), but with regard to some of the cold-tolerant arctic–alpine species, Webb may have been right.

The Pleistocene epoch commenced approximately 2.6 mya. Oxygen isotope records from deep marine sediments provide evidence that large-scale fluctuations in global climate have occurred since the onset of the epoch. There were several glacial cycles, each with a duration of approximately 100,000 years, interrupted by shorter, warmer periods, known as interglacials (Howard-William, 2013). Peter Coxon and Stephen Waldren (1995) found that the cold periods varied in both magnitude and frequency in north-west Europe, with earlier cycles apparently more frequent and less extreme than those in the later stages of the epoch. They suggested that cold–warm–cold sequences happened on average over a 41,000-year cycle in the period before 0.9 mya, whereas more recently, from 0.9 mya to the present, the cycle lasted 100,000 years.

Radiocarbon evidence from the last major interglacial suggests that Ireland was largely ice-free 32,000 BP, after which the last Irish ice sheet (the Midlandian) thickened to cover all mountain summits (Ballantyne & Ó Cofaigh, 2017). The current consensus is that this ice sheet covered the whole island and extended to the edge of the Atlantic shelf in the west and far out onto the Celtic Sea shelf in the south. It was then confluent with ice from Scotland, north-west England and Wales. However, not all experts agree on the extent of ice cover (see below).

For much of the early twentieth century, it was believed that the Midlandian ice sheet in Ireland stopped at a point termed the southern Ireland end moraine (SIEM), which ran through parts of Co. Kilkenny and across the country, with the province of Munster remaining free of ice. However, the SIEM is now interpreted as being a deglacial moraine (i.e. not marking the end limit of an ice sheet, but deposited during its retreat). What this means is that many features originally ascribed to the earlier Munsterian glaciation (374,000–130,000 BP) in the older literature are actually now interpreted as Midlandian. Most surface evidence of the Munsterian glaciation in Ireland was, in fact, obliterated by the Midlandian (K. Craven, pers. comm., 2017).

TABLE 1. Irish glacial and interglacial periods compared with those of Great Britain and the Continent. Derived from a table prepared by Don Cotton (2015), with dates from Kim Cohen and Philip Gibbard (2011).

	Ireland	*Great Britain*	*European Continent*	*Dates ('000 years BP)*
Interglacial	Littletonian	Flandrian	Holocene	11.7–present
Glacial	Midlandian	Devensian	Würm/Weichselian	114–11.7
Interglacial	[not yet found]	Ipswichian	Eemian	130–114
Glacial	Munsterian	Wolstonian/Gipping	Riss/Saale	374–130
Interglacial	Gortian	Hoxnian	Holsteinian	424–374
Glacial	Pre-Gortian	Anglian	Mindel	478–424

Recent research on the global sea-levels suggests that the LGM lasted from approximately 265,000 BP to approximately 19,000 BP (Clark *et al.*, 2009), with thick ice cover and maximum extent occurring at different times in different sectors. However, the British and Irish ice sheet was climatically sensitive and therefore extremely dynamic. We do not know the impact of the North Atlantic Drift at this time, for example, which may have been modified or even intensified by the ice mass. Consequently, there are real uncertainties in the reconstruction of the complex history, thickness and lateral extent of the ice sheets. Along the west of Ireland (i.e. the Burren region), however, there is evidence of ice-free conditions only since 19,100 BP (Ballantyne & Ó Cofaigh, 2017). At Loop Head,

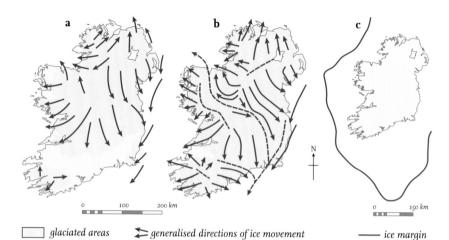

a   b   c

N

0   100   200 km   0   150 km

☐ glaciated areas   ⮴ generalised directions of ice movement   —— ice margin

FIG 100. Models of the extent of glacier ice on Ireland and the adjacent shelf during the last glacial maximum (LGM): a. the traditional model of restricted ice cover, as depicted in McCabe (1987); b. revised model showing offshore extension of ice cover in all sectors apart from an unglaciated enclave in the south-west, as depicted by Warren and Ashley (1994): c. model of LGM ice cover extending to the shelf edge west of Ireland and southwards across the Celtic Sea, as depicted by Sejrup *et al.* (2005). From Ballantyne and Ó Cofaigh (2016).

Co. Clare, deglaciation seems to have occurred by 20,300 BP, based on chlorine-36 ages (although note that these are only approximate). Further north, in the Nephin Beg mountain range, Co. Mayo, there is evidence of the emergence of nunataks at 19,500 BP.

Offshore moraines and related evidence indicate that the ice sheet extended westward to the Atlantic shelf, eastwards into the Irish Sea Basin and southwards to the Celtic Sea shelf edge. The presence of north–south-trending moraines on the Atlantic shelf was first documented by Haflidi Haflidason and others in 1997. The outermost moraines detected occur parallel to the shelf edge at nearly 300 metres below modern ordnance datum (OD), and are thought to mark the maximum westward extent of the Irish ice sheet. Subsequently, the moraines (or moraine complexes) were found to extend to the shelf break and across the Porcupine Bank west of Ireland (Clark *et al.*, 2009). Westwards truncation of the outermost moraines and the presence of iceberg scours on the outer side of these moraines indicate that expansion of the Irish ice sheet was halted by increasing water warmth. The outermost moraines along the western margin are the largest, suggesting that the ice margin may have remained at, or near, the shelf edge for millennia (Clark *et al.*, 2012).

As noted, there are differing views on the extent of ice coverage across Ireland during the last glaciation, which is a critical issue when considering the survival or extinction of earlier interglacial flora and fauna. In addition, the thickness of ice is acutely relevant to the exposure of possible nunataks. There is some information from the level of glacial trimlines on the height of the ice in Irish mountains. Trimlines were initially interpreted as indicating the maximum altitude of moving ice, although there is some evidence to suggest that they define the minimum rather than maximum altitude of former ice cover at the LGM (Ballantyne & Ó Cofaigh, 2017). The surface of the ice sheet was definitely not below this altitude and was probably even higher, because the trimlines so far found in western Ireland are not high enough to sustain a sloping ice sheet extending to the western shelf edge.

Trimline evidence from across Ireland indicates that up to 76 individual nunataks stood up above the ice sheet during the last ice age, cumulatively exposing a maximum of 111.7 square kilometres (it is likely that this figure will be reduced if the trimline adjustment is accepted). The largest of these ice-free areas was across the Comeragh Mountains, Co. Waterford, where just less than 18 square kilometres were exposed. Other extensive ice-free areas included patches of the Galtee Mountains in Co. Tipperary and Co. Limerick, and MacGillycuddy's Reeks and Mangerton Mountain in Co. Kerry. The highest trimline was around Lugnaquilla, Co. Wicklow, where the ice was a minimum of 850 metres thick, while the lowest was across Knocknadobar in Co. Kerry, where the ice trimline occurred at only 310 metres above OD. In all, ice-free areas may have occurred in 11 counties during the last glaciation: Carlow, Cork, Donegal, Dublin, Kerry, Limerick, Mayo, Tipperary, Waterford, Wexford and Wicklow (Meehan, 2016). Note that Co. Clare does not figure in this list.

So, did some of the arctic–alpine species, and possibly others, survive *in situ* from one interglacial to the next, and especially from the last interglacial through the final Midlandian glacial period? Evidence of the presence of either Fringed Sandwort (*Arenaria ciliata*) and/or Arctic Sandwort (these species are difficult to separate by their fossilised seeds) from 30,500 BP in Derryvree Lough, Co. Fermanagh, provides support for Webb's *in situ* survival hypothesis (Colhoun *et al.*, 1972). The site is 50 kilometres from the Benbulbin mountain range in Co. Sligo, where Fringed Sandwort still grows. In addition, DNA analysis of this species points to its much earlier presence in Ireland, up to 150,000 BP (Howard-Williams, 2013) (see below).

The climate during the warm interglacial periods was similar to that today, while in the cold glacial periods, temperatures were 10–25 °C lower. These climatic changes resulted in large redistributions of the flora and fauna.

FIG 101. Benbulbin, Co. Sligo. At 526 metres above sea-level, this mountain range may have acted as a refugium or nunatak for several arctic–alpine plant species that are found locally today. (David Cabot)

Arctic–alpine plants would have moved to higher, colder latitudes as the climate warmed up during the interglacials, and then south again with the approach of another cold phase.

To envisage the landscape at the onset of the final glacial advance, one should also consider the mammalian fauna. Some of these species would have had an impact on the flora at that time, not only the herbivores devouring the tundra vegetation, but also the birds distributing seeds, facilitating the spread of plants. Among the mammals, the herbivores included Giant Deer (*Megaloceros giganteus*), Reindeer (*Rangifer tarandus*), Red Deer (*Cervus elaphus*), Wild Horse (*Equus ferus*), Woolly Mammoth (*Mammuthus primigenius*), Collared Lemming (*Dicrostonyx torquatus*), Norwegian Lemming (*Lemmus lemmus*) and Mountain Hare (*Lepus timidus*). The carnivores were Spotted Hyena (*Crocuta crocuta*), Arctic Fox (*Vulpes lagopus*) and Stoat (*Mustela erminea*), while the Brown Bear (*Ursus arctos*) was also present (Monaghan, 2017). Most of these mammals became extinct during glaciation, but there is good evidence (from late-glacial deposits) that the Stoat,

Mountain Hare and Red Deer reinvaded by themselves. it is less convincing for the rest of the modern fauna. Much of the interglacial flora suffered a similar fate, but we now know that a few hardy arctic–alpine plants, and possibly other species, survived in refugia.

FIG 102. Woolly Mammoth (*Mammuthus primigenius*) and Wild Horses (*Equus ferus*) grazing on open tundra some 40,000 years ago, during the interglacial period before the onset of the Midlandian glacial phases around 35,000 years ago. (Peter Snowball)

Survival of arctic–alpine flora

Among the flora present just before the final, Midlandian glaciation were Fringed Sandwort and possibly Arctic Sandwort, together with Mountain Avens, Moss Campion (*Silene acaulis*), Purple Saxifrage (*Saxifraga oppositifolia*) and hypnoid saxifrages (including seeds ascribed to Irish Saxifrage), and Dwarf Willow (*Salix herbacea*). The evidence for the presence of these species came from the sediments of Derryvree Lough in Co. Fermanagh, as pollen or fossils of leaves, shoots or seeds, and they show open tundra vegetation growing in a periglacial regime (Colhoun *et al.*, 1972).

All the above arctic–alpine species are found in the mountains of Ireland today, albeit in a few disjunct locations. Fringed Sandwort has its only station in Ireland and Britain on Benbulbin (526 metres above sea-level) in Co. Sligo. A study carried out by Peter Coxon in 1988 suggests that the peak of neighbouring Truskmore (647 metres) was above the level of the ice sheet as a nunatak, and so could have acted as a refuge for this and other arctic–alpine species. Nearly 20 years earlier, Francis Synge (1969) also argued that the summit surfaces of the high plateaus in northern Co. Sligo, especially the Benbulbin massif, were also glacier-free, standing above the general level of the ice sheet of the Late Midlandian glaciation. There is also some evidence that two hill summits in the north-west Burren – Slieve Elva and Knockauns Mountain – both acted as nunataks (p. 78), where arctic–alpine flora could also have survived during the final glacial period (Finch, 1966).

It seems likely that this general area was home to the Fringed Sandwort and other species recorded in the investigations into the Derryvree Lough pre-Midlandian flora. As the ice advanced they were confined to a nunatak, and when it retreated they would have spread widely into the adjacent lowland areas. Later, interglacial warming and the spread of trees has confined them to their current mountain cliffs.

If Fringed Sandwort has really been in Ireland for 150,000 years, as the genetic analysis suggests, it must have been growing during the interglacial between the two main phases of ice advance, perhaps again on mountain cliffs (Coxon & Waldren, 1995). The lowland flora of a middle Pleistocene deposit in Kilkenny includes coniferous and broadleaved trees, some of which are no longer present in Europe, banished by the ice to the Caucasus Mountains. Firs (*Abies* spp.), spruces (*Picea* spp.), yews (*Taxus* spp.), elms (*Ulmus* spp.), alders (*Alnus* spp.), oaks and hornbeams (*Carpinus* spp.) were present, as were wing-nuts (*Pterocarya* spp.), which are now absent from Europe. Smaller species below the canopy consisted of grasses, heathers and Rhododendron (*Rhododendron ponticum*), the latter another Caucasus plant though widely introduced. A study of even earlier sediments, dating back about 350,000 years, in Cork Harbour revealed fossil pollen from the Gortian interglacial (Table 1, p. 123), named for a site just east of the Burren. A vegetation very similar to that seen today can be guessed at from the pollen record, which included pines, alders, oaks, birches (*Betula* spp.), willows (*Salix* spp.), and junipers (*Juniperus* spp.). Sea buckthorns (*Hippophae* spp.) and box (*Buxus* spp.) were also present; the latter requires a July temperature of at least 17–18 °C, a degree or two warmer than today. The presence of Holly (*Ilex aquifolium*) and ivy (*Hedera* spp.) indicates mild winters, while yews require relatively wet summers. The climate was damp and oceanic, as illustrated by the presence of filmy ferns (Hymenophyllaceae spp.), while Alder Buckthorn (*Frangula alnus*) and ash (*Fraxinus* spp.) grew in a wetland community. So, if the climate was similar to today, the arctic–alpines may have had a similar distribution and, without the attentions of grazing sheep, grown in much greater abundance on the mountains.

The LGM occurred about 26,000 BP, when the depth of the ice sheets in the Irish Midlands was 1–1.2 kilometres. From about 22,000 BP to 20,000 BP there was a comparatively rapid decline in the ice, so that by about 16,000 BP most of the landscape was once again ice-free and warming up. At this time, the land bridges connecting Ireland to Britain were submerged by the rising sea-levels and simple reinvasion by plants ceased (Edwards & Brooks, 2008). This severance meant that about 180 species that could grow in Ireland were not able to reach the island.

Late-glacial July temperatures during the period 15,000–10,000 BP have been reconstructed from chironomid assemblages in a core taken from a lake basin at Lough Nadourcan in north-west Ireland (Watson *et al.*, 2010). Cold, glacial conditions in Ireland seem to have been terminated abruptly by a July temperature rise of about 5 °C, with the warming occurring over a period of about 300 years. Summer temperatures reached about 13 °C early on, and then oscillated around the 11–12 °C level for most of the following two millennia. It is unlikely that glacier ice survived any of this warming, so we can say that in all likelihood low ground was completely ice-free by about 15,000 BP (Ballantyne & Ó Cofaigh, 2017).

At ice melt, any plant species that had survived the LGM – including Mountain Avens, Fringed Sandwort, Moss Campion and, possibly, others – were then free to flourish and expand out of their refugia, wherever these were. It is unlikely that the more temperature-sensitive Mediterranean–Atlantic species could have survived the LGM. No Mediterranean species were recorded in the Derryvree Lough flora inventory, nor have any been specifically identified from earlier interglacial sites. Some botanists, including David Webb (1983), thought that land off the west coast of Ireland could have acted as refugia for some of the Mediterranean species. However, in view of recent research into the westward movement and duration of ice throughout the LGM, this now seems unlikely (Clark *et al.*, 2009).

Arrival of the Mediterranean–Atlantic flora

Both the Maidenhair Fern and Dense-flowered Orchid may have first arrived in Ireland during an interglacial when conditions were similar or warmer than today, but they are likely to have died out as the cold returned. Then, in the postglacial period, they reinvaded, possibly by overland migration via the land bridges that existed when sea-levels were some 120 metres below their present level. The lightness of the propagules of both species would have hastened any migration, allowing them to be carried over water by southerly gales. In the absence from Britain of most of the Irish-based Lusitanian species, the evidence points towards a southern route into Ireland from Spain and Biscay.

The possible methods of entry for species can be overland migration by growth and seed fall, transport by wind and water, or attached to the muddy feet of birds, especially waterfowl on their northerly migration. Nor should the role of early humans in the mesolithic and neolithic periods be discounted. New evidence indicates that man arrived in Ireland approximately 12,500 BP, some 2,500 years earlier than previously thought (Dowd, 2016), allowing a greater time period over which accidental introductions of flora and fauna via human migrations may have taken place. Whenever Mediterranean and southern species reached the Burren, climatic conditions and soil fertility must have been sufficiently attractive for them to become established. A proportion also made it to the north-west corner of Ireland and to the Hebrides.

As a Mediterranean species, the Dense-flowered Orchid may have followed the same route into Ireland as St Dabeoc's Heath, a member of the Lusitanian flora from north Spain and Portugal. This plant has a long history in Ireland – macrofossils dating to the Gortian interglacial, 424,000–374,000 BP (Table 1, p. 123) have been found. Macrofossils and pollen were also found at Kilbeg, Co. Waterford, where the interglacial deposits were probably earlier, dating back some 130,000–400,000 years (Watts & Ross, 1959). The climate was distinctly cold then, and there was a well-developed subarctic flora. Analysis of the diatoms present indicated that the climate was comparable to that of central Scandinavia today. St Dabeoc's Heath is more cold-tolerant than often realised, and in its Spanish mountain habitats it is sometimes covered by snow. It was probably extinguished in Ireland during subsequent glaciations, but its earlier presence shows that migration into Ireland from the south has been a continuous phenomenon.

Detailed genetic studies of St Dabeoc's Heath have indicated that the plant probably survived the LGM in two places, off the western coast of Galicia in Spain and in the Bay of Biscay. The western Spanish population showed higher genetic diversity than the eastern Spanish or the Irish populations for all three genetic loci examined. It is thought that the Irish plant was at the leading edge of postglacial spread from the Biscay refugium, which was then truncated by the rising seas. Each end of the former Biscay range was isolated in the postglacial period. The divergence time between the Irish and Spanish populations associated with the putative Biscay refugium was estimated as 3,333–32,000 years. Subsequently, the whole Spanish range was reunited by spread along the coastal mountains. This scenario seems more likely than *in situ* survival of the species off the west coast of Ireland.

The invasion of Ireland by Lusitanian and Mediterranean species via land connections seems possible or even likely, but some recent work has cast doubt

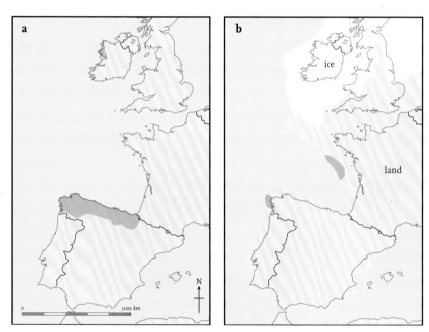

FIG 103. Distribution of St Dabeoc's Heath (*Daboecia cantabrica*): (a) present-day distribution (orange; based on Webb (1983)); (b) palaeodistribution at the last glacial maximum (LGM) around 21,000 years ago (orange), along with the limits of the British–Irish ice sheet (white; after Sejrup *et al.* (2005)) and areas of dry land (beige) at the LGM. From Edwards and Brooks (2008).

on this theory. Computer-simulated models used to produce palaeogeographic reconstructions suggest that the land bridge between Ireland and Britain was drowned some 16,000 BP, several thousands of years earlier than previously thought, when the climate was still cold and some ice caps remained (Edwards & Brooks, 2008). This would have inhibited the spread of warmth-demanding species northwards and throws the debate open once more.

The conclusion is that there is still no actual evidence for the migration of temperate flora and fauna across a land bridge into Ireland during the Holocene (Edwards & Brooks, 2008).

SPECIAL BURREN INVERTEBRATES

Lepidoptera

Several invertebrates are exclusive to the Burren or occur here more abundantly than anywhere else. These have their geographic centres of distribution either in southern or northern Europe and they parallel the disjunct distributions of the plants. Again, it is uncertain why this should be so. The presence of these invertebrates certainly adds to the uniqueness of the area. Taking the butterflies, for example, the Burren is home to more species than any other region in Ireland.

Perhaps the most celebrated butterfly in Ireland is the Pearl-bordered Fritillary (*Boloria euphrosyne*), which is one of its rarest and most endangered species. It is confined to the Burren, the Aran Islands and a karst outlier area in south-east Co. Galway (p. 50). Once widespread throughout much of Britain,

FIG 104. Pearl-bordered Fritillary (*Boloria euphrosyne*), May. (Philip Strickland)

FIG 105. Brown Hairstreak (*Thecla betulae*). Friars Island, near Moycullen, Co. Galway, August. (Colin Stanley)

the species has suffered a severe and rapid decline in England and Wales over the last 50 years, and is now locally extinct in most of Wales, and in central and eastern England. The butterfly is widespread throughout Europe, from northern Spain to Scandinavia, and reaches as far east as Russia. The adults feed on Bugle (*Ajuga reptans*), buttercups (Ranunculaceae spp.) and Bird's-foot Trefoil (*Lotus corniculatus*), while the favoured larval foodplant is the Common Dog-violet (*Viola riviniana*), an extremely regular species in the Burren (Harding, 2008).

The Burren is also the headquarters for the Brown Hairstreak (*Thecla betulae*), whose larvae feed on the fresh young leaves of Blackthorn (*Prunus spinosa*), which is particularly frequent in the Burren and on western limestones. The insect is also found on the shores of Lough Corrib and in small, isolated populations by Lough Derg in Co. Tipperary. In Britain, it has undergone a significant decline, due to hedgerow removal and annual hedge cutting, which destroys the eggs.

FIG 106. Wood White (*Leptidea sinapis*), May. The range of this habitat specialist is restricted to the karst areas of Co. Clare and Co. Galway. (Philip Strickland)

The third butterfly for which the Burren is particularly renowned is the Wood White (*Leptidea sinapis*). This species appears identical to Réal's Wood White (*L. reali*), and the two cannot be distinguished unless their genital characteristics are examined minutely under magnification. They can, however, be separated by

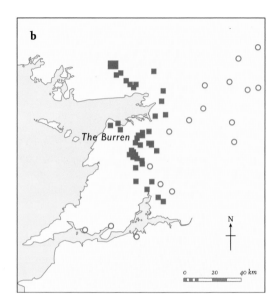

a

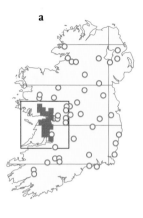

■ Wood White
○ Réal's Wood White

FIG 107. Distribution of the identified Wood White (*Leptidea sinapis*) and Réal's Wood White (*L. reali*) specimens: a. in the whole of Ireland; b. in the zone of potential overlap in western Ireland.The Irish distribution is shown by occurrence in hectads. The grid lines shown are the 100 kilometre squares of the Irish national grid. The symbols on the inset map are plotted on the centroid of the grid references of the sampled specimens, and show those only with a resolution of 1 kilometre or less. From Nelson *et al.* (2011).

their specific habitat preferences. Following an examination of a small sample of museum and field-collected specimens from Britain and Ireland, Brian Nelson, Maurice Hughes and Ken Bond concluded that the distribution of the two species did not overlap in Ireland (Nelson *et al.*, 2011). The Wood White is restricted to the karst areas of Co. Clare and Co. Galway, whereas Réal's Wood White is widespread in Ireland and a habitat generalist, but is not found in the Burren or other karst areas in Clare or Galway. The probable reason why the Wood White is restricted to the Burren and other karst areas is that it is xerothermic, adapted to relatively warm and dry conditions that are provided by the microclimate of these areas. The species is one of the most range-restricted butterflies in Ireland. Its larvae feed on Meadow Vetchling (*Lathryus pratensis*), Bird's-foot Trefoil and vetches (*Vicia* spp.), all of which occur commonly in the Burren (Dublin Naturalists' Field Club, 2017).

It is moths, however, that are the real lepidopteran stars of the Burren. The Transparent Burnet (*Zygaena purpuralis sabulosa*) is one of the more conspicuous

Burren moths, with almost transparent blackish wings that bear red patches. Within Ireland, it is largely confined to the Burren, with extremely localised populations on limestone outcrops in Co. Limerick and beside Lough Corrib and Lough Mask. In Britain, another subspecies is known only from the Inner Hebrides and a few sites on the west coast of mainland Scotland, and from old records (up to 1962) in north Wales. The larvae in the Burren feed on the abundant Wild Thyme.

The discovery of the Burren Green (*Calamia tridens*) near Ballyvaghan in August 1949 came as a major surprise, as the moth had never before been recorded in Ireland or Britain. Was this moth a stray, a migrant or a resident? Eggs and larvae were found in 1950, establishing its Burren residency. The larvae feed principally on Blue Moor-grass, the dominant grass over most of the Burren. The Burren moth differs from the nominate Continental race and is now recognised as a separate geographical subspecies, subsequently named *C. t. occidentalis* (Cockayne, 1954). The European form is widely spread over the Continent; a single individual trapped in 2014 in Essex was initially mistakenly identified as the Burren Green, gaining some fame as a result (Urquhart, 2015).

The Irish Annulet (*Gnophos dumetata*) is found only as a resident in a limited area of the Burren and nowhere else in Ireland or Britain. The Irish subspecies is *G. d. hibernica*, distinct from the nominate race, which is found in central and southern Europe. Eggs and larvae were found in the summer of 1992, and a follow-up study by Brian Elliott and Bernard Skinner in 1995 established that the Irish Annulet was a Burren resident and not a migrant. The larvae feed on Buckthorn (*Rhamnus cathartica*), which is locally frequent on limestone pavement (Nelson *et al.*, 2011).

FIG 108. The Irish subspecies of the Annulet (*Gnophos dumetata* ssp. *hibernica*), August. (Philip Strickland)

A study of the nearest Continental subspecies of the Irish Annulet (of which there are a number), showed that ssp. *hibernica* is closest to ssp. *margaritatus* from Aragon in Spain (Elliott & Skinner, 1995), although there is some doubt about this. In considering the Irish moth's likely origin, Elliott and Skinner presumed it was another interglacial relict from the late Pleistocene and evidence of a lingering Mediterranean fauna in the Burren, although this is refuted by indications that the vegetation died out at this time, as discussed above. In its relationships, the Irish Annulet resembles the Burren Green, whose closest affinities appear to be with specimens from the Iberian Peninsula and not to those in nearby mainland Europe, according to Cockayne (1954). Elliott and Skinner wondered what other relict species, as yet undiscovered, were on the west coast of Ireland.

Another moth that was assumed to be a glacial relict, this time by the eminent entomologist and geneticist Edmund Brisco Ford (1901–88), is the Least Minor (*Photedes captiuncula*); the subspecies *P. c. tincta* is brighter and more strongly marked than the nominate race, and is found only in the Burren and on similar limestone areas in Co. Galway and at Lough Mask in Co. Mayo (Ford, 1955). One specimen has also been found outside the Burren, at Carnoneen in east Co. Galway, but in the same habitat (Ken Bond, pers. comm., 2016, 2017). Ford believed, over a half century ago, that a thorough investigation of the moths of the Burren was clearly necessary. How right he was! This was the inspiration for a series of visits to the Burren by John Bradley, Teddy Pelham-Clinton and Robin Mere (1909–67), in which many species of Lepidoptera new to the Burren, or to Ireland, were discovered – most of their records are included in a comprehensive paper (Bradley & Pelham-Clinton, 1967).

FIG 109. The local subspecies of the Least Minor (*Photedes captiuncula* ssp. *tincta*), August. (Philip Strickland)

Lepidoptera are well known for their genetic plasticity, developing varieties or subspecies according to different environmental or geographic circumstances, and providing a field day for taxonomic splitters. While these varieties are interesting in themselves, they are not of such startling biogeographic interest, apart from probably demonstrating evolution in action. However, when there appear to be more than the usual number found in a relatively restricted area such as the Burren, questions arise as to the reasons for their concentration here. Does it reflect the intensity of lepidopteran collectors or, perhaps, the unusual ecological conditions of the Burren?

Over the years, lepidopterists visiting the Burren have identified Irish subspecies of a further five moths (the nomenclature in *An Annotated Checklist of the Irish Butterflies and Moths (Lepidoptera)* by Ken Bond and James O'Connor (2012) has been followed):

1. Latticed Heath (*Chiasmia clathrata hugginsi*).
2. Straw Belle (*Aspitates gilvaria burrenensis*).

FIG 110. Latticed Heath (*Chiasmia clathrata hugginsi*), June. (Philip Strickland)

3. The plume moth *Merrifielda tridactyla* (as *Alucita icterodactyla phillipsi*).
4. Common Pug (*Eupithecia vulgata clarensis*).
5. Mere's Pug, a form of Freyer's Pug (*Eupithecia intricata hibernica*).

Among butterflies, the Irish subspecies of the Dingy Skipper (*Erynnis tages baynesi*) and Grayling (*Hipparchia semele clarensis*) also occur in the Burren.

In addition to the five subspecies listed above and the local races of the Burren Green, Irish Annulet and Least Minor, there is a group of nine moths that either are restricted to the Burren or have been recorded only a few times outside the area (Allen *et al.*, 2016):

1. Royal Mantle (*Catarhoe cuculata*).
2. Heath Rivulet (*Perizoma minorata*).
3. Violet Cosmet (*Pancalia schwarzella*). Nearly all records for this rare species are from the Burren. A single specimen was recorded at the Crabtree Reserve, Lullybeg, Co. Kildare, in 2014.
4. Pale Twist (*Clepsis rurinana*). There have been only eight records of this moth in Ireland, all but one occurring in the Burren. The non-Burren record was from the Crabtree Reserve in Lullybeg, Co. Kildare.
5. Buff Button (*Acleris permutana*). This rare species is found only in the Burren in Ireland, where it has been recorded from just two 10 kilometre squares.
6. Reddish Light Arches (*Apamea sublustris*). Restricted to the Burren and some coastal sites between Co. Louth and Co. Wexford.
7. Small Argent and Sable (*Epirrhoe tristata*). Found principally in the Burren, with six other scattered records in Ireland (2000–12).
8. Tissue (*Triphosa dubitata*). Found principally and widely in the Burren and in 10 other scattered sites throughout Ireland (2000–12).
9. Scarce Crimson and Gold (*Pyrausta sanguinalis*). Common in the Burren, but present in very few other Irish sites. It is found on one site in the Isle of Man but is now extinct in Britain.

The micro moth *Hypercallia citrinalis* is also restricted to the Burren in Ireland and Britain, like the Burren Green. This species used to occur in a colony in Kent,

OPPOSITE: **FIG 111.** a. The plume moth *Merrifieldia tridactyla*, June. b. Common Pug (*Eupithecia vulgata* ssp. *clarensis*), June. c. Violet Cosmet (*Pancalia schwarzella*), May. d. Pale Twist (*Clepsis rurinana*), August. e. Buff Button (*Acleris permutana*), August. f. Reddish Light Arches (*Apamea sublustris*), June. g. Tissue (*Triphosa dubitata*), August. h. Scarce Crimson and Gold (*Pyrausta sanguinalis*), July. (all photographs by Philip Strickland)

FIG 112. The stunning micro moth *Hypercallia citrinalis*, seen here in July, is confined to the Burren. Its larvae feed on Common Milkwort (*Polygala vulgaris*) and Heath Milkwort (*P. serpyllifolia*), both locally common plants. (Philip Strickland)

England, until about 1975, and previous to that in Durham and Essex. Elsewhere, it is now found only in mainland Europe, Asia Minor and central Asia. The reasons for its occurrence in the Burren are unknown. Here, it is found typically in dry calcareous grassland, where the larval foodplant is Common Milkwort (*Polygala vulgaris*).

Odonata

Dragonflies and damselflies require water, so most are found in the south-east of the Burren where the permanent waterbodies occur together with some turloughs. In a brief review published 18 years ago, Brian Nelson pointed out that nearly half of all the known sites in Ireland for the Robust Spreadwing or Scarce Emerald Damselfly (*Lestes dryas*) are found in north Co. Clare, where the best-recorded site is at Ballyvelaghan Lough in Burren village (Nelson, 2000). Although numbers may be reduced here (none were seen in 2010; Nelson *et al*, 2011). Recent increases have been noted in the south-east Burren (Cham *et al*, 2014). In Britain, it is restricted to two small areas in the Norfolk Breckland and in south Essex and north Kent. It was formerly more widespread, although limited in its distribution. Another damselfly, the Ruddy Darter (*Sympetrum sanguineum*), can also be found at most Burren turloughs. The Burren's permanent lakes, such as Lough Bunny and Loch Gealáin, support a wider range of species, including the Variable Damselfly (*Coenagrion pulchellum*), Black-tailed Skimmer (*Orthetrum cancellatum*) and Hairy Dragonfly (*Brachytron pratense*) (Chapter 10).

Other invertebrates

Relatively few invertebrate specialists have worked in the Burren, but their discoveries – quite often made on fleeting visits – illustrate the interest of the area and may be assumed to be only the tip of the iceberg.

Insects

Three insects new to Ireland and Britain were found in the limestone grikes near Fermoyle, Fanore, in 1959 (Richards, 1961). Two of these were of Lusitanian origin: the leaf-hopper *Rhopalopyx monticola*, elsewhere found only in the Pyrenees; and the weevil *Apion dentirostre* (now a synonym of *Apion* (*Ceratapion*) *carduorum*), which when found was known only from southern Spain but has since been recorded from Britain and other European countries. The third was a dipteran flesh fly, *Sarcophaga soror*, not found in Britain but occurring in central Europe, Sweden, Denmark and Finland.

A collection of weevils and other insects gathered from 10 sites in the Burren in June 1965 turned up another weevil, *Otiorhynchus arcticus*, elsewhere known only from Scotland and Scandinavia (Morris, 1967). It has also been found as a late Pleistocene fossil in Britain, at Upton Warren in Worcestershire (Fossilworks, 2017). The weevil is the most abundant and widespread of all beetles in southern Greenland, and in Europe it can be said to have a low arctic–boreal–alpine distribution (Böcher *et al.*, 2015). Another weevil, also new to Ireland, is *Bagous brevis*, which was discovered in some moss at the fen at Connell's Ford, near Mullagh More, in June 1970 (Morris, 1985). It was recorded more recently by Julian Reynolds at Lough Gealáin, Mullagh More, in 2007 (Foster *et al.*, 2009). The geographical range of this weevil extends throughout Europe, from Britain to Germany, Denmark, Norway, Sweden, Finland and northern France, but it is very rare in these countries.

The freshwater beetle *Ochthebius nilssoni* was first discovered at Lough Briskeen, south-east Co. Galway, in August 2006, and later found in two marl lakes – Lough Cooloorta and Lough Gealáin, Mullagh More – in the Burren. The species was new to Ireland, and is absent from Britain, Europe and the rest of the world apart from a few lakes in Västerbotten province, northern Sweden, and an unconfirmed record from the Alps. It is considered to be a glacial relict in Ireland that was probably more widespread in the late-glacial or early-postglacial periods (O'Callaghan *et al.*, 2009). It has been estimated that the Irish and Swedish populations have been isolated from one another for at least 12,000 years (Wikman, 2010).

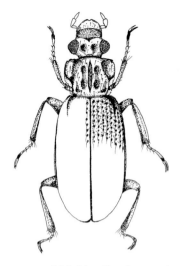

The Wasp Hoverfly (*Doros profuges*) resembles a wasp, especially the solitary wasp *Argogorytes mystaceus*, but has only two wings, not four. In Ireland it is almost exclusively confined to the Burren, but it also occurs in Britain and most western European countries. The Burren provides the ideal habitat of calcareous grasslands for the insect because of its association with the black ant *Lasius fuliginosus*. The latter species 'farms' aphids for their sweet secretions and the hoverfly larvae eat the aphids.

Crustaceans and echinoids

The fairy shrimp *Tanymastix stagnalis* was originally discovered in 1974 in Rahasane Turlough, Co. Galway, 19 kilometres north-east of the Burren (Young, 1976). It was then found two years later in a pool

FIG 113. *Ochthebius nilssoni,* measuring less than 2 mm in length. From Wilkman (2010).

south-east of Slievecarran and has now been recorded from six other locations in Ireland, all of them shallow pools that dry out in the summer. The animal survives as resting eggs that hatch when wet conditions return, and goes through a series of stages before it becomes a fully formed adult shrimp. *Tanymastix stagnalis* is absent from Britain but is widespread throughout Europe, from the Iberian Peninsula to south-west Russia; Sweden and Norway are the northerly limit. One theory for the species' presence in Ireland is that it was transported

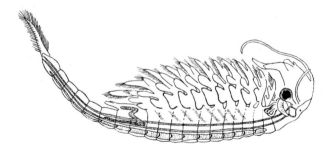

FIG 114. The fairy shrimp *Tanymastix stagnalis,* the only species of Anostraca in Ireland, may reach up to 20 mm in some locations, and has stalked eyes and 11 pairs of bristly, flattened thoracic appendages. It swims upside down and filters food particles from the water.

here on the feet of migratory waterfowl or waders from Europe. But why then is it absent from Britain and other locations in Ireland that are visited by the same birds? Could it instead be a glacial relict that survived the last glaciation? This theory was suggested for populations of the shrimp in Norway recently, and the authors concluded that the species probably survived at least the last glaciation on nunataks or coastal refuges in central north-western Norway (Arukwe & Langeland, 2013). This may also have been the case in Ireland.

One of the more interesting larger marine invertebrates in the Burren is the Purple Sea Urchin (*Paracentrotus lividus*), which is the only member of its genus. Its centre of distribution is the Mediterranean Sea, but it also occurs from the Azores, through the Canaries to Portugal, with some records from south-west Britain and a few in scattered locations in Scotland. While common along parts of the western coasts of Ireland, the species is nowhere near as numerous and abundant as it is in the Burren, its headquarters outside the Mediterranean. Here, it flourishes in an average sea temperature of 9.7 °C in winter, whereas in the southern part of its distribution the range is 10–15 °C. Under favourable

FIG 115. Purple Sea Urchin (*Paracentrotus lividus*), removed from its protective hollow in a tidal pool. Poulsallagh, May. (Fiona Guinness)

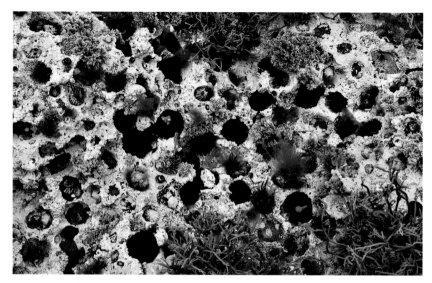

FIG 116. Intertidal rock pool, with Purple Sea Urchins (*Paracentrotus lividus*) in protective hollows they have burrowed out. Note that not all hollows are occupied by the sea urchins. Poulsallagh, May. (David Cabot)

FIG 117. Typical biokarst intertidal rocky coastline near Black Head, along which numerous rock pools harbour the Purple Sea Urchin (*Paracentrotus lividus*). (David Cabot)

circumstances, the animal's shell, or test, grows to a diameter of 7 cm when it is mature and sports sharply pointed purple spines. The specimens in the Burren, however, are much smaller, more in the range of 4–5 cm diameter, probably due the colder conditions in which they live. The largest specimens of Purple Sea Urchin are found in the warm waters of the Mediterranean and Aegean seas, where they live scattered in the open in lower densities in shallow waters. They appear gigantic in comparison to their considerably smaller Burren cousins.

In the Burren, Purple Sea Urchins are most numerous in the coastal strip between Black Head and Poulsallagh, where they live in excavated circular hollows in the soft limestone rock. Up to 100 individuals can be packed into a small-sized rock pool that is exposed during low tides. Their hollowed-out homes protect them from the often severe impact of waves, as well as from predators – mainly starfish and gulls. They are one of the principal agents in the formation of the jagged biokarst that characterises this part of the coastline. As the urchin grows in size, it continues to enlarge its protective hollow, using its mouth and spines as instruments of excavation.

Molluscs
One of the most curious molluscs found in the Burren is the Land Winkle (*Pomatias elegans*; Fig. 119). It is small – only 11–15 mm in length at maturity – and closely related to marine periwinkles. In Ireland, the species is found only in the Burren, where a colony was discovered in 1976 on the limestone pavements just south of the Flaggy Shore near New Quay (Platts, 1977). In Britain, the winkle is uncommon on calcareous soils in southern England, with a few scattered colonies as far north as Lincolnshire and Yorkshire. And in mainland Europe, it is common in the south (Fig. 120). It is thought to be temperature-sensitive, apparently limited by a January isotherm of about 2 °C.

The origin of the Land Winkle in the Burren 'can only be by conjecture' (Platts, 1977). The ecological conditions for its survival include calcareous soils and the absence of winter frosts. Ideal microhabitats include grikes, which provide shelter and high humidity, and rocks scattered on limestone pavements, with intermittent patches of vegetation that are used for shelter.

In a follow-up survey of the extent and abundance of the Flaggy Shore population in 2000, live specimens were found in an area of approximately 21 hectares of limestone pavement between the coastal road and a low hill (Platts *et al.*, 2003). The minimum size of the population was calculated at 247,000 individuals using a mark-recapture programme, so the species is by no means rare.

The species was introduced to Ireland on several occasions by experimenting Victorian naturalists, without apparent success. One wonders

FIG 118. Limestone pavement south of the road running along the Flaggy Shore, habitat of the Land Winkle (*Pomatias elegans*). (David Cabot)

FIG 119. A dead Land Winkle (*Pomatias elegans*) found on the limestone karst south of the road along the Flaggy Shore. (Fiona Guinness)

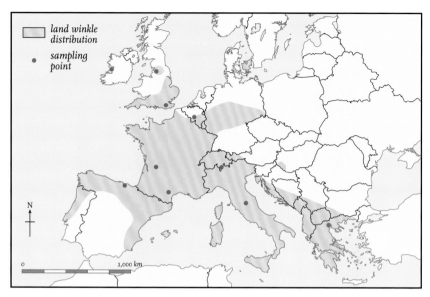

FIG 120. European distribution of the Land Winkle (*Pomatias elegans*; shaded areas), including sampling points used in a genetic and morphological study by Jordaens *et al.* (2001).

about the motivations for such introductions – the animal offers insignificant food rewards compared to marine periwinkles and is of novelty value only. The Irish naturalist William Thompson (1805–52) initially considered that the species had been introduced to Co. Sligo, but later changed his mind (Thompson, 1840, 1842). However, no live Land Winkles have been found in the county since then.

Whether the Burren Land Winkles were introduced or native can be examined by genetic comparison with other populations. This was done in a 2001 study in the context of conserving distinct groupings in Europe that have few links to their neighbours (Jordaens *et al.*, 2001). While there were genetic differences through the range, these were not coupled with any structural changes that could prove isolation for a long enough period to be indicative of a native origin. The results for the Irish population were inconclusive, so the question as to its origin remains open. The lack of fossil evidence and the population's extreme isolation in the west of Ireland may indicate that the winkle has a recent origin.

Other Burren molluscs of particular interest include the estuarine snail *Hydrobia acuta neglecta*, which is found on Inishmore in the Aran Islands, and at one site on Strangford Lough in Co. Down and another on the Mullet in Co.

Mayo. The marine Ear Snail (*Otina ovata*) is extremely localised in Ireland, mostly confined to the Burren but with seven other scattered sites, while the Looping Snail (*Truncatella subcylindrica*) has its only Irish stations on the northern gravelly foreshores of the Burren, extending east into Galway Bay (Byren *et al.*, 2009).

CHAPTER 6

Limestone Pavement

W HEN THEY HEAR THE WORD 'BURREN', most people think of
limestone pavement. They recall the coast road at Poulsallagh or at
Black Head, the lowlands of Coolorta and Roo, or the layered hills
around Mullagh More, without knowing that such bare rock covers only about
20 per cent of the total surface area; the rest is grassland, scrub and heath. However,
it is the pavement that stands out as the rarest and most distinctive habitat.

FIG 121. Limestone pavement and grassland between Black Head and Poulsallagh.
(Steve O'Reilly)

FIG 122. Limestone pavement with the Aran Islands in the distance. (Fiona Guinness)

FIG 123. Clints, grikes and limestone glacial erratics. The sharp edges to the rocks imply 'recent' exposure. (Fiona Guinness)

FIG 124. Well-worn limestone with runnels and solution features, Deelin More. The rounded edges to the rocks imply exposure over a long period. (Fiona Guinness)

FIG 125. Wet solution hollow in limestone pavement with many lichen colonies on the vertical rock surfaces. (David Cabot)

The pavement consists of bare rock (clints) and the joints or cracks between them (grikes), both of them an integral part of the environment. Plants may be restricted to one or the other, but more often they are rooted in the more congenial conditions of the grikes, spreading out onto the surface to a greater or lesser extent. Animal life alternates between the two, taking cover in the cooler conditions of the cracks and venturing out when it is safe to do so. Wood Mouse (*Apodemus sylvaticus*) heaven is found in the interlocking grikes, with their buds and bark, occasional falls of Hazel (*Corylus avellana*) nuts and perhaps a caterpillar or bristletail.

CLIMATE AND MICROCLIMATE

The environment on the open rock surface of the Burren is difficult for plant life. There is rapid wetting and drying year-round, along with almost constant wind, run-off by water and occasional damage by hooves or feet. In summer, there may be no rain for 30 days and a temperature difference each day of up to 15 °C between clints and grikes. Studies have shown that when the air temperature above the rock reaches 25 °C, as it can do occasionally in May or June, the grikes remain at 15 °C while the actual rock surface is at 33–37 °C depending on the angle to the sun. A black lichen (*Verrucaria* sp.) on the surface has to withstand 44 °C, while a white one, by reflecting the heat, reaches only 27 °C (McCarthy, 1981).

These extremes are rare, however, and the feature of the climate that stands out is its oceanicity. The mean January temperature of the coastal regions is 6–7 °C and that of July 15 °C (Met Éireann, 2018). The wind speed averages at 22–25 kilometres per hour, while the annual rainfall is about 1,600 mm, although it varies between 1,100 mm and 2,100 mm. Mean rainfall for December and January normally totals 356 mm, but in the winter of 1994/95 the Burren experienced 642 mm in these months, resulting in extensive flooding in the lowlands. The number of rain days is particularly important to vegetation on limestone pavement, as run-off and soil drainage take away much of what falls. In winter, the number of days per month with 1 mm or more of rain is 21, whereas in summer it is 17.

LIFE ON THE ROCKS

There are no major nutrients in pure limestone except calcium; nitrogen is lacking entirely and phosphate remains unavailable because of the high alkalinity. But life here does not have to wait for physical weathering to break down the limestone. Anywhere one sees rock in the Burren, one sees cyanobacteria, single-celled or minute filamentous organisms that used to be called blue-green algae but are in fact much more primitive than algae. They had a long heyday 2,500–570 mya and have come down to us almost unchanged. Having evolved at a time of higher radiation levels, when genetic material was vulnerable to mutation, they have a gene-repair system second to none but no mechanisms for sexual reproduction (and thereby evolution) (Potts, 2000). This endows the group with longevity, but means they can no longer adapt to changing conditions. They have (probably) seen it all before!

Walking over the pavement after rain, one soon finds colonies of *Nostoc* in the kamenitzas (solution hollows). These are the largest of the cyanobacteria, forming colonies that are brown like seaweed but consist of gelatinous balls and blobs a few centimetres across. They are not restricted to bedrock, and are sometimes seen on limestone-pebble paths in gardens and on felted roofs. The gelatinous blobs were described in Europe 500 years ago, when they were believed to be something magical that fell from the sky – they appear rapidly after rain, swelling from blackened crusts. At this time, before the advent of fertilisers, colonies were more common in fields and grassland. The genus name is derived from the German and Old English words for nostril, owing to the resemblance of the bacterial colonies to snot or, to be technical, 'extracellular polysaccharide' (Potts, 2000). *Nostoc commune* is a form common throughout the world in hot and cold climates from Spitzbergen to Ecuador; it is even eaten in some places. It is tolerant of frequent drying and wetting, and can dry out completely in five hours.

FIG 126. *Nostoc* cyanobacteria in active growth, along with a scatter of colonising sedges. (Roger Goodwillie)

In this state it is black and twisted, resembling humus, and can last a long time before breaking its dormancy – the record is currently 87 years for a museum specimen.

Many cyanobacteria, including *Nostoc*, fix nitrogen from the air, and so can live in very poor soils or on rock. Some also get trapped by fungal hyphae and, together, they grow into lichens. These cyanobacteria are taken in as single cells rather than filaments, so that *Nostoc*, if involved, does not get a chance to form its independent colonies. The first colonising lichens are minute and so sunken into the rock that they are invisible until they reproduce, shooting their spores through tiny holes or pores. These are endolithic lichens (i.e. inside the rock; there are also endolithic cyanobacteria) and they have a pioneer role in creating minute hollows and features that other plants can then take advantage of. Any cracks and ridges allow mature lichen bodies to develop, when they can more easily be identified. In his 1981 ecological study in coastal western Ireland, Pat McCarthy found species of *Verrucaria*, *Petractis* and *Arthopyrenia* lichen to be often the most abundant at the outset, and that there are basically six species on a flat clint surface. Once a little roughness appears, larger lichens may gain a foothold on the surface (epilithic) and become much more noticeable – perhaps a *Lepraria* or *Acrocordia*. Other species choose the higher-moisture surroundings of runnels in the rock or in the grikes between them. In these cases, the photosynthetic partner to the fungus is more often an alga rather than a cyanobacterium, especially in the yellow or orange *Trentepohlia* that covers natural rocks and tree bark all over Ireland, as well as concrete walls and motorway barriers.

The stone walls that cross the limestone pavements in the Burren add considerably to the amount of lichen habitat available, but in general they carry

FIG 127. Encrusting *Caloplaca* lichen on limestone, Black Head. (David Cabot)

FIG 128. The maritime *Ramalina siliquosa* lichen on a granite erratic. (David Cabot)

the same species as the clints. Siliceous boulders found as erratics or where chert beds break the surface add a hundred or so lichen species to the Burren's total of 349, whereas boulders used as bird perches contribute 50 more. The orange *Xanthoria* and *Caloplaca* are particularly noticeable here.

Algae are more palatable than lichens, and one of the larger grazers one meets on limestone pavements is the bristletail *Petrobius brevistylus*, a 30 mm-long cylinder of an insect with three tails behind and two long antennae in front. It is a member of the Archaeognatha, one of the most primitive groups of insects and a relative of the silverfish (*Zygentoma* spp.) seen in (some) houses. Bristletails scurry, making sudden runs followed by inactivity as they feed or watch for danger. One noticeable feature is their large eyes, which practically meet on top of their head, but these are only for self-preservation; the insects do not hunt but instead feed on green algae that they scrape off the rock surfaces, taking cover in grikes or under flakes of rock when threatened. For an insect, they lose water rapidly, and so have evolved special membranes on their underside to absorb it directly from the surface. They therefore tend to live in damp surroundings, although they do come out to warm up in the sunshine.

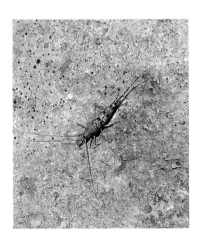

FIG 129. A bristletail (*Petrobius* sp.), on lichens. The Flaggy Shore, May. (Fiona Guinness)

Lichens decay slowly and are eaten by relatively few invertebrates because they contain antimicrobial and antiherbivory chemicals manufactured by the fungal symbiont. These phenolic compounds also seem to prevent the germination of moss spores on the lichen thalli, which would otherwise pose a shading risk to them (Lawrey, 1986). Some lichens are, however, grazed by caterpillars such as the Dew Moth (*Setina irrorella*) and by snails. The white lichen *Aspicilia calcarea*, characteristic of limestone surfaces, is taken by several species, as are some of the terrestrial *Verrucaria* species. Different types of snail have also been found to concentrate on different lichen species, partly because of their mouth structure (Froberg *et al.*, 1993). Typical small snails seen on the Burren's limestone pavements and under flakes of rock are *Pyramidula* species: *P. umbilicata*, which has a Lusitanian distribution from Spain and Portugal to France, Britain and Ireland; and *P. pusilla*. These are tiny species, measuring 2–3 mm, smaller than the hairy snail *Trochulus hispidus* and the plainer *Vitraea* species.

DEVELOPMENT OF VEGETATION

Kamenitzas have long fascinated people, and papers and theses have been written about their creation, evolution and micro-ecology. No particular cyanobacterium or lichen has been singled out as responsible for their initial formation; rather, it seems that slight differences in the rock structure allow the first raindrops to linger, so that their full burden of carbonic acid can act on the rock and cause deepening or, later, expansion. There is usually a thick growth of cyanobacteria in the hollow, many of which fix nitrogen. A 2014 study found a high content of the cyanobacterial pigment scytonemin in the black material at the base of the hollows, as well as a considerable amount of mineral matter (53 per cent of total dry mass), presumably blown in as dust (Doddy & Roden, 2014). One pool in 20 contained a pink rather than a black deposit, which was dominated by the green alga *Haematococcus pluvialis*, implying nutrient enrichment.

A *Collema* lichen is often found in the centre of the hollow, with other lichens on drier ridges and bumps around it (McCarthy 1981). In fact, six different species are prominent. All these organisms produce carbonic acid from respiration at night. When they decay, their dead parts form humic acids, and these further drive the deepening and evolution of the hollow. The acids also release minute quantities of phosphate from the limestone, which is often the limiting element for other life.

Once *Nostoc* and lichen species are growing strongly in a solution hollow, there is a gradual accumulation of organic material. The first higher plant to arrive is often Jointed Rush (*Juncus articulatus*), which is happy to be in water for a large part of its life. The later succession of species depends on where the solution hollow is. If it is beside the sea, perhaps at Poulsallagh, maritime species are available to colonise it and can withstand occasional dowsings by wave spray. Sea Plantain (*Plantago maritima*) is often found, as are Glaucous Sedge (*Carex flacca*), Distant Sedge (*C. distans*), Thrift (*Armeria maritima*) and Red Fescue (*Festuca*

ABOVE: **FIG 130.** Thrift (*Armeria maritima*) and Common Scurvygrass (*Cochlearia officinalis*) growing in limestone hollows. New Quay, May. (Fiona Guinness)

FIG 131. Colonisation of a solution hollow by Glaucous Sedge (*Carex flacca*) and Mountain Everlasting (*Antennaria dioica*), May. (Fiona Guinness)

rubra). It is not known how long it takes for a solution hollow to develop and grow vegetation, but a period of several hundred years would be required before the results of differential weathering in the rock are seen. Seaside and intertidal hollows are wetter and so erode faster.

Away from the sea, the colonisation of a kamenitza can proceed differently, and other sedges are usually present, especially Flea Sedge (*Carex pulicaris*). Grey Willows (*Salix cinerea*) also arrive by wind, although their tenure is limited by occasional drought years. As the 'soil' dries out, *Nostoc* is replaced by mosses, especially the large *Neckera crispa*, which grows among the grassy cover of Blue Moor-grass (*Sesleria caerulea*) or Purple Moor-grass (*Molinia caerulea*), Devil's-bit Scabious (*Succisa pratensis*) and Wild Thyme (*Thymus polytrichus*) (Ivimey-Cook, 1965). The low-nutrient conditions also allow Heather (*Calluna vulgaris*) and Tormentil (*Potentilla erecta*) into the stand, and sometimes Common Butterwort (*Pinguicula vulgaris*) or Black Bog-rush (*Schoenus nigricans*). One of the Burren's speciality mosses, *Breutelia chrysocoma*, is often prominent; it is also found in bogs and moorland all over the west coasts of Ireland and Britain. One can already see the parallels between kamenitzas and moorland: decay is slow in both and peat can build up to form pockets for further plant colonisation.

The tiny semi-aquatic kamenitza habitats are scattered across the open limestone pavements and seem like little islands of hospitality in the desert of dry rock. However, this is misleading; for much of the year there is abundant moisture, as would be expected on the west coast of Ireland (or Britain). As we have seen, rain falls on half the days of summer and two-thirds those of winter, but fog and cloud occur frequently. The trouble is that the water drains away from the open pavement in the absence of a soil cover, and the plant species that succeed are likely to have a definite adaptation. Some resist drought where they are, like the only succulent in the flora, Biting Stonecrop (*Sedum acre*). (White Stonecrop, *Sedum album*, also occurs but is frowned upon for being an introduction, despite having quite a wide distribution.) Other plants grow when the substrate is damp but shrivel in drought, or they grow long roots to reach water. Still others have an annual lifestyle that sees them as seeds in dry conditions. These winter annuals include Whitlow Grass (*Erophila verna*) and Rue-leaved Saxifrage (*Saxifraga tridactylites*), ubiquitous tiny plants that germinate in autumn and flower in February and March. The saxifrage is more conspicuous, as its leaves are often red when exposed to the sun. It also persists later into the year as leafless stems holding its few seed capsules aloft. A summer annual is seen in a few solution hollows; this is the Mudwort (*Limosella aquatica*), which occurs at Poulsallagh, away from its more usual habitats of turlough and lakeshore.

FIG 132. The annual Rue-leaved Saxifrage (*Saxifraga tridactylites*) at the end of the flowering season, May. (Fiona Guinness)

The tendency of plants to grow during the autumn and winter months is pronounced in every community in the Burren. A visit in September or October, when the rains have come, reveals new shoots of all sorts extending over the rock surfaces. Bird's-foot Trefoil (*Lotus corniculatus*), Wild Thyme and the Hoary Rock-rose (*Helianthemum oelandicum*) produce a tracery of tiny leaves that will survive the winter and allow rapid growth the following year. Ivy (*Hedera helix* and *H. hibernica*) extend their shoots even as they are flowering, while Wild Madder (*Rubia peregrina*) sprouts from lower down the stem. This growth is encouraged by the relatively warm temperatures in winter, which is the main reason why Mediterranean species do so well here. Only in January and February do average soil temperatures fall below 5 °C (the point at which growth effectively stops), although there are many days at this time when the daytime surface temperature is above this.

FIG 133. Mudwort (*Limosella aquatica*) in a solution hollow, benefitting from the nitrogen-fixing ability of *Nostoc* cyanobacteria, September. (Roger Goodwillie)

FIG 134. Common Ivy (*Hedera* sp.) flourishing in the absence of goats. The plant is an Atlantic species that would spread widely across the Burren if grazing were to cease. (Fiona Guinness)

PLANT SPECIES OF LIMESTONE PAVEMENT

The commonest plant on limestone pavement is Herb Robert (*Geranium robertianum*). Although it does not cover the greatest area, it is ubiquitous – you can almost guarantee that there will be Herb Robert within 5 metres of any spot, like rats in a city. It says a lot about the human brain that no one ever commented on this until a study was made of the actual vegetation of limestone pavement in 2013 by Sue Wilson and Fernando Fernández. They classified pavement into five types based on the vegetation. The types share about 40 species, including mosses, so are difficult to distinguish in the field without carrying out finer analysis. Each is named for so-called 'faithful' species that are less common than most and are assumed to require distinct environmental conditions.

Herb Robert is so frequent on limestone pavement that it disappears from notice when one is looking at its showy neighbours. It is not showy itself – a humble pink flower is carried over hairy, dissected leaves – and is perhaps most noticeable in autumn, when odd leaves turn such a bright red that they have to be seen. It is normally a plant of hedges, walls and woodland, and is one of the

FIG 135. Herb Robert (*Geranium robertianum*), a ubiquitous plant in limestone pavement, May. (David Cabot)

BELOW: **FIG 136.** Primrose (*Primula vulgaris*) grows through much of the pavement and grassland, in the open or in shade, May. (Fiona Guinness)

woodland coterie that is so prevalent in the Burren, including Primrose (*Primula vulgaris*), Early-purple Orchid (*Orchis mascula*), Broad-leaved Willowherb (*Epilobium montanum*), Wood Sage (*Teucrium scorodonia*) and Tutsan (*Hypericum androsaemum*).

In terms of frequency on the pavement, Herb Robert is followed by Blue Moor-grass, Wall-rue (*Asplenium ruta-muraria*), Wood Sage and *Tortella tortuosa*, a cushion moss whose sickle-like leaves are curved in several different directions, justifying its specific name. Two other mosses are also easily seen: the tiny curving *Ctenidium molluscum* and the large, crinkle-leaved *Neckera crispa*.

FIG 137. Blue Moor-grass (*Sesleria caerulea*), the most successful grass on limestone, in flower in May. (Fiona Guinness)

Blue Moor-grass is one of the most distinctive plants of the open pavement, although it does spread into grassland habitats (Chapter 7). It is a prime example of the fact that not all grasses look the same: it has folded leaves with boat-shaped tips and a stubby bluish flower head that appears early, when spring is still a distant hope. The plant does not tolerate bad drainage or competition, so limestone pavement suits it well. It has a large root system – four or five times larger than most ordinary grasses – and in the Alps and elsewhere it stabilises screes. It can also cope with the very low fertility found in a highly calcareous soil where phosphate is scarcely available. The early flowering habit certainly suits it in limestone pavement, where summer drought is a significant factor and the soil may hold only two days' supply of water. The grass withstands a level of drought that would damage most species, its leaves folding inwards when conditions are stressful.

FIG 138. Purple form of Wild Thyme (*Thymus polytrichus*) growing near the sea in May. It is frequent everywhere in pavement and grassland. (Fiona Guinness)

FIG 139. The unmistakable flowers of Bloody Crane's-bill (*Geranium sanguineum*) appear through spring and summer. May. (Fiona Guinness)

It is hard to go anywhere on the pavement without seeing Bloody Crane's-bill (*Geranium sanguineum*). It needs high light levels, flowering little in shade, and its large flowers attract insects into the inhospitable wastes of rock. It is one of those plants that everyone sees and is consequently widely depicted in tourist memorabilia. Its bright magenta flowers shine among the limestone from May to September. Bloody Crane's-bill is a Continental species, occurring from Portugal to southern Scandinavia and east to the Caucasus and Urals. With this climatic background, it can afford to spread its flowering over several months and makes use of a whole season of pollinating insects – hoverflies, bumblebees and butterflies all come for nectar and incidentally collect pollen. The flowers open mostly in the afternoon, at the warmest time of day. When their nectar has been sucked up by an insect, they take time to replenish it – one can watch bumblebees or hoverflies searching for a flower worth entering rather than trying them all.

FIG 140. Bloody Crane's-bill growing with Hemp-agrimony (*Eupatorium cannabinum*) and Hazel (*Corylus avellana*), May. Note the divided leaves of the crane's-bill. (Fiona Guinness)

Both of these insect groups are probably sensitive to the scent marks of others and do not waste time on flowers that bear such traces (Pearce *et al.*, 2017). Bloody Crane's-bill has evolved without the potential of sudden shocks of cold. By contrast, most of the arctic–alpine species – including Spring Gentian (*Gentiana verna*), Mountain Avens (*Dryas octopetala*) and saxifrages (*Saxifraga* spp.) – have evolved to flower immediately the snow melts. They surge into growth, saturating the landscape with flowers so that insects cannot avoid coming across them.

Wild Madder is the complete opposite of the crane's-bill. Discreet to the point of disappearance, it has tiny yellowish flowers and whorls of evergreen leaves that resemble a large Goosegrass (*Galium aparine*), to which it is related. Its dark green stems clamber and scramble over the pavement, and have backward-pointing bristles on all parts. The young shoots are often bronze in colour and show an unlikely luxuriance for a plant so close to its northern limit in Europe, although they may later become battered by wind. Wild Madder is a Mediterranean plant that is equally at home in the Holm Oak (*Quercus ilex*) forests of France or Spain as in the open pavement of Co. Clare. It needs a mild winter and is coastal in most of Britain and Ireland. Despite appearances, its flowers attract sufficient insects to make a full crop of black berries, which are

FIG 141. New shoot of Wild Madder (*Rubia peregrina*), with bronzed leaves, growing from a grike. (Fiona Guinness)

dispersed by birds. The flowers are arranged in cymes, one of the more attractive forms of symmetry in flowering plants. Each shoot ends in a flower, so side branches below develop the next flowers and further side branches the next. Wild Madder roots produce a natural dye that was used in the past when Dyer's Madder (*Rubia tinctorum*) was in short supply. It gives a pink colour rather than the brighter red of the latter, but it is debatable whether any plants were dug out of the rocks of the Burren for this purpose.

Juniper (*Juniperus communis*) drapes itself over the rock surfaces in many places, its spiny leaves folded against the stem. It is a dioecious plant, meaning that males and females are separate individuals, so only half of the bushes bear fruit – although when one is looking for berries, that proportion seems to be much smaller than half. In spring, the males release clouds of windborne pollen that blow to the female flowers; these then go on to develop green and, finally, bluish berries. The berries take two to three years to ripen, so several stages may be found on the one bush. They are dispersed by migrant thrushes such as Fieldfare (*Turdus pilaris*), Redwing (*T. iliacus*) and Mistle Thrush (*T. viscivorus*), which are seen quite frequently in autumn on the open rocks.

Juniper shows remarkable variation in its shape. There are both prostrate and erect forms, these often dictated by the environment. On the Aran Islands, the flat form predominates, but in sheltered sites in the Burren the plant may stand 3 metres tall. Many people have tried to assign these different forms to subspecies, but there seems very little genetic basis for this; the seed may just as well turn into carpeting plants or erect ones when grown in garden surroundings. In the Burren, it is often hard to tell which the plants are, because the exposure and grazing to which they are subjected makes the ground-hugging

FIG 142. Juniper (*Juniperus communis*), with full-sized berries aged two to three years. May. (Fiona Guinness)

form the most common. Regardless of the shape of the bushes, many insects are associated with the species, as with all native trees. Three species of micro moth are known to feed exclusively upon it in Ireland: Juniper Pug (*Eupithecia pusillata*), Juniper Carpet (*Thera juniperata*) and Chestnut-coloured Carpet (*T. cognata*).

When it reached the Burren in late-glacial times, Juniper was ubiquitous and its pollen so abundant that today it is used as a marker in vegetation history. Interbreeding would have occurred all over Ireland. Nowadays, however, the plant is much more scattered, occurring only locally in the western counties. A recent study found that there was considerable genetic differentiation among the different Irish populations (Cooper *et al.*, 2012). Junipers from Co. Clare and Co. Galway were found to be genetically distinct from those in Co. Donegal or Co. Kerry, and therefore on the way to forming varieties at some distant date in the future. This is a common experience for plants isolated by climate or habitat change, and the rate of genetic drift often gives some indication as to when they became separated.

PLANT SPECIES OF SHATTERED PAVEMENT

As mentioned in Chapter 3, limestone pavement can be shattered rather than remaining intact depending on the underlying lithology. Three herbaceous plants and one shrub characterise such sites: Wall Lettuce (*Mycelis muralis*), Hairy Rock-cress (*Arabis hirsuta*), Dark-red Helleborine (*Epipactis atrorubens*) and Burnet Rose (*Rosa spinosissima*). Each grows in seemingly impossible places, but their presence reveals that the environment is not always as it seems – there is water and soil at depth.

Wall Lettuce is, in fact, an introduction; it was not mentioned by the early explorers of the Burren, although records of it were scattered through Midland Ireland. It wasn't until the 1930s that it was reported from 'between Ballyvaughan and Black Head' (Praeger, 1939), and it has become widespread since then. It has tiny, pale yellow composite flowers carried in an airy panicle. The plant is frequently pigmented red, either under the leaves or on both surfaces – a feature that disappears when it grows in shade. The redness is caused by anthocyanins and may have multiple functions, including reducing insect attention and giving some photo-protection from damaging ultraviolet (UV) light (Hatier & Gould, 2008). Red leaves reflect less green and UV light, and may appear senescent to aphids and flying insects looking for somewhere to lay their eggs. The redness is a feature shared by other species such as saxifrages and Herb Robert, and even the new shoots of Hazel are red in a rather bristly way.

FIG 143. Shattered pavement with an Irish Whitebeam (*Sorbus hibernica*) seedling and Herb Robert (*Geranium robertianum*) to the left. (Fiona Guinness)

FIG 144. One of the most distinctive of the eight or so hawkweeds found in the Burren – *Hieracium hypochaeroides*, with its purple-spotted leaves. Mullagh More, May. (Roger Goodwillie)

FIG 145. Hairy Rock-cress (*Arabis hirsuta*), May. The narrow leaves extend up to the base of the flowers. (David Cabot)

How Wall Lettuce copes with its rocky home is a study in itself. It usually germinates in winter, so has the advantage of damp growing conditions when small. It also has a mycorrhizal association, a symbiotic relationship with a fungus in which the latter scavenges for nutrients and water while taking carbohydrates from the plant. The plant does the opposite, obtaining water and nutrient ions from the fungus while producing carbohydrates. This co-evolution is highly advantageous to both sides and is very common in higher plants growing in difficult habitats.

Hairy Rock-cress is not difficult to spot in the rocky wastes where it grows, but it is unexpected and gets full marks for persistence. It is a small, vertical plant with leaves that fold around the stem and a little head of white flowers at the tip. It grows from a leaf rosette that lasts for two years, and has rather characteristic forked hairs, each like a tiny catapult. Hairy Rock-cress is one of the few species in the vegetation that does not have a mycorrhizal association, a feature it shares with most of its close relatives, the crucifers or members of the cabbage family (Brassicaceae). The family is characteristic of open, nutrient-rich habitats where there is no competition for root space. It may be significant that Hairy Rock-cress can establish itself only in open habitats without existing plants as it cannot become integrated into the mycorrhizal network (Francis *et al.*, 1986).

Hairy Rock-cress self-pollinates regularly and so can form micro-species and races that differ slightly from one another if they remain isolated. It has an interesting origin, being originally formed as a hybrid between two other species of *Arabis* during the glacial period, when its parents were pushed together by the conditions and could interbreed (Titz, 1972). The resulting hybrid has double the number of chromosomes of the parental species, which may give it a greater tolerance of different conditions. The Burren plant is certainly more widespread than its putative parents and in places it grows on walls, as well as in wild habitats of sand dune and rock.

Dark-red Helleborine also grows in challenging places. Poking out of seemingly dry rocks with its two rows of leaves, it flowers in midsummer with a column of hanging rosy-purple blooms that remain rather bell-shaped. It seems to choose the most difficult environment of any Burren orchid, but it must find congenial conditions in rocky sites and is not adapted for competition in a more complete vegetation cover. Again, it has a mycorrhizal association, and this may be a selective factor that limits it to rocky places. Elsewhere in Europe it occurs in open pinewoods, so it is tempting to see it as a relic of those forests, responding to the fungal community that persists in cracks in the rock. It certainly grows on the hills where pines must have lingered the longest. The helleborine can be seen as a doughty survivor in more ways than one. If no insects visit its flowers to effect cross-pollination, three-quarters of them still form seeds. And seeds are what orchids need to produce in quantity – dust-like and requiring a fungus even to germinate, they must only ever develop to maturity in minute numbers. Another strategy of this species is to spread vegetatively via extra shoots from the long root system. This is unusual for an Irish orchid and results in a clonal population of plants, all with a similar genetic background.

Burnet Rose is a plant associated with dry limestone land and sand dunes, and is ubiquitous in the Burren, often growing among shattered pavement and loose rocks. It is well named as its stems are densely covered by prickles and bristles. It was first identified in the late sixteenth century and was called *Rosa spinosissima* by Carl Linnaeus in 1753. Six years later, he described a similar species, which differed in having smooth rather than hairy flower stalks. This he called *R. pimpinellifolia*, but that was rejected when it was eventually discovered that the two were actually the same species. The plant is adapted to its rocky habitat by suckering from the root, so that large colonies can develop, formed of leafy shoots 20–30 cm high. It can grow everywhere in the Burren since it survives pH values between 5.8 and 7.6 (Mayland-Quellhorst *et al.*, 2012). The plant is eaten by goats, but cattle are dissuaded by the thorns. A broad variety of phytophagous insects also feed on the plant, including beetles and fly and moth larvae. The moths include the Belted Beauty (*Lycia zonaria*), while two species of gall wasp make the growths that are known as robin's pincushions.

FIG 146. Burnet Rose (*Rosa spinosissima*) is often found in shattered pavement, May. (Fiona Guinness)

FIG 147. Pink-flowered variety of Burnet Rose (*Rosa spinosissima*), May. (Fiona Guinness)

Between May and July, Burnet Rose produces relatively large flowers, up to 50–70 mm across, which are generally cream-coloured. These are 'morning' flowers, shedding most of the day's pollen early and therefore being visited by insects at that time. Later, in August and September, oversized black hips develop, which are widely eaten by birds and Wood Mice.

PLANT SPECIES OF GRIKES

Conditions in the grikes are completely different from those of the open pavement. Shelter, high humidity, even temperatures and shade predominate, and, depending on the grike, there may be a soil made up of traces of loess or glacial till, enriched by leaves. When the sky is overcast, the microclimate of a grike is not that far removed from a woodland situation, although light levels are generally higher. Even in full sun, the temperature in a grike may be 10 °C or more lower and the humidity eight or ten times higher than on the surface (Dickinson *et al.*, 1964). This has obvious repercussions on plant growth and allows many woodland species to thrive regardless of the vegetation cover. Mosses grow with particular luxuriance and, seen in winter or in damp weather, they seem to blanket everything, rock and twig alike.

Among higher plants, Tutsan, Wood Sanicle (*Sanicula europaea*), Wood Sage, Common Dog-violet (*Viola riviniana*) and Wood False-brome (*Brachypodium sylvaticum*) are especially common in grikes. Hemp-agrimony (*Eupatorium cannabinum*), which is normally a plant of riverbanks and wet places, is more local and found mainly in the western half of the Burren. Purple Moor-grass and Bracken (*Pteridium aquilinum*) also favour grikes, as does the rare grass

FIG 148. Hart's-tongue ferns (*Asplenium scolopendrium*) spreading along a grike, May. (Fiona Guinness)

Wood Small-reed (*Calamagrostis epigejos*), which occurs only on the Aran Islands. The last species is one of the lesser-known puzzles of the region. How did the plant get here? It is almost unknown in the rest of Ireland, despite its broad ecological requirements elsewhere, growing from the 'water-logged littoral of fishponds to dry steppic grasslands, from deep shaded understory of coniferous forests to fully insolated clearings, and from oligotrophic sandy soils to eutrophic humus-rich habitats' (Gloser & Gloser, 1999). However, as with a few other species in the Aran Islands, it is quite common in the Inner Hebrides and the western coast of Scotland.

PLANT SPECIES OF OTHER NICHES

While limestone pavement may sound like a uniform rocky habitat, there are niches other than the open rock, the solution hollows and the grikes. These include small stony or gravelly areas formed by weathering or physical abrasion. Irish Saxifrage (*Saxifraga rosacea*) grows in such sites, and along the coast road and in the Aran Islands it creates a fine show in May and June with its cushions of red leaves and large white flowers. It is part of a complex group of species and forms from all over Europe that may have met up and interbred during the glacial period but are now more isolated. In Ireland, the plant is scattered in several mountain ranges as well as on the west coast, but it is never as common as in the Burren. In Britain, it has been found wild only on Snowdon, although it hasn't been seen there since 1978 despite later searches (Preston *et al.*, 2002). It has three- or five-pointed leaves, sometimes with a little spine on the end. When the plant is found on cliff tops, mixed with Thrift or Sea Plantain, it is hard not to see it as a relic from glacial times, driven to places that could not support a tree

FIG 149. Irish Saxifrage (*Saxifraga rosacea*) in a solution hollow on limestone pavement, May. (Fiona Guinness)

FIG 150. Dried-up plant of Arctic Sandwort (*Arenaria norvegica*) near Black Head, May. (Fiona Guinness)

cover because of exposure. Surviving there, it was later unable to spread further inland even when the trees were removed. There are equivalent sites on the west-facing cliffs of the Aran Islands, where any tree cover would have been dwarfed and open.

Such exposed sites can be seen as refugia for the glacial flora, holding other plants like Arctic Sandwort (*Arenaria norvegica*) and Pyramidal Bugle (*Ajuga pyramidalis*), which are arctic or montane plants and now occur only close to the coast road. The sandwort was famously 'lost' for 47 years, having been recorded in 1961 but not seen again until 2008 (Chapter 2). At that time, the population numbered 249 individuals and was associated with shallow depressions on the

FIG 151. Flowers of Spring Sandwort (*Minuartia verna*), May. The leaves are long and narrow. Inset: Flower detail. Note the glandular hairs on the sepals. (both photographs Fiona Guinness)

pavement holding a skeletal soil of gravel and fine stones. This habitat can be described as 'open, xerophytic conditions with high pH and low fertility', and is exactly what is found after ice has melted from a landscape (Walker *et al.*, 2013). Neighbouring plants were Sheep's Fescue (*Festuca ovina*), Common Bent (*Agrostis capillaris*) and Spring Sandwort (*Minuartia verna*), the last a species characteristic of central but not arctic Europe. Arctic Sandwort has succulent leaves and white flowers less than 1 cm across. It was first found on a path, showing that it can spread to new habitats from its original home on limestone pavement.

Hoary Rock-rose is another plant of stony areas; it has evergreen leaves and lemon-yellow flowers that open in sunshine but not in dull weather. The flowers are distinctive in that the petals open below the horizontal, giving maximum exposure to the fuzz of stamens, which move in response to bee (or pencil) contact. In the Burren, Hoary Rock-rose grows luxuriantly on the crests of rocky ridges on the coast road from Black Head to Poulsallagh, on the glacial erratics there and on the shelves of Mullagh More, where the soil is thin. It also has a station on the coast at Bell Harbour. This pattern again implies a limited distribution during the period of heavy tree cover, from which the plant is still recovering. The species was widespread in the late-glacial period, and leaves have been found in deposits dating to this time in Co. Tipperary and Co. Louth (Proctor, 1956). One can envisage a retreat to open slopes and cliffs with good drainage when the tree cover first arrived, and a slow spread since the subsequent clearance, perhaps limited by a drainage factor. The species is characteristically abundant where it occurs, but it has been found to succumb to winter damp when in cultivation (Proctor, 1956). In the field, it does not maintain itself in closed grassland or screes. Mycorrhizae are important to the species, particularly the soil fungus *Cenococcum*, which is notably drought-resistant and may be outcompeted in damper soil. It is interesting that this fungus is also common on roots of pine (*Pinus* spp.) and birch (*Betula* spp.) trees in quite acid soils.

FIG 152. Above: habitat of Hoary Rock-rose (*Helianthemum oelandicum*) on the west coast, May. (Roger Goodwillie) Left: Hoary Rock-rose is usually abundant where it occurs, May. (Fiona Guinness)

FIG 153. Details of Hoary Rock-rose (*Helianthemum oelandicum*) flowers and leaf, May. (Fiona Guinness)

Pyramidal Bugle seems even more reluctant to spread and still occurs on the site where it was discovered in 1862 – grassy ledges in the coastal cliff at Poulsallagh beside the road. It was found subsequently in other places on the Aran Islands and across Galway Bay, always close to the sea. What restricts it to these sites is an open question. It is generally a mountain plant outside Britain and Ireland, growing to an altitude of 2,700 metres in the Alps, and so is unlikely to require warm winters. It can reproduce both by rhizomes and seeds, and individuals may live for at least eight years. However, it flowers quite rarely and studies have shown that only 10–15 per cent of plants produce flowers in any year (Walker, 2015). A quirk of its make-up is that its habitat needs to be kept open

FIG 154. Pyramidal
Bugle (*Ajuga
pyramidalis*) at the
end of the flowering
season, May.
(Fiona Guinness)

by grazing but it is itself quite palatable. This may be its undoing in the context
of the Burren, and it may only now be recovering from the onslaught of sheep
grazing that occurred in the first half of the nineteenth century.

WOODY SPECIES

Woody plants frequently attract attention on the Burren's limestone pavement,
but calling them trees gives a wrong impression. More often than not, they are
seen as forlorn, twiggy shoots pressed closely to the rock and rooted in a grike. In
summer, they are leafy mounds, often with the marks of grazing clearly seen and

FIG 155. Holly (*Ilex
aquifolium*), probably
grazed by goats, which
spend a lot of time on
pavement.
(Fiona Guinness)

FIG 156. Honeysuckle (*Lonicera periclymenum*), rooted in rough ground and thereby avoiding grazing animals, May. (Fiona Guinness)

the droppings of goats nearby. Most Irish species are here, but oaks (*Quercus* spp.) are very few and birches are found only locally. The most frequent trees are Hazel, Blackthorn (*Prunus spinosa*), Hawthorn (*Crataegus monogyna*), Holly (*Ilex aquifolium*) and Ash (*Fraxinus excelsior*), with Spindle (*Euonymus europaeus*), Rowan (*Sorbus aucuparia*) and Yew (*Taxus baccata*) also widespread. Four willows may be found: commonly Grey Willow; less often Eared Willow (*Salix aurita*) and Goat Willow (*Salix caprea*); and, only rarely, Creeping Willow (*Salix repens*). Irish Whitebeam (*Sorbus hibernica*) and Common Whitebeam (*Sorbus aria*) are occasional, as are Crab Apple (*Malus sylvestris*) and Aspen (*Populus tremula*). Yew is grazed just like the other species, although it contains poisonous alkaloids; it grows tall only on hillside cliffs out of reach of goats. The species can be fatal when animals are confined and eat too much of it, but nibbled as a mix with other forage it seems harmless. Goats and sheep are particularly suited to grazing it because they keep moving, always on the lookout for something better.

Growing with these 'trees' are Honeysuckle (*Lonicera periclymenum*), ivy (already noted as regular species in pavement) and Bramble (*Rubus fruticosus*). The Burren boasts an additional species of *Rubus*, the Stone Bramble (*R. saxatilis*). It has slightly prickly green shoots that grow from a perennial root and last for two years, flowering in the second. They may root at the tips and form a colony of identical, clonal individuals. This is just as well, for the species cannot pollinate itself and spreads by seed only rarely. The flowers have tiny petals, seldom longer than the sepals, and prominent stamens. The 'berries' are bright red clusters of two to five drupelets, more separate than in a comparable raspberry. A study in Sweden found that less than 10 per cent of Stone Bramble flowers formed fruit,

FIG 157. The fresh new leaves of Stone Bramble (*Rubus saxatilis*) with its white flower, May. (Fiona Guinness)

the proportion highest when they were closer to a neighbouring colony (Eriksson & Bremer, 1993). Insects are obviously more likely to visit flowers in the same colony than in a different one, which is important if the species is scattered or becomes rarer. As isolated colonies get further from each other, the chances of good pollination and fruit-set are reduced. The plant persists vegetatively but cannot fruit or recruit new genetic combinations that might be useful in a changing environment. Such a set-up is called a metapopulation – a population of different but interdependent colonies with unoccupied space in between. Each depends on another for seed production.

FERNS

Ferns are seen on the Burren's limestone pavement at all times but are especially noticeable out of the flowering season. The smaller ones are usually spleenworts, growing directly on the rock. Small and tough, they can survive many days of drought and die back to the rhizome if necessary. Limestone is their natural habitat, but the use of lime mortar for walls in the past allowed them to multiply and spread into acidic regions where they could not grow on natural rock faces. They are all familiar for this reason. Maidenhair Spleenwort (*Asplenium trichomanes*), Rustyback Fern (*A. ceterach*) and Wall-rue are everywhere in the Burren, although they may not thrust themselves into view at all times. Polypody ferns are widespread also, particularly Southern Polypody (*Polypodium cambricum*), which has broader, more triangular fronds than the other species, their segments

FIG 158. Maidenhair Spleenwort (*Asplenium trichomanes*) and Rusty-back Fern (*A. ceterach*), May. (Fiona Guinness)

FIG 159. Wall-rue (*Asplenium ruta-muraria*) growing among lichens in a rocky cleft, May. (Fiona Guinness)

with serrated edges. It is a Mediterranean form that is found all over Ireland but only in the west in Britain. Growing preferentially on limestone, it produces its new fronds in autumn and winter when the rains have come, usually sporing early in the new year.

A more delicate species is the Brittle Bladder-fern (*Cystopteris fragilis*), which grows similarly from rock but usually in dampness or in shade. It is associated with the northern coastal area of the Burren but occurs occasionally throughout. The genus name refers to the indusium (covering) of the spores, which in this case is cup-shaped rather than flat (*Cystis* is Greek for 'bladder' or 'pouch').

More visible are the larger fern species, including shield Ferns (*Polystichum* spp.), Male Fern (*Dryopteris filix-mas*) and Hart's-tongue (*Asplenium scolopendrium*). They are rooted in the grikes and grow fully exposed to the weather, reminding one of the local climatic features that allow for this. Ferns need dampness and water to reproduce – they have aquatic gametes that swim from male to female, unlike the dry pollen of higher plants – so the moist, if not wet, habitat in the grikes encourages them. Hard Shield-fern (*P. aculeatum*) is a Burren speciality, perhaps because the warm winters favour its evergreen fronds more than those of its relation, Soft Shield-fern (*P. setiferum*). Both species grow frequently in the grikes, but Hard Shield-fern is more common than might be expected. Generally, it is associated with wet, open conditions with a relatively high pH, and is most frequent in the north-west of Ireland, and in Scotland and Wales. Like Hairy Rock-cress, discussed above, it evolved as a hybrid species between two others, Soft Shield-fern and Holly Fern (*P. lonchitis*), and grows in rather intermediate conditions. Its frond is narrower and less divided than that of its common cousin.

One cannot leave the pavement without looking for, or looking at, the Maidenhair Fern (*Adiantum capillus-veneris*). Hidden in grikes and never growing above the surface, it is a plant that needs time and good eyes to find but rewards the effort with its delicate sprays of fan-shaped leaflets on blackish stems. It is well known in cultivation but seldom looks its best in the wild, battered by wind into the adjacent rocks and browned by age or cold. Its tenacity depends on firm attachment of its rhizome, which lives for many years. The Burren's coastal limestone pavement seems to support most colonies of the species, and in some places it could even be called common. But there is nowhere in the Burren or on the Aran Islands where the fern would be a complete surprise, and

FIG 160. Maidenhair Fern (*Adiantum capillus-veneris*) at Black Head, May. Weathered leaves can be seen in the shade below. (Fiona Guinness)

its occurrence might best be described as sporadic. The fern is often considered a Mediterranean species, but it also occurs worldwide in warm temperate regions, including Japan, Australia, Africa and the Americas. It reaches its northern limit in Europe in Co. Donegal, where it grows even closer to the sea than in the Burren. On the world scale it requires a high moisture content and is equally at home in cliffs or forests, given enough light. The genus name, *Adiantum*, is from the Greek *adiantos*, meaning 'unwetted', as the leaves of this plant repel water and never look wet.

SURVIVING THE PAVEMENT ENVIRONMENT

The available flora in the west of Ireland is not large in comparison to that of much of Europe, and often there are few enough plant species available that are really suited to growing on the limestone pavement here. Other species therefore have to manage the task, even if they are not fully adapted to the conditions. Several of these are woodland plants, and three in particular have

been studied in terms of their efficiency in photosynthesis (Osborne *et al.*, 2003). Wood Sage, Wall Lettuce and Hazel were chosen for the study as they grow frequently on rocks in the Burren but are woodland species in the rest of Europe. The authors found that none of these plants can keep up with the amount of light available for growth because of a lack of water. If there is enough light and water, the stomata on plant leaves open to the maximum, with a corresponding increase in photosynthesis. However, the three species in the Burren were found to cut back on photosynthesis as water levels declined in their shoots and leaves. A shower of rain soon righted the situation, showing that the capacity of the pavement soil for water storage is tiny. So, although rain occurs in the Burren on 250 days a year on average, this does not mean that water is always available for plant growth.

FIG 161. The rough-looking leaves of Wood Sage (*Teucrium scorodonia*) are a common sight before its greenish flowers appear, May. (Fiona Guinness)

FIG 162. Red Valerian (*Centranthus ruber*) colonising the verge of the coast road, May. (Fiona Guinness)

Die-back is regularly seen in Hazel shoots in summer. Although sometimes caused by insect attack, it is so widespread that desiccation during summer gales is a much more likely reason. The young shoots have no strength at first, and wilt before new supplies of water arrive. This happens regularly on the exposed pavement, but also sometimes in the shelter of the Burren's eastern valleys. In the long term, it inhibits plant succession, when one might otherwise expect the Hazel to grow taller each year and create conditions for Ash or oaks (or even Wych Elm, *Ulmus glabra*) to appear in its shelter and then grow into taller woodland.

A new species that seems designed for pavement living is Red Valerian (*Centranthus ruber*), a familiar and long-introduced plant that is now spreading in many places, particularly in the eastern Burren under Slievecarran. Brought originally from the Mediterranean as a garden plant, it has been established on walls since the nineteenth century and more recently has again become fashionable in gardening circles. It is now spreading along roadsides in the Burren and also over the open pavement, thanks to its light, windblown seeds. As the plant is tolerant of drought and exposure, there is little to prevent it from following the example of Wall Lettuce and spreading throughout the Burren. Now that it has so many inocula in the region to initiate wild colonies, it will have a much more harmful impact on the rest of the flora.

FAUNA

The productivity of limestone pavement is obviously low compared to grassland or woodland, but there is enough to support a healthy population of Wood Mouse. In fact, this species is central to animal life in the pavement. It is a consumer of seeds, bark and young shoots, a predator of caterpillars and other insects, particularly in spring and summer, and a hoarder of Hazel nuts for use over the winter. It is also the prey of Kestrel (*Falco tinnunculus*), Red Fox (*Vulpes vulpes*) and Pine Marten (*Martes martes*). If Wood Mouse numbers decline, more Hazel seedlings may appear but fewer young Kestrels may be reared. If Hazel produces more nuts because of good pollination or fewer insect consumers, Wood Mouse survival will be greater over the winter and the population higher for the rest of the year. But all this activity is hidden in the grikes. In one study using the mark-recapture method, the population was estimated at 22 per hectare at its peak (November), which is equivalent to two mice over the area of a tennis court (Gallagher & Fairley, 1979). Within this zone, the adults were found to have a range of 30–65 metres in any direction, or a territory of about 0.3–1.4 hectares. The males moved greater distances in the breeding season, some of them over an area of 2.4 hectares.

The Bank Vole (*Myodes glareolus*) arrived in the Burren very recently in its gradual spread from its introduction to Co. Limerick in the 1920s. It is very likely to have some impact on the Wood Mouse if it can adapt to living in grikes rather than its more general habitat of thick grass cover. It eats much of the same food, including Hazel nuts, but tends to concentrate on bark and vegetable matter. Unlike the mouse, it eats dead leaves, having microorganisms in its gut that can digest cellulose, helped perhaps by nitrogen-fixing bacteria (Manaeva *et al.*, 2012). This gives it a definite advantage in times of food scarcity, such as in early spring. Elsewhere in Ireland, the Bank Vole has had considerable negative effects on Wood Mouse (and Pygmy Shrew, *Sorex minutus*) populations, effects that seem to be compounded if another introduction, the Greater White-toothed Shrew (*Crocidura russula*), is also present, although this consumes different food (Montgomery *et al.*, 2012).

If there is one butterfly particularly associated with limestone pavement, it is the Grayling (*Hipparchia semele*), a large, quick-flying species that folds its wings immediately it lands and seems to disappear. Unlike most other butterflies, it never suns itself with wings flat; they are always closed and held in a vertical position. Close observation will reveal that Graylings angle their wings at right angles to the sun, not to the ground, so they often look off balance. The underside of the wing – which the butterfly shows – is mottled with brown and grey, resulting in perfect camouflage. The Burren version of the Grayling is paler than those in the rest of Ireland and in Britain, and is recognised as the separate race *H. s. clarensis*. It is a 'late' butterfly, being most numerous in August, with some flying in July and September. It lays single eggs on grasses including Wood False-brome and Red Fescue, and the larva feeds for about two months and then hibernates until April, when it wakes up and continues to feed.

The Pearl-bordered Fritillary (*Boloria euphrosyne*) is also suited to the habitat of the limestone pavement, as it feeds as a caterpillar on violets (*Viola* spp.) and lays its eggs on Bracken or nearby Hazel. It resembles a small version of its relatives, the Silver-washed Fritillary (*Argynnis paphia*) and Dark Green Fritillary (*A. aglaia*), and speeds over the pavement on red-brown wings with black markings. It flies from late April to June, and in cool temperatures it comes to ground occasionally to sunbathe on flat rocks. Apparently most frequent in the eastern lowland Burren, the species has nevertheless been seen at Black Head and on the Aran Islands. It was discovered almost a hundred years ago, in 1922 (Phillips, 1923).

We have met with tiny snails on lichen-covered rocks earlier in this chapter but there are other, larger species that are seen everywhere. Large striped snails are probably familiar from gardens but the Brown-lipped Snail (*Cepaea nemoralis*) has a particular affinity for exposed limestone, in which it forms tunnels for refuge in dry weather. The tunnels are smooth vertical tubes, about 20 mm across, which are hidden beneath overhanging rocks where there is grass or other vegetation. The tunnels develop slowly over a succession of snail generations, through a mixture of rasping and perhaps acid secretions. They take about 10 years to extend 1.5 mm but can be 10 cm deep or more, showing use for 600 or 700 years (Stanton, 1986).

The Brown-lipped Snail feeds feed mostly on dying or dead leaves. Indeed, the habit of feeding on debris and leaf litter is highly developed among all the

FIG 163. Brown-lipped Snail (*Cepaea nemoralis*) gliding over a bed of Sea Plantain (*Plantago maritima*), Black Head. (David Cabot)

invertebrates that live in the grikes. Woodlice, millipedes, springtails, fly larvae and worms are listed as detritivores by Owain Richards in his 1959 survey, with predators including species of centipedes, spiders and harvestmen (Richards, 1961). The entomologist noted few species other than Lepidoptera larvae that feed directly on green plants. One of the most famous snails of the Burren, the Land Winkle (*Pomatias elegans*), favours dead leaves and leaf litter, biting off tiny pieces with a specially adapted tongue. The species is named for its general winkle-like shape and, like a marine winkle, it has an operculum, a disc that is attached to its foot (sole) and closes the shell in dry conditions. It is much smaller than its seashore relatives, measuring about 1.5 cm in length when fully grown, and has a ridged shell. The Land Winkle was discovered in the Burren as recently as 1976, adding Ireland to its Mediterranean range, which also includes south-eastern Britain and isolated colonies in Lancashire and Yorkshire (p. 145). It was surveyed in 2000 and found in an area of about 20 hectares at Flaggy Shore near New Quay (Platts *et al.*, 2003). Shells were found only on the limestone pavement here and not on the grassy paths or cattle-grazed areas in between, and it seems that the animals may take refuge in the grikes during dry periods rather than burrowing into loose soil as they do elsewhere in their range. The total population was estimated at 247,000. Unusually for a mollusc, the Land Winkle is not hermaphrodite and the females are a little larger than the males. Being southern in origin, it is sensitive to frost, and the high winter temperature and high humidity of the area suit it well. During the survey, it was found only up to 43 metres above sea-level. In the absence of fossil shells, the jury is still out on whether the species reached the Burren under its own steam or not. Finding additional colonies in the area would go a long way to answering this question.

An animal that is highly suited to the pavement habitat is the Common or Viviparous Lizard (*Zootoca vivipara*), Ireland's only native species of reptile.

FIG 164. Empty shells of Land Winkle (*Pomatias elegans*; left) and Heath Snail (*Helicella itala*), both measuring about 15 mm. Flaggy Shore. (Fiona Guinness)

FIG 165. Common or Viviparous Lizard (*Zootoca vivipara*), ready to scuttle away to cover. (David Cabot)

However, the speed at which they take refuge in the grikes makes the lizards very difficult to see. Studying the reptile in other habitats is helped by the provision of basking platforms, especially useful in moorland or bogs but these are largely redundant in the Burren, where every south-facing rock is a suitable sunbed. The consequence is that nobody knows how numerous the Burren population may be or any peculiarities the lizards have for living in the habitat.

Limestone pavement is seldom birdless, but there is only one species that actively seeks it out. This is the Wheatear (*Oenanthe oenanthe*), which nests in grikes and stone walls and does not like cover of any sort. It is thought to be declining in much of Britain and Ireland, but continues to nest in the coastal strip of the Burren and the Aran Islands, where it is the first migrant to arrive in spring (Balmer *et al.*, 2013).

FIG 166. Male Wheatear (*Oenanthe oenanthe*) bringing food to its nest in limestone pavement, May. (Fiona Guinness)

CHAPTER 7

Calcareous Grassland and Heath

DEVELOPMENT AND DEFINITIONS

The type of vegetation that spreads from grike and solution hollow onto the rock or develops directly on the surface depends on a combination of factors: soil, exposure, water, grazing and history.

FIG 167. Grassland in the western Burren with an abundance of Bird's-foot Trefoil (*Lotus corniculatus*) in May. (David Cabot)

While we might think that all soils in the Burren would be derived from the limestone that underlies them, this is far from the case. Other constituents are also important, including loess (windblown silt), till (glacial drift) and peat. Wind is frequent around any melting ice sheet and often blows downhill to the surrounding lowland owing to density differences in the air. Any silt recently dumped during glacial thaw will be picked up and carried to less windy places, there to be deposited as loess. In some places, the deposits may be thick enough to form hill-like sand dunes, which today are quarried to provide the perfect surface material for horse gallops and playing fields. In the Burren, the loess

FIG 168. Autumn colours remind one how prevalent Purple Moor-grass (*Molinia caerulea*) may be in some Burren grassland. This is a stony desert below Abbey Hill, with Juniper (*Juniperus communis*) and Yew (*Taxus baccata*) in the foreground, October. (Roger Goodwillie)

FIG 169. Lousewort (*Pedicularis sylvatica*), a sign of slightly acid conditions in the grassland. Poulavallan, May. (Roger Goodwillie)

is more of a shallow layer, being derived from the shorter local glaciation that occurred in Connemara. It is recognisable by being stone-free, unlike glacial till. The material originates from Galway granite and is rich in quartz and silica, which means it is less alkaline than soil formed from limestone. Soils in the Burren's limestone grassland may contain a proportion of loess, which reduces the pH to 5–6.5, making the soil relatively acid (Jeffrey, 2003). As these soils became mobile after ancient deforestation, some of the loess and till accumulated in the grikes, thereby offering a habitat for such acidophile species as Wood Sage (*Teucrium scorodonia*).

Low pH levels also occur in three other situations: on the remains of the shale or sandstone cap that once covered the limestone; on till deposits that have been leached or decalcified by rainfall *in situ*; and on chert, which occurs as narrow beds through the limestone. Chert is made of silica, like flint, and in the high pressures of rock formation it tends to dissolve and accumulate as a layer between limestone beds. It is a black, often contorted, material that resists erosion when exposed and so stands out from the underlying rock. A particularly prominent layer can be seen beside the lake under Mullagh More, where a solid band of Heather (*Calluna vulgaris*) draws attention to it (Fig. 170).

Peat is also common in the Burren and can even accumulate directly on the limestone in solution hollows (Chapter 6). It builds up in conditions of high rainfall and low evaporation, where decomposition is prevented by waterlogging and an absence of nutrients. In solution hollows or around the Burren's few springs, waterlogging itself is the main factor, but on the open rock a peat may develop on knolls and ridges that do not receive groundwater flushing. The same

FIG 170. A band of chert standing out above the limestone and covered by Heather (*Calluna vulgaris*), May. (Roger Goodwillie)

situation occurs on limestone mountains in Co. Sligo and Co. Fermanagh, and so is not the sole prerogative of the Burren. In fact, on these other hills the peat is distinctly thicker, hiding the rock almost completely by up to a metre. There is *Sphagnum* peat on the higher hills, but in the Burren the plant remains are of other mosses, sedges and heathers, and the accumulation of material has most likely been slower.

Against this background, it is no wonder that the vegetation of the Burren is variable by nature. Here, the definition of a community as grassland or heath is difficult, and the description of a whole hillside impossible. The first step on a hill may be grassland, the next limestone debris, the next limestone pavement and the next limestone heath. In many places the vegetation can best be called a mosaic, as the individual patches are so small. The advantage is that one sees a multitude of species in a short time; the disadvantage is that the ecologist cannot work out why the plants grow where they do.

Faced with such a quandary, the phytosociologist looks for communities in the vegetation – combinations of species that recur in different places and are most likely caused by similar environmental factors. To this end, he or she records species and coverage in a unit area suitable for the habitat, then sorts the results to put the most similar together. Hopefully, this shows up units that can be named. The actual process of naming the units then poses a further problem, as the common species (the most tolerant ones) often overlap and occur in multiple communities, for example Blue Moor-grass (*Sesleria caerulea*). Instead, the phytosociologist looks for rarer species (called indicator or faithful species) that occur only in certain of the groups and presumably are responding to environmental factors that may otherwise be unclear. Working on a broad European basis, there is Festuco–Brometalia grassland, a community dominated by fescue (*Festuca*) and brome (*Bromus*) species. However, such general names do not fit well with the Burren communities, where there are no bromes. Consequently, botanists now tend to use local names as these are more useful in discussing the specific vegetation and bring to mind something of its appearance. A fine synopsis of Burren plant communities was produced by Sharon Parr and others in 2009, later broadened out with a countrywide survey of limestone pavement (Parr *et al.*, 2009; Wilson & Fernández, 2013). Local community names, such as *Briza media–Thymus polytrichus* (Quaking Grass–Wild Thyme) grassland or *Dryas octopetala–Carex flacca* (Mountain Avens–Glaucous Sedge) heath, are certainly more user-friendly than some of the classical schemes. Their use also gets around the problem of finding Spring Gentians (*Gentiana verna*) and Dense-flowered Orchids (*Neotinea maculata*) in the same community – something that does not happen in the rest of Europe.

FIG 171. Pavement or grassland? Common Milkwort (*Polygala vulgaris*) with Early-purple Orchid (*Orchis mascula*) and mosses, May. (Fiona Guinness)

The division between grassland and heath lies probably at pH 6.5 (Jeffrey, 2003). At levels above this, lime-loving species like Kidney Vetch (*Anthyllis vulneraria*), Carline Thistle (*Carlina vulgaris*) and Yellow-wort (*Blackstonia perfoliata*) occur in the vegetation, while below it are found Bitter-vetch (*Lathyrus*

FIG 172. Kidney Vetch (*Anthyllis vulneraria*) is common both in grassland and on limestone pavement, particularly in coastal areas. It is the foodplant of the Small Blue butterfly (*Cupido minimus*). May. (Fiona Guinness)

FIG 173. Goldenrod (*Solidago virgaurea*) on limestone, May. This low-nutrient-demander usually grows in acid surroundings. (Fiona Guinness)

FIG 174. Mountain Everlasting (*Antennaria dioica*) flowers and leaves (note the white edges), growing with Bird's-foot Trefoil (*Lotus corniculatus*) and Bloody Crane's-bill (*Geranium sanguineum*), May. (Fiona Guinness)

linifolius), Goldenrod (*Solidago virgaurea*) and Heath Speedwell (*Veronica officinalis*). However, the distinction is not this clear; there are many stands of vegetation in which all species are mixed together and it would seem that climate and nutrients are the main agents. There are also some species that straddle the divide, like Mountain Everlasting (*Antennaria dioica*). If consistent rainfall washes the calcium away and the soil is nutrient-poor, there may be much less influence of soil pH than would be expected. In such areas, the distinction between calcicole and calcifuge species declines, and nutrient-rich and nutrient-poor becomes a more obvious difference.

THE FLOWERING SEASON

The Burren has a long flowering season, from the earliest Spring Gentians that appear in April to the last of the Harebells in September. There is a rush in the flowering of alpine and arctic species in the spring, together with woodland plants, then the more gradual appearance of continental flowers like Bloody Crane's-bill (*Geranium sanguineum*), Yellow-wort and Pyramidal Orchid (*Anacamptis pyramidalis*) over the summer. In July, Harebell, Lady's Bedstraw (*Galium verum*) and Wild Carrot (*Daucus carota*) tend to dominate, whereas in later summer, Heather, Bell Heather (*Erica cinerea*) and Devil's-bit Scabious take over. Wild Carrot is replaced by Burnet Saxifrage (*Pimpinella saxifraga*), and Autumn Hawkbit (*Scorzoneroides autumnalis*) and Knotted Pearlwort (*Sagina nodosa*) continue well into October.

But spare a thought for the grasses and other wind-pollinated plants. Doing without insect pollinators as a group, they can afford to spread their flowering over a longer period. Sedges (*Carex* spp.) flower notoriously early, from March onwards, at the same time as Blue Moor-grass. The grass-like Field Wood-rush (*Luzula campestris*) comes out in April and May, while the majority of grasses flower in June. Yellow Oat-grass (*Trisetum flavescens*) and the bent grasses (*Agrostis* spp.) are often the latest grasses to flower, in July. It is as important for these plants to have a flowering season as it is for insect-pollinated ones, to give the pollen the maximum chance of being carried to a female flower of the right species.

Salad Burnet (*Sanguisorba minor*) is usually an early plant and occurs in the south-eastern quarter of Ireland and the Burren. After overwintering as a clump of pinnate leaves, it produces a globular flower head in May or June. But this flower head is like no other in the vegetation.

FIG 175. Pyramidal Orchid (*Anacamptis pyramidalis*), ubiquitous in calcareous grassland and dunes from mid-June onwards. (Fiona Guinness)

The flowers are wind-pollinated, none has petals, and most are either male or female. The females occur on top and appear first, showing as a cluster of frilly red stigmas,

FIG 176. A tale of two grasslands. In the foreground is unimproved and unfertilised grassland on limestone, while in the distance is reclaimed rocky land with reseeded grassland that has been fertilised. May. (Fiona Guinness)

before the males extend their stamens. This arrangement of flowering from the top downwards is somewhat unusual, as one would expect the oldest parts of the flower head (at the base) to develop first, but it ensures cross-pollination.

Field Mouse-ear (*Cerastium arvense*) may be seen in flower from April to August. It is a plant of the drier, eastern side of Britain and of the driest part of Ireland (around Dublin), but it is also very abundant on the well-drained soils of the Burren and Aran Islands. It straggles through other plants and can be confused with Common Mouse-ear (*C. fontanum*) before it produces its centimetre-wide flowers with neatly divided petals. The overall effect is marred later in the year by the persistence of dead leaves on the stems.

PLANT SPECIES ON SHALLOW GRASSLAND SOILS

Grassland on the thinnest soils is usually dominated by Blue Moor-grass and, when one includes mosses and lichens, has the most diverse flora of any habitat.

Blue Moor-grass is particularly well adapted to poor limestone soils because of its drought resistance and ability to grow with very little phosphate. Other, taller, species that would blanket it out in better soils are curtailed, although a little Quaking Grass is often present, along with a fescue or Crested Hair-grass (*Koeleria macrantha* and the ubiquitous Glaucous Sedge. The soil is often no thicker than 5 cm and the plant cover is definitely mossy, with *Dicranum scoparium, Breutelia*

FIG 177. Blue Moor-grass (*Sesleria caerulea*) in May; the flowers are steely blue. (Fiona Guinness)

BELOW: FIG 178. Spring Gentian (*Gentiana verna*) showing the long tube below the petals and the small extra petal forming a fringe to the tube, May. (Fiona Guinness)

FIG 179. A clump of Spring Gentian (*Gentiana verna*) growing in context, with Mountain Avens (*Dryas octopetala*), Bird's-foot Trefoil (*Lotus corniculatus*) and a milkwort (*Polygala* sp.), May. (Fiona Guinness)

chrysocoma and *Ctenidium molluscum* all common (Wilson & Fernández, 2013). In addition, it contains multiple flowering plants, including yellow Bird's-foot Trefoil (*Lotus corniculatus*) and Kidney Vetch, a white scattering of Fairy Flax (*Linum catharticum*) or Squinancywort (*Asperula cynanchica*), mauve Wild Thyme, and blue Devil's-bit Scabious (*Succisa pratensis*), Common Milkwort (*Polygala vulgaris*) and – in pride of place – Spring Gentian.

Spring Gentian is the jewel that draws many people to the Burren, and it certainly does not disappoint. Gentian blue is bright and is so striking against the rest of the vegetation that any person – or flying insect – is automatically drawn to it. The flower is huge compared to the rest of the plant – as befits a native of the Alps, where insects are few – and it is larger than the leaves or stems would suggest. Indeed, it is hard to find the plant at all in the absence of flowers or seedpods. The flower has a long tube, suggesting it evolved to be pollinated by long-tongued insects such as butterflies. However, bumblebees (*Bombus* spp.) have discovered that they can enter the flower by biting through the tube to reach the nectar. This does not directly pollinate the flower but allows in ants, which may have the same effect. The result is that every flower seems to develop characteristic tall seed capsules. The seeds have an elaiosome (a parcel of fat and protein fixed to the outside), and when they fall they are collected by ants, which feed on the elaiosome and in turn disperse the seeds.

Spring Gentians flower in April and May, with a few just reaching into June. The blue Harebell (*Campanula rotundifolia*) waits until July to put on a display and is at its best in August and September, when its dainty bells blow in the wind over most of the Burren. It is quite rare in Ireland despite its frequency in Britain, and in fact the Irish plant is a hexaploid form, with six sets of chromosomes, a feature it shares with plants in the Celtic fringe of Cornwall, Wales and Scotland (Stevens *et al.*, 2012). Presumably this genetic constitution benefits the plant in areas with high rainfall or low sunshine hours, although its presence in such locations could simply be an accident of postglacial immigration

The Harebell is a retiring species, with small heart-shaped leaves that are difficult to see among other vegetation unless it is in flower. It does not compete successfully with other plants and never forms pure stands. Quite often it will be at the edge of a clump of vegetation against a rock surface. The bell-shaped flower is obviously insect-pollinated, and small bees and even ants may sometimes be seen visiting it. It is self-sterile; the flower produces pollen before its stigma becomes receptive, and if no insect removes it, the stigma ripens more slowly and the pollen falls out of the flower. Although perennial, the species seems to lose vigour after a year or two and seedlings outdo established plants (Stevens *et al.*, 2012). Mixing in new blood in this way is an advantage to a plant that is self-sterile, as it brings in a genetic mix of individuals rather than remaining as a clone derived from a single individual that has spread vegetatively.

FIG 180. Harebells (*Campanula rotundifolia*) in grassland with Pyramidal Orchid (*Anacamptis pyramidalis*) and Cat's-ear (*Hypochaeris radicata*). Inishmaan, July. (Roger Goodwillie)

FIG 181. Bird's-foot Trefoil (*Lotus corniculatus*) is one the most widespread plants in grassland and heath. It is called Cuckoo Flower in the Aran Islands because of the timing of its first flowers. May. (Fiona Guinness)

FIG 182. Cathair an Aird Rois, with a former mass house and shebeen (pub) on the Burren Way between Formoyle and Feenagh, May. There is extensive growth of Mossy Saxifrage (*Saxifraga hypnoides*). (Fiona Guinness)

The grassland on the skeletal Burren soils is floriferous, partly because intermittent drought and physical damage create spaces for additional herbs to become established. On the stoniest sites, Sea Plantain (*Plantago maritima*), Carline Thistle, the moss *Neckera crispa* and the lichen *Cladonia rangiformis* are frequently seen. One might think that Sea Plantain is a curious species to combine a life in saltmarshes with one on shallow limestone soils. It is common around all the coasts of Britain and Ireland, coming inland only in western Scotland and Ireland. During the postglacial period, when soils were skeletal, both Sea Plantain and Thrift (*Armeria maritima*) were widespread away from the coast. In many places they were banished to coastal regions by the spread of forest, but a lake core taken for pollen analysis by Bill Watts showed the continuous presence of Sea Plantain (and Hoary Rock-rose, *Helianthemum oelandicum*) close to Mullagh More from the late glacial to the present day (Sheehy Skeffington & Jeffrey, 1985). This suggests that the question we should be asking is why the plant is now confined to saltmarshes, rather than why its range extends inland in the west. It may be that it is adapted to a high mineral content in the

FIG 183. Mossy Saxifrage (*Saxifraga hypnoides*), May. The species is widespread in the Burren, usually in the shelter of walls of rocks, but not in the Aran Islands. (Fiona Guinness)

FIG 184. Old heads of Carline Thistle (*Carlina vulgaris*) the year after flowering, May. (Fiona Guinness)

FIG 185. Female flowers of Mountain Everlasting (*Antennaria dioica*), May. The stems are woolly with white hairs. (Fiona Guinness)

soil, either from bare rock debris or from sea water. It also illustrates the plasticity of the many plant species that can be grown in a garden soil, regardless of their natural habitat. Working out the competitive forces that restrict them to their current niche in the wild is one of the many things that makes ecology interesting.

The Carline Thistle is the neatest of the thistles, at just 20 cm high. It has whitish-green leaves, trimmed with spines, and in its second year produces a flowering stem with five or more whitish flower heads. Although usually biennial, it is, in fact, monocarpic and in the poorest conditions may take more years to come to flowering size. The flowers are central in each head and are purplish, but they are surrounded by yellowish bracts that make an effective attraction to insects. The bracts are hygroscopic, opening and closing in response to humidity. Even in the depths of winter, Carline Thistles persist as skeletons in the vegetation and so the species feels almost ever-present in the Burren.

Mountain Everlasting draws attention to itself in May and June, when colonies of short, 10 cm stems hold their furry flowers aloft and merit the plant's alternative common names of Catsfoot and Pussytoes. Six to eight little daisy flowers are held together, and they often have pink or red scales. Looking at the flowers, one sees that some are male with stamens and some female without, but that these are on different plants. More of the white plants are male and more of the females are pink, although this is not an absolute rule.

Studies have shown that there is no significant difference in breeding success of either colour, so there is no selective pressure to be either pink or white (von Euler, 2006–07). The leaves, which can be seen at all times of year, have white rims and undersides, and form neat rosettes among the ground cover. They are impervious to both frost and drought. Mountain Everlasting has a misleading name where Ireland is concerned, as it grows through the central plain as well as on the coast in Co. Donegal. In Britain, it is now a hill plant, although in the past it did have lowland stations.

More calcicole species grow in some of this thin Burren grassland, like Kidney Vetch, Rough Hawkbit (*Leontodon hispidus*) and Mouse-eared Hawkweed (*Pilosella officinarum*), whereas Heather is an obvious partner elsewhere, suggesting a transition to limestone heath and small patches of peat. Self-heal (*Prunella vulgaris*) and Common Milkwort are everywhere, as is Water Avens (*Geum rivale*), with its distinctive nodding pinkish flower or head of hooked achenes. Normally a plant of wet places, streamsides, clayey woods and tall herb meadows, it grows in limestone grassland and heath in the Burren, and also in woodland, regardless of shade. It is sensitive to grazing, however, and sheep may keep it in a dwarf non-flowering state, when it can persist only vegetatively. Reduce the level of grazing, and the plant can spread rapidly thanks to its excellent dispersal mechanism. In the autumn, it retains some green leaves, allowing it to take advantage of growing temperatures when they occur.

Three small white-flowered plants sparkle in the green patches on shallow soils: Fairy Flax, Irish Eyebright (*Euphrasia salisburgensis*) and Limestone Bedstraw (*Galium sterneri*). The eyebright is a speciality of western Ireland, and particularly of the Burren. It is one of the few species that does not also occur in Britain,

FIG 186. Water Avens (*Geum rivale*), with hanging flowers and heads of achenes, May. (Fiona Guinness)

FIG 187. Irish Eyebright (*Euphrasia salisburgensis*) has sharply pointed bronzed leaves and is restricted to western Ireland. May. (Fiona Guinness)

FIG 188. Limestone Bedstraw (*Galium sterneri*) threading its way through moss in the thin grassland, May. (Fiona Guinness)

its nearest stations being in the Pyrenees and Alps. The plant is small and has sharply toothed bronze-coloured leaves. Its flowers are predominantly white with purple veins and the yellow spot in the throat that is found in all eyebrights. The Irish form is distinct from others in Europe and has been isolated long enough to evolve into a separate subspecies, no doubt helped by its habit of self-pollination. Like all eyebrights, it is a hemiparasite, getting some of its nutrients from the roots of other species, in this case Wild Thyme. When the plant germinates in spring, the infant root attaches itself to a root of Wild Thyme and obtains most of its mineral nutrients and water in this way. Although the eyebright can grow in cultivation without a host, it remains smaller, indicating that wild plants must also get some carbohydrate from the host.

Yellow Rattle (*Rhinanthus minor*) has a similar habit, in this case gaining its nutrition from up to 20 different host species (Westbury, 2004). These are mostly grasses, and their resulting poor performance is sometimes clearly visible if the rattle is common enough. Some broadleaved herbs, including Ox-eye Daisy (*Leucanthemum vulgare*) and Ribwort Plantain (*Plantago lanceolata*), resist the hemiparasite's approaches by walling up or preventing access to their vascular tissue, making the invading root harmless (Cameron *et al.*, 2006). Other species are vulnerable, however, among them Tufted Vetch (*Vicia cracca*).

Wild Thyme must have few defences as it is also preyed upon by the parasitic Thyme Broomrape (*Orobanche alba*) and Dodder (*Cuscuta epithymum*), the latter more generally found on sand dunes in Ireland. Thyme Broomrape is well behaved, confining itself to one species of native plant, but other members of the genus are troublesome weeds of crops (the name *Orobanche* comes from the Greek for 'pea strangler'). Broomrapes are total parasites; they never produce their own chlorophyll for photosynthesis. The seeds are tiny, like those of an orchid, so can be widely dispersed. However, in order to grow they need to land within 2–3 mm of a host root so that this can be reached by the first weak attachment thread. This

FIG 189. Wild Thyme (*Thymus polytrichus*), the host of Thyme Broomrape (*Orobanche alba*) and Dodder (*Cuscuta epithymum*). May. (Fiona Guinness)

clamps onto the host root and thereafter the broomrape extracts all its needs via this connection. It forms a swelling from which the red flower stem rises, bearing eight to 10 tubular flowers. Thyme Broomrape behaves as an annual, but some seeds germinate in autumn in the mild Burren climate and are able to produce flowers early, during May. The further north one goes in Scotland and Northern Ireland, the more coastal the plant becomes, seeking the warmth of the sea.

Dodder resembles a tangle of fine red or yellow string on the surface of the ground. Looking at it more closely, one can see little suction pads (called haustoria) attached to the underside of the plant stems, and in late summer there may be clusters of small flowers, the most normal-looking part of the whole plant. Despite its Latin specific epithet (*epi*- means 'over' and -*thymum* means 'thyme'), the plant parasitises many different species; in the Burren and Aran Islands, it has been found on 48 different species (Doyle, 1993). The most common of these are Yarrow (*Achillea millefolium*), Bird's-foot Trefoil and Wild Thyme, while Red Fescue (*Festuca rubra*), Tormentil (*Potentilla erecta*), Red Clover (*Trifolium pratense*), Mountain Avens and Heather also fall victim. Indeed, it is tempting to ask why Dodder is so rare if it can grow on so many different hosts.

One of the reasons is that Dodder is a definite southern species, commonest around the Mediterranean. In Britain, it is widespread in the south-east but becomes coastal north of the latitude of Bristol. Its optimum germination temperature in greenhouse conditions has been found to be 15 °C, which in the Burren is reached only in May or June. A consequence of this is that it flowers late and may not always make seeds – which it needs as an annual. As a safeguard, not all the seeds germinate the following season; only one-third did so in the

FIG 190. The yellow variety of Dodder (*Cuscuta epithymum*), July. (David Cabot)

first year in one study, resulting in the build-up of a seed bank (Meulebrouck *et al.*, 2008). When the seed starts to grow, it makes a small attachment root and a thin shoot searches for a suitable host (called foraging). It is drawn towards green plants by the light reflected from them, and can reach to about 10 cm if it has to. When it makes contact, it twines around the stem before producing haustoria. Although Dodder is essentially an annual, it can also regrow from the suckers it used for feeding in the first year. This is a further safety measure for survival if seedlings fail. It also makes it all the more likely that the plant will be in the same place year after year.

PLANT SPECIES ON RICHER GRASSLAND SOILS

In places where the soil is thicker, Blue Moor-grass loses some ground, although not all, to other grasses – Sweet Vernal Grass (*Anthoxanthum odoratum*), Crested Dog's-tail (*Cynosurus cristatus*) and some Cock's-foot (*Dactylis glomerata*). Self-heal and Devil's-bit Scabious come into their own, and 'ordinary' grassland species such as Red Clover and White Clover (*Trifolium repens*) appear in numbers, along with Lady's Bedstraw, Yellow Rattle and perhaps Greater Knapweed (*Centaurea scabiosa*) (Wilson & Fernández, 2013). The last is a large plant up to a metre high; it is most frequent in the Aran Islands and eastern Burren, and less so on the western hills. It is a species that almost always appears in rude health, its large lobed leaves and strong stem supporting a thistle-like purple flower. It is one of the best examples to use for appreciating the evolution of a composite flower (those in the daisy family, Asteraceae), since it has a ring of long but sterile single flowers around the outside and a boss of fertile ones in the centre. In this way it is intermediate between a thistle and a daisy, and has a similarly intermediate visual impact. The head of flowers is supported on a ball of bristly bracts – used in the past

FIG 191. Yellow Rattle (*Rhinanthus minor*) in Burren grassland, May. It is a hemiparasite, taking some sustenance from adjacent grasses. (Fiona Guinness)

FIG 192. Greater Knapweed (*Centaurea scabiosa*) is commonest on the limestones of the Burren, and scattered elsewhere in the country. July. (Roger Goodwillie)

FIG 193. Dropwort (*Filipendula vulgaris*) flowers are pollinated by both insects and the wind. May. (Fiona Guinness)

by weavers to raise the nap on fabrics. Both this species and the Common Knapweed (*Centaurea nigra*) offer a food source to several fruit flies, which burrow inside as larvae and eat the developing seeds.

Greater Knapweed shows an understandable distribution in the Burren, growing on relatively deep patches of soil that are either natural or, on the Aran Islands, man-made (sand and seaweed were historically used in the islands to cover the limestone). This cannot be said for Dropwort (*Filipendula vulgaris*), which apart from a single station on the high Burren is otherwise restricted to the eastern side of the region, from the lowlands beside Slievecarran and Lough Bunny eastwards for about 30 kilometres. In Britain, it is widespread in non-acidic grasslands over most of England and some of Wales. Where it is present, it is quite common and its flowering stems stand clear of the leaves, making them available to flies and other insects. A certain degree of wind pollination seems to occur also (Weidema *et al.*, 2000).

GRASSLAND ORCHIDS

Orchids are a well-known part of the grassland vegetation, and in fact the Burren boasts 23 species. The Early-purple Orchid (*Orchis mascula*) is the most conspicuous early in the season, replaced later by the Common Spotted-orchid (*Dactylorhiza fuchsii*) and the Fragrant Orchid (*Gymnadenia conopsea*). The Bee Orchid (*Ophrys apifera*) and Frog Orchid (*Coeloglossum viride*) are rarer, or at least are seen less often. Finally, in August, Autumn Lady's-tresses (*Spiranthes spiralis*) appears, its minute flowering stems seldom reaching 10 cm.

FIG 194. Early-purple Orchids (*Orchis mascula*), the most frequent orchid species seen early in the year, May. (Fiona Guinness)

FIG 195. A single Early-purple Orchid (*Orchis mascula*) in typical open habitat with Hoary Rock-rose (*Helianthemum oelandicum*) behind, May. (Fiona Guinness)

FIG 196. The Early-purple Orchid (*Orchis mascula*) varies relatively little but is occasionally white. May. (Fiona Guinness)

FIG 197. Fragrant Orchid (*Gymnadenia conopsea*) has a neat flowering spike, with horizontal 'handlebar' petals. June. (Carl Wright)

FIG 198. Common Twayblade (*Neottia ovata*), May. Formerly placed in the genus *Listera*, this orchid is found in grasslands and scrub throughout the Burren. It normally has a straight stem up to 40 cm in height. (Fiona Guinness)

The orchids have suffered hugely from renaming, that practice beloved of taxonomists all over the world. We were brought up to believe that scientific or Latin names were immutable, and that everyone could recognise this, no matter what their native language. However, that ignores the holy grail of precedence and, latterly, of genetics. If an older name is found in some dusty archive by a taxonomist, that one takes priority, as long as it was published. One can see the logic of this process, as it must eventually cease to operate and the Latin names stay the same. But now the science of genetics has taken up the mantle, with genetic sequencing showing up some hitherto improbable relationships. In the earliest books, almost every orchid started off life as *Orchis*, whereas nowadays there is a rejigging of many of the genera. For example, the Twayblade genus

Listera has gone and its two former species are now placed in *Neottia*, along with the Bird's-nest Orchid (*N. nidus-avis*) and other relatives.

The Early-purple Orchid is ubiquitous in the Burren grasslands. It grows as individuals; it has little capacity to spread vegetatively, so one never sees clumps of shoots. It multiplies only by seed and, unlike several other orchid species, needs a pollinator to make the seeds. However, it attracts insects by deceit as it does not produce nectar. It is one of the best orchids to examine for the special arrangement of pollen, called pollinia, that characterises the family. If you push a ballpoint pen into the mouth of the flower, the chances are that you will come away with two little club-like pollinia stuck onto the pen with naturally quick-setting glue. This is what the bumblebee gets for its trouble of landing on the flower, and after a few minutes the pollinia droop so they are in just the right position to catch on the stigma and fertilise a flower visited by the insect later on. Often there are several flowers on each spike that do not develop further; some people have seen this as the plant's just rewards for its deceit. Each year, after their initial enthusiasm, the insects learn that there is no point in visiting these flowers, so do not continue doing so to the end of the flowering period.

A relatively long-lived species, the Early-purple takes 10 months to appear above ground, four years to flower and then continues for a further eight years (Jacquemyn *et al.*, 2009). It grows early in the year, its spotted leaves first appearing in late winter and elongating in March and April. It is quite palatable to grazing animals, and in fact a drink called salep was made from its tubers in the past, before the advent of coffee and tea. This is still made in Greece and Turkey, and is also turned into an ice cream. As in all orchids, the Early-purple requires a soil fungus to activate seed germination and growth, and this mycorrhizal association persists over its whole life, with the fungus scavenging for water and nutrients, and in turn feeding off starch or other products produced by the plant. The fungus associated with Early-purple is a woodland species and the orchid also favours woodland in much of its range. Perhaps its abundance in the Burren reflects the past cover of trees here.

Early-purple Orchids are a feature of the flora in May, while later in the year the Common Spotted-orchid comes into its own. The species epithet of this plant, *fuchsii*, commemorates Leonhart Fuchs (1501–66), a German herbalist who is also remembered by the genus name *Fuchsia*. The orchid is a variable plant, usually with spotted leaves and pale pink flowers that are patterned with purple lines and dots. The lower lip of the flower has three equal divisions in contrast to its relative, the Heath Spotted-orchid (*Dactylorhiza maculata*), which is almost as frequent in the Burren. On both flowers the arrangement of the lines and spots directs an insect's attention to the spur at the back, where there is nectar.

The white form, O'Kelly's Spotted-orchid (*D. fuchsii okellyi*), has none of these purple lines and was named from specimens collected in the Burren. It is by no means rare here and also occurs in Co. Galway, Co. Fermanagh and the south-west Scottish islands. It has pale green leaves that lack spots and a rather short flowering spike (see Fig. 32).

The way Common Spotted-orchid appears on newly exposed subsoil on roadsides, quarries and even in gardens shows how widespread its seeds must be. It has been estimated that each capsule contains several thousand seeds (rather few by orchid standards), resulting in a total per plant of 60,000 or so. The seeds themselves are dust-like, with a wafer-thin profile that suits wind dispersal. They

FIG 199. Common Spotted-orchid (*Dactylorhiza fuchsii*) at a young stage, May. The three similar lobes on the bottom petal (the labellum) are diagnostic. (Fiona Guinness)

FIG 200. Heath Spotted-orchid (*Dactylorhiza maculata*) in heathy grassland with Devil's-bit Scabious (*Succisa pratensis*) and Tormentil (*Potentilla erecta*), May. Note the labellum, with its small central lobe. (Fiona Guinness)

get away without food reserves since they parasitise soil fungi once they land. This is a successful ploy common to the Orchidaceae that has allowed the family to develop the second-largest number of species (after the daisy family).

The Bee Orchid is the one species that causes an intake of breath in the wandering botanist, no matter where it is seen. It is a relatively large orchid, certainly in terms of its flowers, but is surprisingly inconspicuous and is often come upon with little warning. It has just the right amount of rarity for a plant – not seen every day, but found at least once a year. The flowers are pink and a centimetre or so across, and they have the bee part in the centre – the labellum or landing stage. This is chocolate brown and yellow with bristly sides, so simulates a type of solitary bee, an *Andrena* species. There is even an iridescent patch that gives the same reflection as the insect's wings. Going further, the plant has developed an odour resembling that of the sex pheromone of the female bee.

FIG 201. Bee Orchid (*Ophrys apifera*) flower showing mimicry of a bee. The pollinia hang from the apex of the flower and stick onto visiting bees. May. (Carl Wright)

The Frog Orchid is even more inconspicuous but relatively frequent in both the Burren and Aran Islands. It has hooded reddish flowers over a strap-like green labellum that is longer than expected. Its growing conditions vary from place to place, ranging from dry pastures to grassland over rock or the edges of marshes and fens. It also has the ability to persist underground without any visible leaves for a year or more, such is its hold on the mycorrhizal fungus. This could be seen as the first stage in the evolution of an alternative way of life to photosynthesis, as has happened in a few species like the Bird's-nest Orchid and Ghost Orchid (*Epipogium aphyllum*).

Autumn Lady's-tresses is unusual in several ways. For a start, its flowers spiral around the stem, with some facing every direction. It is also a wintergreen plant, the tuber producing leaf rosettes in August that grow through the winter before withering in spring. Then, having survived a few months of potential drought, it wakes up again in August or September, and either flowers or sends out more vegetative shoots, especially if grazed by an animal or mown in a lawn. A pair of patient botanists found that it takes 11 years for a seed to produce an above-ground Autumn Lady's-tresses plant and 13–14 years for this to flower (Jacquemyn & Hutchings, 2010). It is a southern species, common in the Mediterranean and reaching its northern limit in Co. Sligo. It seems to require good drainage and both spring and autumn rains, which it gets in the Mediterranean and also in the Burren. In addition, it needs a level of grazing, as its diminutive size bars it from taller grassland – drought-limited calcareous grasslands thus suit it admirably. Once one of these orchids has been spotted, there will be more nearby as they are highly gregarious. They spread vegetatively from budding, and their seeds fall relatively close to the parent (for an orchid), allowing rapid colonisation within a suitable habitat. The amount of flowering varies from year to year, and plants normally take a break, resting for one year after flowering.

This selection of orchids has so far omitted the Dense-flowered Orchid, the most famous Mediterranean plant in the Burren along with the Maidenhair Fern. It is not easy to see, being tiny and green, but it is present above ground for seven months. Two leaves appear in October and the plant builds up reserves over the winter to produce a flower stalk in April, about 10 cm high. This adds a few extra leaves and the yellowish flowers, which do not open fully and are always self-pollinated (even in the Mediterranean). Although the Irish plants are in an evolutionary dead end, they still contain as much genetic diversity as those in the Mediterranean, so their isolation has not yet harmed their survival (Duffy *et al.*, 2009). A spring visit to the Burren usually rewards botanists with a withered flower spike poking through a low vegetation of Blue Moor-grass or Wild

FIG 202. Single flowering stem of Autumn Lady's-tresses (*Spiranthes spiralis*) in August, appearing without leaves (which are present in autumn and spring). (David Cabot)

FIG 203. Dense-flowered Orchid (*Neotinea maculata*) growing typically in a thin soil near rocks. The flowers never open wide and are self-pollinated. May. (Fiona Guinness)

Thyme. It is an anticlimax of a plant at this time, despite its exotic origins. In the Mediterranean, it grows in woodland (like Wild Madder), and even in the Burren it is sometimes found under trees and bushes (Doyle, 1985).

The Moonwort (*Botrychium lunaria*) may be included as an 'honorary orchid' although it is in fact a fern. When it comes up in spring it is the size of a Dense-flowered Orchid, and it may cause a double take when its cluster of sporangia appear like flower buds. Later on, it opens a single pinnate leaf with paired fan-shaped leaflets. The spores germinate as a prothallus (or gametophyte), but only when in darkness (Whittier, 1981). They develop a mycorrhizal association immediately and grow deeper into the soil, where eventually the gametes are produced. After fertilisation, the new fern (the sporophyte generation) also spends a significant amount of time below ground, redeveloping a mycorrhizal association. Species of Moonwort elsewhere have been found to benefit from the carbohydrate of green plants nearby as they are both connected to the same *Glomus* fungus (Winther & Friedman, 2007). With this arrangement dependent on so many other factors, it is no wonder that the plant is seldom seen and seems to be getting rarer. In some years it may spend all its time underground.

FIG 204. Brightly coloured waxcaps occur in short grassland as well as in woodland in the Burren. This Scarlet Waxcap (*Hygrocybe coccinea*) seems to avoid mycorrhizal plants such as Mountain Avens (*Dryas octopetala*) and Heather (*Calluna vulgaris*), succeeding in mossy grassland rather than in heath. September. (Carl Wright)

HEATH

Heaths are open habitats on poor soils that are usually acidic. They often form in areas where felling has removed the tree layer and where soil impoverishment or leaching, drought or wind maintain a community of dwarf shrubs and herbs.

In the Burren, the tree layer has certainly been removed and, in the absence of leaf recycling, the nutrient content of the remaining soil has declined. What little organic matter is produced blows into the grikes, where it is consumed by invertebrates, or accumulates on the rock surface as peat. Plant material is found between fragments of limestone as a rendzina soil, or directly on the rock, where intermediary mosses protect it from nutrient flushing. Peat appears in hollows where drainage (and oxygen) is impeded by the shape of the rock, by a layer of chert or by fine leached till. One meets such hollows on Cappanawalla or Gleninagh Mountain above the northern coast, where they have yielded valuable information on the past vegetation from the preserved pollen grains within (Feeser & O'Connell, 2009). Even Round-leaved Sundew (*Drosera rotundifolia*) grows in one peat hollow above Black Head.

Whatever the initial cause of the accumulation of organic material, the pH becomes slightly acid and the vegetation is dominated either by Mountain Avens or Heather, usually with Slender St John's-wort (*Hypericum pulchrum*), Devil's-bit Scabious, Blue Moor-grass and Wild Thyme (Hanrahan & Sheehy Skeffington, 2015). Heath retains a number of the grassland species, for example Bird's-foot Trefoil, Bloody Crane's-bill and Harebell, and occasionally supports the pale yellow Lesser Butterfly-orchid (*Platanthera bifolia*). In many places with peaty

FIG 205. A small depression with (trampled) peat on top of Cappanawalla, April. (Roger Goodwillie)

FIG 206. Layer of peat on the sandstones of Slieve Elva, south of the Burren. This rock lacks the drainage system of karstic limestone, so supports much deeper peat. (Fiona Guinness)

soils, certainly higher in the hills, the dark-red liverwort *Fruillania tamarisci* is characteristically supported. Also here are Bearberry (*Arctostaphylos uva-ursi*) and Crowberry (*Empetrum nigrum*), creeping among the stems of the Mountain Avens. These three sub-shrubs also occur together in open pine forests in Scandinavia and may have persisted *in situ* on the high ground of the Burren. The Bearberry communities that include Crowberry grow on the purer peat, with 75 per cent organic matter and a pH of 6.8. On rockier sites, Bearberry is mixed with Bell Heather and Bitter-vetch, and the substrate is more acidic, averaging pH 6.2. Sometimes these shrubs descend to lower levels – around Black Head, Crowberry grows practically within the splash zone of the waves. Bearberry also grows in the Caher River valley and eastern Burren, along with Juniper (*Juniperus communis*).

FIG 207. Lesser Butterfly-orchid (*Platanthera bifolia*) in company with Heather (*Calluna vulgaris*), Sweet Vernal Grass (*Anthoxanthum odoratum*) and Tormentil (*Potentilla erecta*). May. (Fiona Guinness)

Mountain Avens is both beautiful and tough. It is circumpolar, growing close to the ice sheets of Greenland and Alaska, and in mountain locations south to Italy. It has excellent drought resistance, and can be frozen in winter and desiccated by summer winds. The whole plant hugs the ground, and its leaves are leathery and unlikely to wilt, with white undersides coated in fine hairs. It is not difficult to see the plant surviving in harsh glacial conditions and welcoming the exposure of bare rock once the ice sheets melted. The odd thing about Mountain Avens in the Burren is that the plant experiences little competition in this niche nowadays. Vegetative growth of the species occurs in clones, whereby individual plants spread outwards from their parent, rooting as they go. They remain attached through woody branches for a long time and benefit from at least one tap root, up to 2 metres in length. This is a valuable feature given the rapid drainage in fell field and limestone pavement alike. The whole plant (or colony) is a long-lived entity, with colonies up to 500 years old. These have experienced many climatic changes in the past (deWitte *et al.*, 2017), and the older individuals in the Burren are quite likely to have been used for fuel – a practice that Foot (1864) records (Chapter 4).

When in flower, Mountain Avens is overwhelming – the endless terraces of blooms around Black Head are hard to beat as a feature of the area. The flowers themselves are well known for being paraboloid, a shape that concentrates the sun's rays on the central reproductive parts and creates a cosy resting place for insects in cold climates. The flower also tracks the sun's movement during the day. It develops into beautiful seedheads that are almost as noticeable as the flowers, and each seed has a long silky parachute to carry it away to pastures new. Simulations of a warmer climate show that increased growth in spring and autumn results in greater biomass, taller flowers and more seeds, and it is not

FIG 208. Seedhead and flower of Mountain Avens (*Dryas octopetala*), May. Note the leathery leaves. (Fiona Guinness)

FIG 209. Bank of Mountain Avens (*Dryas octopetala*) above Black Head, May. (Fiona Guinness)

hard to believe that Mountain Avens grows better in the Burren than in its more northerly or more alpine outposts (Welker *et al.*, 1997). Its secret is to remain wintergreen; the leaves are green for two years and stay on the plant as long again, so they can respond to any increase in the length of the growing season.

How important seed production is for Mountain Avens is an open question, although probably significant in the Burren. A study in Svalbard found that seed germination in the species was very rare in the wild – only 3 per cent of seed germinated, mostly in moss and none at all in gravel (Müller *et al.*, 2011). More than half of the same seeds grew in controlled (and warmer) conditions in a greenhouse, so it may be that the Burren climate has favoured seedling germination and establishment on a wide scale. Certainly, the onset of equable temperatures in the Burren creates optimal conditions for early flowering and seed-set.

Mountain Avens also occurs quite widely in Scotland and forms a very similar heathy vegetation on the limestones of Skye and Durness, spreading to calcareous shell sand, as it does around Fanore in the Burren. In the massive *British Plant Communities*, John Rodwell (1991) calls this vegetation the Mountain Avens–Glaucous Sedge (*Dryas octopetala–Carex flacca*) heath, and it shares many of the frequent species with the Burren. The community includes Wild Thyme, Sea Plantain, Harebell, Fairy Flax and Bird's-foot Trefoil, but also three Burren specialities: Dark-red Helleborine (*Epipactis atrorubens*), Arctic Sandwort and Hard Shield-fern (*Polystichum aculeatum*). In Scotland, however, it is enriched by several other mountain species that do not occur in Co. Clare, for example Alpine Bistort (*Persicaria vivipara*) and Rock Sedge (*Carex rupestris*). This should be kept in mind by the exploring botanist, as there is no better way to find a plant than by specifically looking for it.

The roots of Mountain Avens are always covered by mycorrhizae. Indeed, the heaths support an unusual fungal community, including many webcaps (*Cortinarius* spp.), which are more often associates of conifers on calcareous soils. In a thorough study of the communities in the Burren, 88 separate macrofungi were recorded and more than a third of these were ectomycorrhizal species (Harrington, 2003). There were 19 webcap species, 17 pinkgills (*Entoloma* spp.) and 10 waxcaps (*Hygrocybe* spp.). Other well-known genera were boletes (*Boletus* spp.), milkcaps (*Lactarius* spp.) and brittlegills (*Russula* spp.). The study was conducted over three years and the numbers of fruiting bodies varied by a factor of four over the years. The most important species – based on the number of fruiting bodies, and their combined weight and frequency – was the Yellow-foot (*Craterellus lutescens*), although it was more local in occurrence than some others. The number of ectomycorrhizal species went up with the amount of Mountain Avens in the vegetation, and was particularly high on the deeper and more organic soils. Non-mycorrhizal species are also more frequent where there is an abundance of Mountain Avens, making use of its abundant leaf litter. As would be expected, the growth of fungi in this community is far greater than in similar stands in the Arctic, and in fact approaches that found in temperate woodlands. Their success is notable, but from all appearances their presence is equally beneficial to Mountain Avens. It is interesting that 39 of the mycorrhizal species are normally associated with pinewoods and in the Burren seem to occur only in these upland heaths, possibly clarifying the history of the communities (Chapter 4) (Harrington & Mitchell, 2005).

Another 'open forest' species in Scandinavia and Scotland is the Intermediate Wintergreen (*Pyrola media*), which grows on the peaty summits of the northern Burren alongside Bearberry and Crowberry. It reaches its southern limit in the Burren – one of the only plants to do so – and reminds one of the disparate origins of the flora. The genus is distinctive: all species have rounded, stalked leaves in a rosette and a single flowering stem with hanging bell-like flowers. Like other heathland plants, it depends on a mycorrhiza – or indeed on a community of fungi, for 49 separate species were isolated from its roots in a Scottish study (Toftegaard *et al.*, 2010). Some of these, however, may be feeding parasitically on the plant rather than contributing to its upkeep.

INSECTS

The diversity of the vegetation in the Burren supports a similar diversity of insects – for example, all but two of the Irish butterflies occur in the area. Some insects feed on only one plant; the Carline Thistle has 12 species of insects restricted to it, while Devil's-bit Scabious attracts different specialists. The most famous of the latter is the Marsh Fritillary butterfly (*Euphydryas aurinia*), the harlequin of the fritillaries, with a mixture of buff, orange and brown panels on its wings. It lays its eggs in clusters on the scabious in May and June, and when they hatch in August, the larvae stay as a group, weaving leaves together with silk for protection and eating inside the tent so made. When they finish on one plant, they move en masse to another – a black line of bristly larvae. They hibernate in a larger web hidden in vegetation, and emerge in early spring to bask in the sun and digest the last of their food. Finally, after 10 months they pupate, emerging as adults six weeks later (Nash *et al.*, 2012).

The Dingy Skipper (*Erynnis tages*) is a common sight in May and June, at the height of the flowering season. It is called a 'skipper' because of its habit of darting up off a rock or plant for a rapid flight and then dropping back onto the ground a few metres away. It looks like a moth in its coloration, its brown and white markings giving it excellent camouflage. When at rest on a grass stem, it also overlaps its wings and holds them flat, not in the vertical way like most butterflies. Its larval foodplant is Bird's-foot Trefoil, and as with Marsh Fritillary on Devil's-bit Scabious, the caterpillar sticks two leaves together to escape attention. It is fully grown in August but then hibernates over winter until it pupates in April (Nash *et al.*, 2012). It appears that Bird's-foot Trefoils contain variable amounts of cyanogenic compounds (those that produce hydrocyanic acid) and that the skipper may be restricted to those populations where this form is rare, such as in the Burren. The Common Blue butterfly (*Polyommatus icarus*), by contrast, is able to denature the chemical, so it does not suffer the same limitations and is therefore much more common. It is seen everywhere in the Burren at hatching time in early June or late August (there are two broods). When at rest, it folds it wings, showing a largely brown surface with scattered black dots ringed in white.

Burnet moths are day flyers and visible for much of the summer. The Six-spot Burnet (*Zygaena filipendulae*) is the common species, and although there is just one generation each year, it can be seen over a long period, in late June, July or August. It has a striking black and yellow larva; like the adult's black and red, this is a warning coloration to predators. It feeds on Bird's-foot Trefoil and is both resistant to any cyanogenic compounds in the plant and also manufactures its own supply, thus making itself distasteful (Jensen *et al.*, 2011). The caterpillar

FIG 210. Rose Chafer (*Cetonia aurata*) on a Hazel (*Corylus avellana*) leaf. The adults fly clumsily around between April and September, typically in sunny weather. They feed on Burnet Rose (*Rosa pimpinellifolia*), as well as other plants. Their brilliant metallic green is a structural colour, caused by the reflection of polarised light. (Fiona Guinness)

FIG 211. A male Common Blue butterfly (*Polyommatus icarus*) on a buttercup. Note the long, curved tongue. (Fiona Guinness)

pupates in June on a grass stem about 25 cm above the ground. Its pupal case remains stuck there until the grass dies and is diagnostic of this species. The Burren is most famous for the Transparent Burnet (*Zygaena purpuralis*), and any burnet flying in June should be looked at since this species hatches earlier than the others. The Transparent has red streaks on its wing rather than the more compact spots of its relative. In this case the foodplant is Wild Thyme. Lastly, there is the Forester (*Adscita statices*), another burnet but less often noticed. It has glossy green wings and is not uncommon in the eastern Burren, where it feeds on Sorrel (*Rumex acetosa*).

FIG 212. Six-spot Burnet (*Zygaena filipendulae*) on Common Spotted-orchid (*Dactylorhiza fuchsii*), May. (Carl Wright)

FIG 213. Transparent Burnet (*Zygaena purpuralis*) has red patches between the veins on its wings. May. (Fiona Guinness)

The choice of foodplant by the burnets is unusually varied for a related group of moths, although it is not absolute and they can survive on other plants. It contrasts with the food of the 'white' butterflies, which is always a crucifer (Brassicaceae), and with most of the fritillaries, which feed on violets (*Viola* spp.). Grasses in general are the foodplants of many of the commoner butterflies, such as Meadow Brown (*Maniola jurtina*), Small Heath (*Coenonympha pamphilus*), Wall Brown (*Lasiommata megera*) and Grayling (*Hipparchia semele*).

Another day-flying moth, although small (a micro moth), is the Scarce Crimson and Gold (*Pyrausta sanguinalis*), with bright purple lines on its yellow wings. It feeds on Wild Thyme, especially the flowers, and may be seen in June. Elsewhere in Ireland it seems restricted to sand dunes, but it is obviously favoured in the Burren thanks to the large expanses of its foodplant. One should not forget about the Burren Green (*Calamia tridens occidentalis*) when visiting grassland, although it flies only by night. It is quite large (almost 2 cm when at rest) and has a typical noctuid shape with a stout body and wings held tent-like when at rest; in this case, the wings are pale green. Other varieties of the moth are

FIG 214. Burren Green (*Calamia tridens occidentalis*), a moth found only in mainland Burren, and absent elsewhere in Ireland and Britain. (Dave Allen)

found in central Europe and western Asia, so it is a mystery why it does not occur in Britain. Its foodplant is Blue Moor-grass, which is relatively common in the northern Pennines and on the Lancashire coast.

Bees and flies may be seen pollinating flowers such as the Carline Thistle and knapweeds, and with luck one may see the Shrill Carder Bee (*Bombus sylvarum*) at a flower. In Ireland, this bumblebee is now restricted to the Burren, although formerly it was more widespread. The name carder bee comes from the insect's habit of combing vegetation together to make an above-ground nest in dense vegetation. The Shrill Carder is a small greyish bee and has a slow

FIG 215. The Shrill Carder Bee (*Bombus sylvarum*) on Devil's-bit Scabious (*Succisa pratensis*), one of the less common bees in the Burren. (John Breen)

FIG 216. The Aran Island Bee (*Bombus muscorum* var. *allenellus*) on Red Clover (*Trifolium pratense*). Note the black abdomen below the chestnut thorax. (Gavin O Sé)

buzzing flight between flowers that, once heard, is not forgotten. Another carder bee, the Moss Carder (*B. muscorum*), also occurs in the Burren and there is a melanic form on the Aran Islands. The latter is sometimes given the subspecific rank *B. m. allenellus* and called the Aran Island Bee. However, recent DNA work suggests that the dark colour of the body, which is also seen in individuals on other islands in Ireland, the Hebrides and Shetland, has arisen through some form of selection and does not represent a true subspecies. The selective pressure that would operate on islands and not on the mainland has not been determined, although light would seem to be a possible factor (Deenihan, 2011).

Ants are also a feature of the Burren grasslands – at least 13 species have been recorded here. The most obvious is the Yellow Meadow Ant (*Lasius flavus*), whose mounded nests sometimes occur in the stoniest places. These show that there is plenty of soil in the grikes with which to build nests, even though there is nothing on the surface except shattered flakes of rock. The ants construct such nests for warmth, and a flat rather than curved surface is often exposed to the south-east to catch the morning sun (Brian, 1977). Once built, the anthill provides a new habitat in the wilderness, especially suited to annual plants because there are new additions of soil in most years. Work on the Co. Limerick limestones showed that about 45 species of plant grow on the anthills, and both there and in

FIG 217. Yellow Meadow Ant (*Lasius flavus*) nest, Glen of Clab. Wild Thyme (*Thymus polytrichus*) is growing on the sides and Procumbent Pearlwort (*Sagina procumbens*) at the summit. (Fiona Guinness)

the Burren, Crested Hair-grass, Silver Hair-grass (*Aira caryophyllea*) and Parsley-piert (*Aphanes arvensis*) are often found on younger parts of the mound (Breen & O'Brien, 1995). After a few years, Wild Thyme, Lady's Bedstraw and fescue grasses take over on the lower parts – or cover the whole nest if it is no longer active. Germander Speedwell (*Veronica chamaedrys*) is often present too.

AGRICULTURAL GRASSLAND

Deeper soils do occur in places on the Burren, notably in the two valleys that penetrate south into the area from Ballyvaghan and Bell Harbour, but also on the slopes above the Carran depression and on the west coast south of Fanore. These

soils can support a more intensive agriculture than the rough grazing of the rocks, the cattle concentrate on them for longer and the nutrient status increases through dunging. The least modified of these areas are the calcareous grasslands, which share such grasses as Crested Dog's-tail, Sweet Vernal Grass and Cock's-foot, as well as the clovers and Common Knapweed. They retain Lady's Bedstraw, Ox-eye Daisy, Yarrow and Cat's-ear (*Hypochaeris radicata*), and in places have Cowslips (*Primula veris*) (Wilson & Fernández, 2013). Slight dampness or overgrazing favours Yorkshire Fog (*Holcus lanatus*) and the buttercups, especially Bulbous Buttercup (*Ranunculus bulbosus*), along with Sorrel and Tufted Vetch. Another 'natural' grassland that occurs in the absence of reseeding includes False Oat-grass (*Arrhenatherum elatius*), Bush Vetch (*Vicia sepium*) and Germander Speedwell. It often has a scrubby component, perhaps with some Blackthorn or the woodland Pignut (*Conopodium majus*).

FIG 218. Cowslips (*Primula veris*) are often a sign of cattle grazing and slightly enriched surroundings. May. (Fiona Guinness)

Cattle and goats are the two most obvious consumers of Burren grassland and heath, although one should not forget the snails and caterpillars, which may take a few per cent of overall production. The larger herbivores are generalists to an extent, while the insects are specialists. Mammals obviously have the greater overall impact, and exclusion studies have suggested that they may take 25 per cent of annual production (O'Donovan, 1987). In their absence,

FIG 219. Germander Speedwell (*Veronica chamaedrys*) grows in grassland on deep soils as well as in woodland. May. (Fiona Guinness)

FIG 220. A young cow after a winter on the hills. (David Cabot)

FIG 221. An unexpected herbivore – llamas – in small fields near Fanore, May.
(Fiona Guinness)

dead vegetation builds up on the surface and annual plants disappear. In one
study over three years, 14 pre-existing species were lost from a protected area
(an exclosure), notably the annual hemiparasites of Irish Eyebright, Red Bartsia
(*Odontites vernus*) and Yellow Rattle, but also Fairy Flax and Common Mouse-ear
(Long, 2011). Red Clover and Self-heal also declined but Bracken (*Pteridium
aquilinum*) increased. The absence of physical damage to emerging Bracken
fronds by cattle is usually taken as the reason for this. Some grass species
increased with protection, as did Tormentil, which can scramble through taller
vegetation. The exclosures favoured snail numbers considerably, as many
species eat dead rather than green vegetation, and they all like cover. The
richest snail fauna before exclosure was in the tallest vegetation, which retained
moisture and was nutritionally the most fertile. One of the characteristic
species is the Heath Snail (*Helicella itala*), which has seen a 60 per cent decline
in distribution in Ireland in the last 30 years, probably due to changes in
agriculture in the midlands (Byrne *et al.*, 2009). It is one of the largest of the
Burren snails, noticeable because it aestivates high up on vegetation.

FIG 222. Empty shells of the Heath Snail (*Helicella itala*), measuring about 20 mm across. (Carl Wright)

In cases where swards have been enclosed for longer, there is an increase in Hazel (*Corylus avellana*) and other shrubby species, sometimes derived from previously suppressed plants. However, the grasses often dominate for many years, forming a thatch on the surface that prevents the seeds of other plants from germinating and the immigration of additional species. Moderate grazing levels by several different animals seem to yield the best results for plant diversity and would have occurred in the past from Red Deer (*Cervus elaphus*) and the Irish Hare (*Lepus timidus hibernicus*), the only native grazers. In one experiment in Britain, grazing by cattle and sheep had a much greater positive effect on species diversity than by either animal alone (Rook *et al.*, 2004). So, goats, cattle and Irish Hares may be the best option for the Burren today.

The abundance of sheep in the early nineteenth century, when they predominated over the other domestic stock, must have left its mark on the vegetation. Sheep are more selective than cattle, and from dung analysis it has been shown that, for example, they take Cock's-foot in preference to fescue grasses and sedges, with Perennial Ryegrass (*Lolium* perenne) towards the bottom of the list (Arnold, 1962). They also alter their food intake over the year, eating more broadleaved species like Bird's-foot Trefoil and clovers in the autumn after the grasses have flowered, and more Heather over the winter when it remains available. Separating the effects of the different herbivores is always difficult and it is an oversimplification to say that goats seem to prefer twigs, sheep herbs and cattle grass. A dung study of the goats in the Burren might be of great interest.

Large animals (including humans) also have a role in seed dispersal, either of seeds carried on the hair or feet, or through the dung. The reappearance of the Arctic Sandwort (*Arenaria norvegica*) on a Burren path in 2008 (Chapter 2) obviously involved some sort of seed transport, perhaps on a walker's boots. Farm animals are mainly responsible for such dispersal, however – in one German study, 8,500 seeds of 85 species were obtained from a single sheep's fleece (Fischer *et al.*, 1996). The seeds of half of the species recorded for the Burren also have the potential to be trapped in a fleece (Good, 1998). Irish Hares and Rabbits (*Oryctolagus cuniculus*) may additionally be important because both have grooming places with short or little vegetation where new seedlings can become established. In contrast, domestic animals do not groom, and their scratching places are frequently full of dung and therefore unsuitable germination sites for many of the grassland plants.

Finally, it is important to remember that when goats die they form a significant part of the diet of the Raven (*Corvus corax*) and Hooded Crow (*C. cornix*), thereby supporting a larger population than might be expected. Thirty Ravens were found feeding on such a corpse near Slievecarran in August 2015 (Clare Birdwatching, 2018).

BIRDLIFE

The Burren would not be complete in summer without the sound of Skylarks (*Alauda arvensis*) and Meadow Pipits (*Anthus pratensis*). Both birds nest on the ground in tussocks of grass, heather or bog sedge, and it would be interesting to

FIG 223. Skylarks (*Alauda arvensis*) are common throughout the Burren and breed on all the Aran Islands. May. (Fiona Guinness)

FIG 224. Meadow Pipit (*Anthus pratensis*), May. One of the most widespread breeding birds of the Burren, it takes over from the darker Rock Pipit (*A. petrosus*) away from the coast. Both species also occur on the Aran Islands. (Fiona Guinness)

know how the spread of scrub is influencing their populations. Skylarks continue to sing into August, with two or three broods of young. They are joined in June and July by the first young Starlings (*Sturnus vulgaris*), which quarter the grassland as small, scurrying groups of brown birds. Linnets (*Carduelis cannabina*) are also seen feeding in the grassland close to scrub areas, but the Blackbird (*Turdus merula*) and Song Thrush (*T. philomelos*) are probably the most visible species.

A feature of a winter walk on hill grassland is likely to be a small flock of Golden Plover (*Pluvialis apricaria*), perhaps 10 or 20 birds running in short bursts through the grass and around the cattle. This species may have nested here in the past; it continues to do so on a small scale in Connemara, across Galway Bay.

CHAPTER 8

Scrub and Woodland

I N THE BURREN, IT IS BEST to think of scrub as an open vegetation of woody plants (almost always Hazel, *Corylus avellana*) where light still reaches the ground and a variety of the pavement plants also occur. This open stage may be maintained by grazing, but where this is insufficient there is an inexorable spread of the leaf canopies and an increase in shade. Once canopy closure occurs, there is such a change in the environment that the stand is called low woodland, even though it is still only 2–5 m in height. When taller trees are present, it can be called full woodland or even forest, although the name seems out of place for the small clumps and pockets of trees that occur. Another factor that may keep the scrub as scrub is exposure.

Hazel scrub may be the climax community for the prevailing conditions on the west coast. No other species has the capacity to resprout repeatedly and maintain itself for hundreds of years. Similar climax scrub occurs in Co. Fermanagh and in western Scotland, although here it is much older, having not been cleared as in the Burren. It can be dignified by the name Atlantic Hazel, and in Scotland individual 'trees' have been estimated as being at least 300 years old (Coppins & Coppins, 2010). As they have more vigorous shoots on the outside of the stool, there is the suggestion that rings of daughters may arise from a central stump when the centre decays, in a pattern just like a fairy ring of fungi. Many of the areas of Hazel scrub in the Burren are comparatively young and still recovering from the intensive grazing of the nineteenth century. Noel Kirby (1981) states that 659 acres (267 hectares) of wood and 2,000 acres (809 hectares) of scrub were recorded in the Barony of Burren in 1655, but by 1840 the Ordnance Survey maps of the region show neither. In the resurvey in 1893, both scattered and dense scrub is mapped while it had increased further by 1915, when actual

FIG 225. Hazel (*Corylus avellana*) scrub spreading below the old fort of Cathair Chomáin. (Fiona Guinness)

woodlands are shown for the first time. While it may be that the mapping criteria changed over this period, it may also be significant that Frederick Foot does not mention Hazel at all in his survey of the Burren in the 1860s (Foot, 1864).

SCRUB COMMUNITY

Plant ecologists sometimes tend to denigrate scrub as neither one thing nor the other. While it may seem to be the first stage of colonising woodland, it often does not fit neatly into a grassland-to-forest transition. It may also be so varied that one cannot find a uniform patch for sampling; it changes from year to year and is difficult (and often prickly) to work in. But all such contentions have to be reviewed in the Burren, where scrub is a major habitat, increasing in area from year to year as winter grazing declines, and covering pavement and grassland to the detriment of some of the flora – but none of the larger fauna. Scrub supports its own mammal and bird community – without it there would be no Cuckoos (*Cuculus canorus*), Willow Warblers (*Phylloscopus trochilus*), Whitethroats (*Sylvia communis*) or Yellowhammers (*Emberiza citrinella*) that so enhance a visit to the area. Scrub has also been the cover that has allowed the Pine Marten (*Martes martes*) to survive a period of intense persecution in the nineteenth and early twentieth centuries. The Burren was one of the centres from which the mustelid has spread to the rest of the country in the last 30 years.

Open scrub is usually based on Hazel and Blackthorn (*Prunus spinosa*), along with Hawthorn (*Crataegus monogyna*), Holly (*Ilex aquifolium*), ivy (*Hedera helix* and *H. hibernica*) and Bramble (*Rubus fruticosus*), plus a little Ash (*Fraxinus excelsior*). Except on the extreme west coast, it is easy to see this as the early successional stage of woodland and to look out for larger and larger trees and tree regeneration as time goes on. Aside from exposure, the succession may be held back by grazing or by a current lack of suitable climax tree species to supply seeds – for example, Scots Pine (*Pinus sylvestris*) or oaks (*Quercus* spp.). Such a stage of arrested development is called a plagioclimax and may be applied to both scrub and woodland.

In the Burren scrub, pavement species continue to occur, including Bloody Crane's-bill (*Geranium sanguineum*), Glaucous Sedge (*Carex flacca*), Wood Sage (*Teucrium scorodonia*) and Blue Moor-grass (*Sesleria caerulea*) (Wilson & Fernández, 2013). The small spleenwort ferns (*Asplenium* spp.) have enough light to grow

FIG 226. Whitebeams (*Sorbus* spp.) are small, bushy trees that establish themselves in scrub but seldom persist into woodland. Both the Irish Whitebeam (*S. hibernica*) and the Common Whitebeam (*S. aria*) occur in the Burren. (Fiona Guinness)

and share the rocky surfaces with the mosses *Ctenidium molluscum, Thuidium tamariscinum* and *Neckera crispa*. Mosses are a fact of life in such areas and may cover up to 90 per cent of the ground. In general, the scrub stands are very rich in plant and animal species – indeed, they are sometimes richer than either woodland or grassland (Long, 2011). Along with the plants, the snail fauna is as diverse in scrub as in woodland, as both habitats provide enough shelter from the elements and from predators.

Two of the Irish butterflies depend on scrub: the Brown Hairstreak (*Thecla betulae*) and the Wood White (*Leptidea sinapis*). The hairstreak is a small brown butterfly with a yellow flash on the forewing (if female). Unusually, the underside of the wings is more colourful than the upperside, being orange with white streaks. This hairstreak is very restricted in Ireland, centred in the Burren and adjacent parts of Co. Galway, although its foodplant (Blackthorn) is much more widespread. It flies late in the year, mainly in August, so is not seen on spring visits unless the eggs or caterpillar are found. This is easier than it sounds, since the single eggs are laid at the base of a Blackthorn side shoot and remain there over winter until April the following year. The caterpillar then finds young,

FIG 227. Brown Hairstreak (*Thecla betulae*), revealing its more brightly coloured underwing. (Colin Stanley)

newly opened leaves to feed on, before pupating in June. When resting, it goes to the underside of the leaf where it attaches itself with a little silk strand. The pupa is often found in ants' nests, possibly choosing these sites for protection from predators. Many other butterfly species, particularly the blues, also have a relationship with ants.

The Wood White is one of a species complex that includes seven similar butterflies in Europe, although only two occur in Britain and Ireland. They are known jointly as a cryptic species and can best be distinguished by dissection of their genitalia or by chromosome analysis. However, they are likely to have behavioural differences so that each species will recognise a mate from a distance. The Wood White of the Burren is restricted to the limestones of this area and nearby Co. Galway, and is replaced by the Cryptic Wood White (*Leptidea juvernica*) in the rest of Ireland (Nash *et al.*, 2012). It is the smallest 'white', with a fluttering flight, and can be identified by the fact that it closes its wings when at rest or when feeding from a flower. The wings themselves are rather narrow and the forewing is not angled as in the other whites, being more rounded. Wood Whites seem to have simple habitat requirements: they like the shelter and warmth provided by scrub, and they feed on nectar from Bird's-foot Trefoil (*Lotus corniculatus*), Bramble and Bugle (*Ajuga reptans*). The eggs are laid on Bird's-foot Trefoil, Meadow Vetchling (*Lathyrus pratensis*) or Bitter-vetch (*Lathyrus linifolius*). Why the butterfly occurs in the Burren is still a mystery, as is the case with the Pearl-bordered Fritillary (*Boloria euphrosyne*; Chapter 5). Is it retreating in the face of a newcomer (the Cryptic), or is it better adapted to the limestone habitat, where Bitter-vetch is unusually frequent?

WOODLAND

As the Hazel trees become larger, so the fringe of mosses on their stems and on the ground grows deeper. The flattened sprays of the moss *Neckera complanata* are particularly distinctive, but several other species also occur. Such a growth appears natural for most of the year, dripping with dew or mist and coming away in the hand of the unwary. Only in prolonged summer droughts would one question its presence.

At a certain stage of canopy closure, some of the light-demanding mosses tend to die out and woodland species such as *Rhytidiadelphus triquetrus*, *Thamnobryum alopecurum* and *Plagiomnium undulatum* become more noticeable. Ash may start to play a more central role in the canopy, standing out above the Hazel if there is enough shelter in the lee of the hills. Somewhat unusually, large

ABOVE: **FIG 228.**
The dappled shade of
a rocky Hazel (*Corylus
avellana*) wood as
the tree leaves are
expanding, May.
(Fiona Guinness)

FIG 229. Swathes
of *Neckera* moss on
Hazel stems, along
with unfurling Soft
Shield-fern (*Polystichum
setiferum*) fronds, May.
(Roger Goodwillie)

trees of Goat Willow (*Salix caprea*) also occur in the few woodlands that exist, and
its presence shows that these stands are relatively young, since the tree cannot
become established in shade. Many of the individual willows in the eastern
woodlands like the Glen of Clab or Slievecarran are old. Some have clearly arisen

FIG 230. A fallen Goat Willow (*Salix caprea*) with three upright stems, now part of the canopy. Also visible are Bluebell (*Hyacinthoides non-scripta*), Wood Anemone (*Anemone nemorosa*), Soft Shield-fern (*Polystichum setiferum*) and Ramsons (*Allium ursinum*). May. (Roger Goodwillie)

as secondary shoots from fallen specimens but now rival the Ash canopy. This corresponds to the appearance of woods in the 1915 Ordnance Survey maps, as each iteration of a Goat Willow tree would cover about 60 years. A fallen tree in the Glen of Clab in 1980 was found to be 70 years old (Kirby, 1981). Young trees are still found in places, but only in open situations.

As mentioned above, the cover of woods and scrub has been vital to the survival of the Pine Marten, and the Burren augments this with grikes and other rock crevices for breeding or resting dens. Outside the breeding season, the animals tend to lie a few metres off the ground, perhaps on a fallen tree or on a rock outcrop. The species is never common and each individual requires a territory of 50–400 hectares, so they are not encountered often (O'Mahony & Turner, 2006). Usually it is the black droppings left on prominent rocks or trees that alerts one to their presence. These may be full of seeds in autumn – the animal is an omnivore and will eat haws, Rowan (*Sorbus aucuparia*) berries and crab apples, as well as any mice, frogs or birds it can catch. Pine Martens are excellent tree climbers – the Irish-language name for the animal, *cat crainn*, means 'tree cat' – and they may take eggs and nestling birds. Their claw structure allows them to run on the ground as well as climb trees. They are active by day in summer, when the tree canopy offers a potential source of food. In winter, they seem to spend more time on the ground and hunt by night.

ABOVE: **FIG 231.** Pine Marten (*Martes martes*) in the Caher River valley. (Carl Wright)

FIG 232. Pine Marten (*Martes martes*) droppings with seeds of fruit, probably ivy (*Hedera* sp.), Deelin More. (Fiona Guinness)

Annual growth cycle

As we saw in Chapter 6, many woodland herbs occur out in the open in the Burren – one should not be surprised to see Primrose (*Primula vulgaris*), Wood Sorrel (*Oxalis acetosella*), Wood Anemone (*Anemone nemorosa*) or Common Dog-violet (*Viola riviniana*) growing in places that are entirely free of tree cover. Although these plants may have been associated with the woodlands that were there in the distant past, they have inbuilt advantages for growth in the Burren of today. They are early leafing species, and the warm winter temperatures mean that they may start into growth in January or February, depending on the year. Desiccating winds are almost unknown in spring and the only significant droughts occur after they have finished flowering.

FIG 233. Bluebell (*Hyacinthoides non-scripta*) usually indicates old woodland, as it is slow to spread into scrub. May. (Fiona Guinness)

FIG 234. The Burren has many woodland plants that grow in open grassland and not necessarily under trees; Wood Sorrel (*Oxalis acetosella*) is one example. May. (Fiona Guinness)

The annual cycle sees about 15 herbaceous species visible in January (including ferns), rising to 20 in March, when the first flowers are seen (Kirby, 1981). Barren Strawberry (*Potentilla sterilis*), Early Dog-violet (*Viola reichenbachiana*), Wood Anemone and Lesser Celandine (*Ficaria verna*) are some of the earliest.

FIG 235. Ramsons (*Allium ursinum*) at its luxuriant best, where damp soil occurs at the base of cliffs. May. (Fiona Guinness)

OPPOSITE: **FIG 236.** a. Common Dog-violet (*Viola riviniana*) is seen everywhere in grassland and woods in May. In this condition it is ready for grazing by the caterpillars of the larger fritillary butterflies. b. The woodland floor in May, with Early Dog-violet (*Viola reichenbachiana*), Lesser Celandine (*Ficaria verna*) and leaves of Barren Strawberry (*Potentilla sterilis*). c. A youngish plant of Male Fern (*Dryopteris filix-mas*), a species that is less common than Soft Shield-fern (*Polystichum setiferum*). May. d. The delicate shoot of Pignut (*Conopodium majus*) is a common sight in woodland. The plant grows from an underground tuber each year and flowers in May. e. The simple inflorescence of Ramsons (*Allium ursinum*) – an umbel of six-petalled flowers. May. (all photographs by Fiona Guinness)

Most are frequent in other woodland situations and there is no special Burren woodland flora. Lords-and-ladies (*Arum maculatum*), Wild Strawberry (*Fragaria vesca*), Common Dog-violet and Germander Speedwell (*Veronica chamaedrys*) flower later. Yellow Pimpernel (*Lysimachia nemorum*), Bluebell (*Hyacinthoides non-scripta*), Ramsons (*Allium ursinum*), Meadowsweet (*Filipendula ulmaria*) and Lady Fern (*Athyrium filix-femina*) are slightly less common, and associated with deeper soil and a greater degree of dampness. In May, just as the tree canopy is expanding, the flowers may be augmented by Enchanter's Nightshade (*Circaea lutetiana*), while June or July bring in Broad-leaved Willowherb (*Epilobium montanum*) and two orchids – the tall, narrow-leaved form of Common Spotted Orchid (*Dactylorhiza fuchsii*) and the Broad-leaved Helleborine (*Epipactis helleborine*). The latter is superficially like the Dark-red Helleborine (*Epipactis atrorubens*) but has broader leaves and a flower that is only vaguely pink. In woodland it is generally green, but it is more colourful in open conditions. The orchid is a widespread plant in Ireland but nowhere as frequent as in the Burren and on the Co. Fermanagh limestones. It turns up in new habitats from time to time, but it also gives this impression by remaining underground in some years. Studies in North America (where it was introduced) showed that only one-third of plants appeared every year and only a third of them flowered (Light & MacConaill, 2006).

This sporadic appearance also characterises the rarest orchid in the Burren, the Bird's-nest Orchid (*Neottia nidus-avis*). It has little if any chlorophyll and was formerly thought to be a saprophyte, obtaining nutrients from decaying leaf litter. More recently, it has been revealed to be parasitic on a fungus that has a mycorrhizal association with trees (Leake, 2005). Occurring on the fringes of the area, near Corrofin and Gort, and largely in planted woods, the species may also reach into the eastern woodlands if the requisite fungal partner, an encrusting *Sebacina*, occurs.

Lords-and-ladies is a remarkable plant for several reasons. For a start, it is well known for its unique flowers, which are enveloped by a leaf-like spathe, with a spadix in the centre for advertisement. In 1778, the French naturalist Jean-Baptiste Lamarck (1744–1829) was the first to report that the spadix generates heat – in fact, it may be up to 15 °C warmer than the surrounding air. This, as well as its smell, attracts small flies, especially owl midges (Psychodidae). The flies are trapped temporarily in the base of the spathe by a ring of hairs and pick up pollen by walking around on the flowers. The hairs then wilt and the insects escape with pollen, when a proportion will be lured to another plant whose female flowers are receptive. The chamber at the base of the flower is in fact a good place to find the hostage midges.

FIG 237. Lords-and-ladies (*Arum maculatum*), showing the central spadix and surrounding spathe, which is swollen at the base around the tiny flowers. May. (David Cabot)

The plant also has the (justified) reputation of being poisonous – the unripe berries (and leaves) contain various poisonous compounds, including calcium oxalate, cyanide and coniine (as in Hemlock, *Conium maculatum*). If unripe berries are eaten, the needle-like crystals of calcium oxalate penetrate the skin cells in the mouth, causing intense irritation. Once the berries ripen, the crystals disappear and they can be eaten with impunity by the brave. Thrushes of all sorts take some of them, as do Pheasants (*Phasianus colchicus*) and Woodpigeons (*Columba palumbus*), so dispersal is effective. The seeds germinate in darkness (i.e. when covered) and the first root grows vertically downwards, forming a tiny tuber at its end that absorbs the rest of the starch from the seed. This remains invisible until the second year, when small leaves appear above ground. In late summer, the tuber grows contractile roots, which pull it deeper into the soil as it grows larger. It ends up about 10 cm down, and the pulling movement can be as much as 5 cm in a week (Meeuse, 1989). The tuber flowers when it is five to seven years old and in later life is often surrounded by a suite of younger plants from previous years' seeds.

The range of chemicals in Lords-and-ladies protects the green parts from being grazed by mammals, although slugs or snails sometimes takes pieces of the flower. Molluscs have an array of enzymes that humans do not, and they also consume fungi that are poisonous to us. The Lords-and-ladies tuber has less in the way of poisons and contains both starch and sugars. It was widely eaten by people in the past after heating – including during the Great Famine – but its peak of usage was in starching the ruffs of the Elizabethan era.

The actual date that growth of plants starts in the spring depends largely on temperature, as might be expected. In one study in Valentia, Co. Kerry, the onset of growth correlated best with the minimum air temperature and actual soil temperature in February, March and April, the temperature in March being especially important (Donnelly *et al.*, 2011). For several species, leaf opening now occurs about three weeks earlier than it did in 1970, and some evidence exists that the hatching of insect consumers is also advancing with the warming climate (Gleeson *et al.*, 2013). However, there is also a sign that the arrival of bird migrants, though earlier than before, is not sufficiently early to take maximum advantage of this food source. If this is the case, resident birds may be at an advantage and changes in the bird community are likely to occur. At the other end of the year, leaf-fall is generally later than it was, but day length and weather conditions each year play a part in causing it.

Above-ground herbaceous species disappear gradually during the summer. The violets (*Viola* spp.) retain their leaves for a long time, which is taken advantage of by the Silver-washed Fritillary (*Argynnis paphia*), a large, strikingly orange butterfly with black spots. It spends a lot of time flying, patrolling a territory in scrub or woodland, but when it lands, the silver washing and

FIG 238. Ash (*Fraxinus excelsior*) and Aspen (*Populus tremula*) are the last trees to come into leaf in spring (mid-May), long after Blackthorn (*Prunus spinosa*; April), Hazel (*Corylus avellana*; May) and Hawthorn (*Crataegus monogyna*; May) – the trees that are in flower in this view. (Fiona Guinness)

streaking is seen on the hindwings. This contrasts with its nearest relation, the Dark Green Fritillary (*A. aglaja*), which is similar but with discrete silver spots. It flies over more open ground and is coastal in much of Ireland. Both fritillaries occur commonly in the eastern Burren, but only the Dark Green is seen on the Aran Islands (Nash *et al.*, 2012). The insects seek out violets in July and August when they are still visible, if not really growing. They lay single eggs beside the plant, either on a tree trunk (Silver-washed) or in moss (Dark Green), and when the egg hatches the larvae immediately go into hibernation. They wake up again in March and April, in violet season, when the plants are at their largest. The caterpillars find nutritious leaf material to eat, pupating in May and June. The Silver-washed suspends its pupa in vegetation, a metre or so off the ground, while the Dark Green is generally on the ground in moss.

In late summer, the woodland floor is a shadow of its former self. A few leaves of most of the herbs can be found, but the Lesser Celandine disappears completely. Only the bright red berries of Lords-and-ladies and dried Bluebell seedheads persist until November, when they are blanketed under the falling Hazel leaves.

The trees
Hazel

Hazel occurs all over the Burren and on each of the Aran Islands. It grows on the rocks in the limestone pavement and in the hedges where agriculture has created fields. It occurs close to the sea, but probably not as close as Blackthorn, which can shelter within the grikes thanks to its branching style of growth.

The bark of young Hazel stems is smooth and pale brown, but pioneer lichens appear on it within two years. *Arthonia* and *Lecidea* species are early colonists, as are *Opegrapha* and *Graphis* species. These are mostly members of the Graphidion community – whitish lichens with dark fruiting bodies in spots or lines like writing. The rare *Pyrenula hibernica* has been likened to blackberries in custard owing to its purple and yellow colour combination. Other, rarer, *Graphis* species are prominent in the Scottish Hazel woods and would doubtless be in the Burren if the Hazel here was of as great an age – perhaps they are already colonising, however, as it is the fungal partner that arrives first from the aerospora. The small lichens are characteristic of the upper, well-lit stems, which lower down have a skirt of mosses. Two liverworts are often included among the lichens, the dark red *Frullania dilatata*, with its curious hooded leaves, and the more straightforward green *Metzgeria furcata* (Kirby, 1981). In time, if the Hazel stems grow larger instead of being constantly replaced, bigger lichens colonise the bark, culminating in the Lobarion community. This comprises large leafy

FIG 239. The dark red liverwort *Frullania dilatata* encrusting a Blackthorn bush (*Prunus spinosa*) in March, with Hazel (*Corylus avellana*) on the left. (Roger Goodwillie)

FIG 240. *Lobaria pulmonaria*, showing the characteristic dimpled surface. The orange apothecia are the sporing structure of the lungwort's fungus partner. December. (David Cabot)

species such as the lungworts – *Lobaria pulmonaria*, with a green algal symbiont, and the bluer *L. scrobicularia*, with a cyanobacterium, *Nostoc* (Chapter 6). These are usually associated with mosses, but probably because the moss canopy raises the humidity at the bark surface for initial colonisation by the lichen. It has been found that young lungwort thalli are particularly vulnerable to snail grazing and that climbing snails determine the lower limit of their occurrence (Asplund & Gauslaa, 2008). *Sticta* lichens, including *S. sylvatica*, occur on drier bark; they are dark grey or brown and often smell of fish when rubbed lightly. This community is rare in the Burren and found only in sheltered woods in the eastern part, such as under Slievecarran.

In the Hazel canopy, one should look for Hazel Gloves (*Hypocreopsis rhododendri*), the hand-like yellow growth of an ascomycete fungus. Although it appears to grow on the Hazel, the species is actually thought to be parasitic on the glue crust fungus, *Hymenochaete corrugata*.

Hazel itself is grazed by animals large and small, from Rabbits (*Oryctolagus cuniculus*) and Irish Hares (*Lepus timidus hibernicus*) to sheep, goats and cattle. It is eaten whether in leaf or not, and it also has its own leaf miner, *Stigmella floslactella*. The seedlings are nipped every year, and there may be quite a number of them in the grassland beside existing scrub. Once they get to a height of 1.5 metres, the trees begin to escape grazing goats as their branches are seldom suitable for climbing. The only grazers left on the tree at this stage are insects, but this pressure may be quite significant. Kirby (1981) recorded the defoliation of about 80 hectares of Hazel in one year by caterpillars of geometrid and tortrix moths. Some of the most abundant species were the Winter Moth (*Operophtera brumata*) and the July Highflyer (*Hydriomena furcata*). The defoliation was a temporary affair and the stems produced new leaves within a month, although they were softer and smaller than the first leaves. Other studies have shown that there is an increase in defensive compounds such as tannins in the tree after defoliation, and Kirby considered that the second leaves were hairier than the first – which could also be a defensive reaction.

One of the main causes of concern for the Burren flora is the spread of Hazel scrub, initiated by seed but more often brought about by a relaxation of grazing pressure. The Wood Mouse (*Apodemus sylvaticus*) and Red Squirrel (*Sciurus vulgaris*) are two of the agents of spread, as they both collect and store the nuts for later feeding. Mice are reluctant to leave cover, usually making their caches at the base of a bush. Squirrels are somewhat braver, but longer-range dispersal takes place by means of Hooded Crows (*Corvus cornix*) and Jackdaws (*C. monedula*), which carry individual nuts to a rock surface and attempt to break them. Jays (*Garrulus glandarius*) are also effective dispersers as they store nuts for the winter, but they are as yet only just spreading into the Burren from the east (Balmer *et al.*, 2013). One study found that most nuts buried beside scrub were eaten over the course of the following winter, whereas those that were buried 70 metres away often survived to germinate (Labarde & Thompson, 2009). Hazel, like oaks,

FIG 241. Clearance of Hazel (*Corylus avellana*) by hand from limestone pavement. (Fiona Guinness)

has a large seed and is particularly effective at germinating in grassland or other non-wooded habitats. It can also regenerate within the stand if grazing permits. Willows (*Salix* spp.), by contrast, carry very little food reserves in their tiny seeds and have to land on open ground to establish themselves.

The Wood Mouse is the major consumer of Hazel nuts in the Burren. Its population density is greater in woodland than on the open pavement, although in the latter the cover of the grikes does replace that of Hazel bushes to some extent. A study in the central Burren showed populations of 46 mice per hectare for woodland and 22 per hectare for open pavement (Gallagher & Fairley, 1979). This corresponds to about 25 per hectare in other Irish woodlands and suggests that it is the exceptional amount of food from the Hazel nut harvest that sustains the greater number of animals. Even in the Burren, there were more Wood Mice per hectare in high-canopy Hazel (which would produce more nuts) than in low canopy. The study also found that the mice were heavier in woodland but that they survived slightly longer in the open habitats. Presumably their predators – the Red Fox (*Vulpes vulpes*) and Stoat (*Mustela erminea*) – were more successful in the cover of woodland, although the high numbers of the animals may also put them under greater stress levels. More mice were active on rainy nights than on clear ones, perhaps in response to their invertebrate prey, which is an important source of food in spring and early summer before the Hazel nuts ripen. Since the study used a capture–recapture method, some indication of movement and home range was calculated. The average movement of individual Wood Mice over the season was about 70 metres, suggesting a home range of 1.5 hectares, although a small proportion of animals moved up to 180 metres.

FIG 242. Hazel (*Corylus avellana*) nuts grow with an enveloping, bract-like involucre. (Carl Wright)

Ash

Ash trees occur in practically all areas of Hazel, in low numbers even on the west coast. Their seeds are dispersed by wind, but as they are relatively heavy they do not travel much more than 150 metres except in exceptional storms. Some seeds develop every year, but the species has mast years, when much larger numbers are produced. This habit is generally thought to be a strategy whereby the seed consumers are so swamped that some seeds will survive to germinate. In the case of Ash, one of the consumers is the small tortrix moth *Pseudargyrotoza conwagana*, which eats the developing fruits and is often seen by day around the trees.

FIG 243. Ash trees (*Fraxinus excelsior*) spread into Hazel (*Corylus avellana*) scrub following a reduction of grazing pressure. June. (Steve O'Reilly)

Ash seeds, or keys, usually germinate in abundance and the seedlings at first have simple, undivided leaves. Although they are shade-tolerant, they may suffer in their early stages from a dense ground flora, perhaps through attacks by damping-off fungi. However, the seedlings are always better off in woodland than in grassland, and it is very seldom one meets a newly established Ash in open ground. As the seedlings develop, they lose their tolerance of shade and require light levels near full sunlight to grow well. While small, they are very susceptible to grazing and trees may be kept at 1.5 metres almost indefinitely. Even when they are large, goats climb on the lower branches if they can, and graze bark and shoots more than 2 metres above ground. Rabbits and Irish Hares also favour the plant, but insects less so. On an older, mossy trunk it is worth looking for a door snail from the family Clausiliidae. The shells of these molluscs are long rather than disc-like, and about a centimetre tall with a small entrance at the base. Their common name comes from the fact that they can close off the shell with a calcareous tongue of material to prevent drying out and to avoid the attentions of predators.

Many Ash woods in England became established after myxomatosis reduced the Rabbit population there in the 1950s (Wardle, 1961). In the Burren, one may presume, the same occurred after the loss of sheep in 1850. But nowhere here does one find veteran trees such as occur sporadically through the rest of Ireland, and the most senior trees seldom seem more than 100 years old.

Ash is one of the fastest deciduous trees for putting on stem growth thanks to the efficiency of its vessel structure in transporting water and photosynthate. However, it is sensitive to spring frosts and is the last of the native trees to break into leaf. Late frosts set the ensuing summer's growth back by damaging the conducting vessels. The species must have been a relative latecomer during the Irish postglacial period, and genetic tests reveal that it came from the south (Iberia) rather than from eastern Europe (Heuertz *et al.*, 2004). Pollen cores tell that it arrived with Wych Elm (*Ulmus glabra*) but after oaks. At first, Ash seems to have been rare, probably occurring only in damp places along streams or lake shores. There was one increase in the Ash population in the Burren about 5,000 years ago, but this was soon overtaken by Yew (*Taxus baccata*), which is a more effective competitor on limestone. Only with the advent of agriculture did Ash increase significantly, reaching a peak in the Iron Age, perhaps 1,500 years ago (Molloy & O'Connell, 2014). So, although it is not a newcomer, Ash has benefitted greatly from man's activities.

Despite the fact that Ash appears in most of the scrub and woodland in the Burren, it often does not grow very well and dead trees are fairly common. On the terraces of Mullagh More, for example, the Ash trees seldom reach old age and appear to have a lifespan of only 50–60 years (Goodwillie, pers. obs.). This probably stems from a lack of water, as the tree is shallow-rooting except for its tap root and a few 'sinkers'. Everywhere, a dry spring reduces Ash growth, and a series of dry springs may so debilitate a Burren tree that it becomes susceptible to fungal attack. It is not difficult to see Scots Pine as a better climax tree on these dry sites (as is suggested by the pollen record). The only healthy Ash trees in the Burren grow at the base of cliffs or on slopes where there is adequate water. One of the oldest trees (150 years or more) grows right next to the Mullagh More cliff, its branches diverging in a way that suggests it was on its own for much of its youth, although it is now surrounded by scrub. Other large trees occur lower down, nearer the adjacent lake.

The flower buds of Ash open in March to April, bearing clusters of purple stamens that fall off relatively soon afterwards. The species is gender fluid, such that the trees may be male, female or hermaphrodite (Thomas, 2016). Flowers on male trees open earliest, then those on hermaphrodite trees and, finally female trees. The leaf buds open in May, although in understorey trees and seedlings this takes place earlier. Ash is said to be determinate in growth: all leaves are formed in the bud, so shoot expansion is rapid and is finished in about three months. The next year's terminal bud is produced in July, and no further branch growth usually occurs for 9–11 months – there is no renewed, or lammas, growth later in the summer, such as occurs in oaks.

FIG 244. Female Bullfinch (*Pyrrhula pyrrhula*), May. Seeds are not plentiful in the spring and these birds feed on dandelion (*Taraxacum* spp.) regularly, although they live in scrub. (Fiona Guinness)

Ash seeds fall from September onwards, but most stay on the tree over winter, dropping to the ground when the growing season starts in April. Once on the ground, most seeds do not germinate for two winters, the seedlings finally appearing from mid-April three years after the flowers that formed them. In this they are assisted by a high content of phenolic chemicals, which protects them from the attentions of small mammals. In a study in England, Ash seeds were found to be the last to be removed from a woodland soil after oak, Beech (*Fagus sylvatica*), Hazel, elm (*Ulmus* spp.) and Yew (Jinks *et al.*, 2012).

FIG 245. Bullfinch (*Pyrrhula pyrrhula*) nest. They lay four to five eggs in a nest made from twigs lined with fine roots. Rockforest, May. (Fiona Guinness)

Ash seeds form a large part of the food of the Bullfinch (*Pyrrhula pyrrhula*) in winter, and the birds may depend on them until the Blackthorn buds start to swell in March. They take large numbers of seeds from certain trees and lesser quantities from those with a high phenolic content. They also drop small seeds when there is an abundance, so help in dispersal, especially as some dropped seeds are split and this reduces the time for germination (Thomas, 2016). In the Burren, Bullfinches are seen in scrub and trees away from the coast, most notably in the northern valleys.

Hawthorn

Hawthorn is ubiquitous as an understorey species in tall woodland and competes as an equal with Hazel in the scrub. Sometimes it is mixed through the other trees, while elsewhere it forms a ring around the Hazel, supplying a protective shield against grazers through preferential survival. It carries its own epiphytic lichens and can sometimes be picked out in winter by the grey-green scales of *Cladonia coniocraea*. Like ivy, it is a pivotal plant in the ecosystem, interacting with many other species through its flowers, leaves and fruit.

The tree flowers later than Blackthorn, and it sustains bees of all sorts, in some years contributing greatly to the honey bee harvest. Up to 30 different Lepidoptera have been recorded feeding on it, including well-known moths such as the Brimstone (*Opisthograptis luteolata*), Magpie (*Abraxas grossulariata*), Broad-bordered Yellow Underwing (*Noctua fimbriata*), Lackey (*Malacosoma neustria*) and Buff-tip (*Phalera bucephala*), the last resembling a broken twig when at rest (Robinson *et al.*, 2010). The Scalloped Hazel (*Odontopera bidentata*) is also on this list; despite its common name, it eats Hawthorn rather than Hazel. We should also not forget the impressive Hawthorn Shieldbug (*Acanthosoma haemorrhoidale*), the largest of the Irish bugs, which feeds on Hawthorn berries in summer and autumn, and the aptly named Hairy Shieldbug (*Dolycoris baccarum*). Shieldbugs have the habit of falling onto unsuspecting naturalists when dislodged from the leaf canopy. Mice depend on Hawthorn berries to varying degrees, and thrushes and Starlings (*Sturnus vulgaris*) are obvious consumers and dispersal agents since they do not digest the seeds.

FIG 246. Flowering Hawthorn (*Crataegus monogyna*) bushes are visited by many bees and flies. (Fiona Guinness)

FIG 247. Two different ages of the Common Green Shield Bug (*Palomena prasina*). The insect grows as a nymph and passes through five moults before it reaches the mature 'shield' shape. (Carl Wright)

Spindle

Spindle (*Euonymus europaeus*) shrubs are a constant presence in the Burren, but are usually so closely grazed that they produce little fruit – certainly not in the quantities seen in hedges elsewhere in limestone country. The species' most distinctive feature is that the young twigs are green, and in fact they produce about 15 per cent of the total photosynthate of the plant (Thomas *et al.*, 2010). This feature may allow the plant to be shade-tolerant, producing a proportion of its food after leaf-fall. But it is also tolerant of sunny places, such as among the rocks of the Burren. The leaves are thin and have a high nitrogen content, so are tasty to grazing animals. Indeed, a loss of 25 per cent of the total leaf area through insect attack has been noted. One of the most noticeable insects is the Spindle Ermine (*Yponomeuta cagnagella*), a micro moth whose tiny caterpillars spin a joint

FIG 248. A Spindle tree (*Euonymus europaeus*) in flower, dwarfed by grazing and with the leaves already eaten by invertebrates. May. (Fiona Guinness)

web over the shoots, devouring almost all greenery inside. The leaves regrow once the moths have developed into adults.

Spindle trees open their small greenish flowers in May, and over the summer they produce three- or four-chambered fruits that assume a distinctive pink hue before splitting to reveal their orange seeds – a colour combination worthy of a plastic manufacturer. The orange tissue is, in fact, an aril, a fleshy structure covering the seed itself, and is widely taken by birds. The aril has the highest food value of any native 'fruit' for its size, being rich in fats and protein. The seed itself is poisonous, and small birds (including tits) and Wood Mice have been seen eating the aril but discarding the seed (Thomas *et al.*, 2010). Thrushes may be able to eat both the aril and seed because of their greater size, which would be to the advantage of the plant. Otherwise, seed dispersal must be very local, although wind does eventually blow the fruits off a tree.

Rowan, birches and elms
After Hazel, Ash, Hawthorn and Spindle, there is a surprising amount of Rowan and some scattered Downy Birch (*Betula pubescens*). Rowan is usually associated with an acidic soil, and in the Burren one of its more notable sites is on the front slope of the hills on the south side of Galway Bay. Here, the pale, upright bushes stand out in the winter sunshine among the dark Heather (*Calluna vulgaris*),

FIG 249. The front slope of the hills above Galway Bay in winter, with Rowan (*Sorbus aucuparia*) growing among Heather (*Calluna vulgaris*). (Roger Goodwillie)

FIG 250. Rowan (*Sorbus aucuparia*) in late summer. The leaves show the effects of insect or fungal attack. (David Cabot)

possibly aided in their growth by the remnants of glacial till from the Galway granite. However, Rowan is also found elsewhere, regardless of soil. Its berries are eaten by thrushes so it is dispersed widely.

Downy Birches are regularly but sparsely scattered throughout the Burren, but are frequent only at Rockforest, Slievecarran and in the topmost parts of the Glen of Clab. In the latter site, the species' presence may be related to a more acid soil than usual, as there are traces of the former sandstone cap in the local soils. The rock type is reflected also in the ground flora of this woodland, where Greater Woodrush (*Luzula sylvatica*) and Bilberry (*Vaccinium myrtillus*) occur, and there is even some Wilson's Filmy Fern (*Hymenophyllum wilsonii*) growing on Ash trunks and rocks in the thick woodland in Poulavallan at the head of the glen.

Wych Elm is an enigmatic tree, rare nowadays owing to Dutch elm disease but once common in the period following the end of the Pleistocene. Even today, it has a wide distribution, from Ireland to the Urals and from northern Norway to Greece. Like Scots Pine, its pollen is dispersed by wind and shows up in all pollen diagrams for the periods deciduous woodland colonised these islands. Then, about 6,000 years ago, the pollen – and therefore the trees – began a catastrophic decline that reduced the species to a rarity in little more than a century (Parker *et al.*, 2002). The decline was almost simultaneous in Britain and Ireland, and annual lake sediments in Norfolk show that 73 per cent of the total reduction occurred in just six years. The decline coincided with the period when the mesolithic way of life – basically hunting and gathering – was being replaced by neolithic agriculture. It was thought previously that the elm trees were reduced by foddering (even in dried form, the leaves are highly palatable) and also by felling, as they were an indication of good soils for growing crops. However, no neolithic people seem capable of achieving such rapid deforestation. The speed of the decline therefore suggests that disease is the only possible cause (Mitchell & Ryan, 1997). This is another occasion for which the geological maxim 'the present is the key to the past' is appropriate.

Dutch elm disease is caused by a fungus and is characterised by blockages of the vessels in the twigs of the tree, first causing wilt and then death. It is spread by bark beetles, which emerge as adults from infected trees and carry the fungal spores to new trees, where they lay eggs. In this way, the disease can progress by 4 kilometres or so a year. Two factors that led to the disease explanation for the mid-Holocene decline in elms being taken more seriously were that bark beetles were found in fossil deposits in Britain, and that infected logs turned up in Denmark, where a similar decline took place earlier (as did the mesolithic–neolithic transition) (Parker *et al.*, 2002). This situation shows that agriculture does indeed have an effect, as bark beetles favour forest-edge habitats that are created by human clearance.

Wych Elm did not die out altogether following its main decline and appears all through the later pollen record in varying amounts. A few trees persist in the woods of the eastern Burren, particularly those west of Mullagh More. Here, they are protected from fungal infestation by being far enough away from other trees – too far for a bark beetle to fly. Wych Elm has a fissured bark with quite a high pH and supports many *Parmelia* and other large leafy lichens – more so than Ash.

Yew

Yew deserves a mention as a Burren tree as its dark colour makes it visible from afar. It is now scattered thinly on cliffs and occasionally in open pavement, although here it usually suffers from regular grazing. Larger trees occur at random in places not frequented by goats, such as south of Ballyeighter Lough. However, for a short time the species formed extensive woodland in the Burren, as it does today at Killarney in Co. Kerry. There it also grows on bare karstic limestone, dying out at the boundary of the adjoining acidic rock. It behaves mainly as a colonising species and if seedlings appear beneath the parent trees they do not survive because of the shade (Perrin *et al.*, 2006). Yew's short burst to prominence in the pollen record followed a secondary decline in elms, and may even have been directly related to the sudden loss of shade when the elm trees were killed. One can imagine flocks of Mistle Thrushes (*Turdus viscivorus*) spreading the seed far and wide as they perched on the dead elm branches over the woodland floor. Yew resembles Holly in being able to regenerate through a canopy of Hazel or other shrubs in a way that birches or willows cannot because of their small seed reserves. The resulting expansion of Yew lasted only a century or two according to the pollen history on Inisheer, and its reduction at that stage (4,900 years ago) may have been caused by agricultural activity (Molloy & O'Connell, 2014).

FIG 251. Yew (*Taxus baccata*) growing tall in the absence of grazing. The evergreen leaves shade out any competition. (Roger Goodwillie)

FIG 252. Holly (*Ilex aquifolium*) is scattered everywhere but becomes more prominent where grazing is reduced – as in this scrub, west of Ballyvaghan. May. (David Cabot)

Scots Pine

One of the more exciting strands of research in recent years has led to the realisation that Scots Pine has had a continuous presence in the Burren throughout the Holocene (see pp. 97–8). Previously, the accepted wisdom was that the pine was at first widespread and common in this period but became extinct in Ireland and Britain – except for Scotland, where native woods persisted. A continuously high level of pine pollen in the Burren was found but questioned by Bill Watts in 1984. More recently, in 2010, fine-grained analysis was carried out on material from a lake beside an open stand of pine at Rockforest by Jenni Roche. She concentrated on the last 2,000 years and found that there was no real decline of pine pollen over this period despite significant agricultural impacts (indicated by the presence of weed and cereal pollen). The clinching evidence was the discovery of a fossil pine leaf dated to about 840 CE, a time when pine pollen had disappeared from all other pollen profiles in Ireland and before reintroduction of the tree elsewhere in the island in the eighteenth century (Roche, 2010). The pollen rain suggests that a relict population of Scots Pine persisted in the Burren to the present day as an open pinewood with abundant Hazel. Later work from a sheltered marshy site at Aughrim Swamp, near Rockforest Lodge (Chapter 4), indicated that here there was a decline in pine pollen levels, corresponding to widespread pine extinction. It also showed that the Rockforest population lived through this, probably aided by the establishment of a walled private estate here (McGeever & Mitchell, 2016).

FIG 253. Scots Pine (*Pinus sylvestris*) flowers in May, releasing tell-tale pollen from the male cones. (Fiona Guinness)

FIG 254. Typical Scots Pines (*Pinus sylvestris*) at Rockforest, with Hazel (*Corylus avellana*) understorey and Heather (*Calluna vulgaris*). May. (Fiona Guinness)

Visiting the Rockforest wood today, one is struck not by venerable old trees but by the integration of Scots Pine into a mixed stand of Hazel, Hawthorn and Brambles. The trees are scattered in twos and threes, they have broad canopies and they overtop the other species. The largest (up to 20 metres tall) grow at the edges of finger-like turloughs, perhaps with a few willow trees, while the smallest spring out of the rocky ribs along with metre-high Hazel scrub. Downy Birch is surprisingly frequent, and there are many Pedunculate Oak (*Quercus robur*) seedlings, derived from several old trees of large size nearer the turlough. Recolonisation after grazing is everywhere apparent. The pines in drier habitats often shelter Holly seedlings and saplings, which have arrived through the agency of bird droppings. In wetter surroundings, Spindle or Blackthorn grow underneath in a more impenetrable scrub. Young Scots Pines seedlings are found on the limestone or among the heathers, Bracken (*Pteridium aquilinum*) and Purple Moor-grass (*Molinia caerulea*). The species is well adapted to life on the rocks. Its leaves are xerophytic in design compared to those of Ash and other broadleaved trees, and it tolerates drying as well as wetting. Comparable stands, derived from planted trees, occur occasionally on other limestones, such as by Lough Carra and Lough Mask in Co. Mayo.

FIG 255. Some of the largest Scots Pines (*Pinus sylvestris*) grow at the edges of a turlough inside Rockforest woodland, where there are wetter conditions. October. (Roger Goodwillie)

FIG 256. Scarlet Elf-cup (*Sarcoscyhpa austriaca*) develops on dead Hazel branches in winter and early spring. (Carl Wright)

FIG 257. Aniseed Funnel (*Clitocybe odora*) is named for its strong scent and taste. It grows in woods with abundant leaf litter and does not form a mycorrhiza. (Carl Wright)

Fungi

Leaf and branch fall towards the end of the year gives the saprophytic fungi of the woodland floor a new pulse of organic material for growth. In some species this initiates a cycle that will end with the production of fruiting bodies in the following months or years. Leaf fungi are usually small and mould-like, and it is only when dead branches and humus are available that larger fungi take over. Many of the obvious species are dead-wood fungi, forming white or yellow fruiting crusts on the surface of fallen branches (e.g. *Stereum, Peniophora*). Others grow a more discrete cup (e.g. Scarlet Elf-cup, *Sarcoscypha coccinea*, on Hazel), or are gelatinous, like the jelly fungi (e.g. *Tremella, Dacrymyces*). In most years there are few toadstool-like species in the Burren woods, although several small *Mycena* are often present. The shaggy brown *Flammulaster muricatus* is one to look out for on dead wood, while the Saffron Milkcap (*Lactarius deliciosus*) occurs below Scots Pine.

All tree species in the Burren have some form of mycorrhizal association. Ash and Yew have an endomycorrhiza within their roots that produces spore clusters in the soil but nothing visible above ground. In Hazel, Hawthorn, Downy Birch and Scots Pine, the more common ectomycorrhizae grow as a sheath around the outside of each root. Many of these are so-called higher fungi – basidiomycetes or ascomycetes. Their fruiting is sporadic, but *Hebeloma* and *Helvella* are regularly found in the woods. Modern work on mycorrhizae using genetic sequencing has shown the relationships to be much more complicated than was thought previously (Smith *et al.*, 2007). Half of the mycorrhizal species may not produce a conspicuous fruiting body, and the fungi that one collects around a tree are unlikely to be a true representation of what is on its roots. Hazel, for example, forms relationships with about 50 different fungi, including, most famously, the Summer Truffle (*Tuber aestivum*). Depending on nutrient fluxes in the soil, some of the mycorrhizal species are annuals, and they also differ in old and young trees.

Animal life

In the Burren's woodland community, production is concentrated in the growth of the woody species, but a good ground flora may photosynthesise almost as quickly over the few weeks that it is fully developed. In autumn, after the trees lose their leaves, Brambles prolong the productive season of the ground flora, being almost evergreen. Their leaves often contain the marks of insect attack, especially of leaf miners, which live between the upper and lower surfaces. These are usually micro moths, for example *Stigmella* species (one of which rejoices in the name *S. splendidissimella*), which produce corridor mines, or *Coptotriche marginea*, which creates blotches on the leaves. Bramble leaves are also a preferred food of goats and deer, and are much taken in winter. Deer have yet to reach the Burren in significant numbers, but Fallow Deer (*Dama dama*) are the most likely to do so as they occur in adjoining parts of Co. Clare. Goats spend time in woodlands in any month of the year when they are left undisturbed, but quickly vacate a site if they see people. They eat ivy to such an extent that it is relatively uncommon in the Burren woodland, especially in the more upland sites. In winter, Hazel scrub in the lowlands north of Mullagh More is peppered with treeborne clumps of ivy, but in the taller eastern woodland there are only a few trees that are affected. In a grazed woodland, the ivy only starts to produce leaves 2 metres up a tree, beyond the reach of grazing animals. Many other woods in Ireland are full of ivy, both on the trees and on the ground, and the species is increasing as domestic animals are more commonly being housed in winter.

The Irish form of ivy is generally Atlantic Ivy (*Hedera hibernica*), although both it and Common Ivy (*H. helix*) are thought to occur all over the Burren. Atlantic Ivy is particularly vigorous in the prevailing climate, as it is in western Britain (but not in Scotland). As a seedling, it competes better on the woodland floor than the nominate race, which is more easily shaded out (McAllister & Rutherford, 1990). It takes any opportunity to climb and may grow at a rate of 2–3 metres a year once established. Both ivies are ecologically important in the woodlands where they occur, creating food for insects, birds and mammals. The leaves and shoots are eaten by deer and by all domestic animals, while in spring the berries constitute an important food supply for birds. The plant contains unpalatable saponins and is seldom eaten by molluscs. There is some evidence that it is grazed by animals more in winter, either because of its chemical content or a lack of alternatives. When the berries are unripe they may be taken by Woodpigeons, but as they ripen they become palatable to more species – thrushes and, to a lesser extent, Blackcaps (*Sylvia atricapilla*). All are active in dispersing the seeds, several of which may grow from the same bird dropping. Clusters of two or three seedlings are relatively common and may grow on together into adulthood, forming a network of stems. The birds may also feed the berries to their nestlings if they are large enough, and sometimes use ivy itself as nesting cover. The plant is particularly useful to birds in Hazel woods, where the branches are rather bare for nesting.

About 80 species of insect and mite are known to feed on ivy in Britain, and perhaps 50 in the Burren, including beetles, plant bugs and Lepidoptera. One of the more celebrated is the Holly Blue butterfly (*Celastrina argiolus*), which has two broods, one appearing in April and May before the Common Blue (*Polyommatus icarus*) is flying, and the second in August, coinciding with the second peak of its relative. The male has a more silvery-blue look than other blues, and the female is similar but with a dark border to the wings. The butterfly flies at a height, often among bushes, scrub and hedgerows. The first brood of adults lay eggs in young Holly flowers and the second in ivy flowers, and the caterpillar eats the flowers and leaves of each plant. The butterfly population is subject to cyclical fluctuations, caused in Britain – and probably in Ireland – by a parasitic wasp that attacks the pupae. Despite this, it has shown an expansion of range in recent years and may be benefitting from a warmer climate (Regan *et al.*, 2010). It is also spreading in other European countries (Asher *et al.*, 2001).

The insects that live in and on woodland ivy provide a considerable food source for bats, several species of which are known to pick the insects off the foliage as well as catch them in the air. The Lesser Horseshoe Bat (*Rhinolophus hipposideros*), a Burren speciality, certainly does this, as does the Brown Long-eared Bat (*Plecotus auritus*). The Lesser Horseshoe occurs only in six western counties in Ireland and is fairly numerous in the eastern part of the Burren. Nowadays, it depends on human structures for both breeding and wintering; only one-tenth of the roosts found in one study were in natural caves, although this was, of course, the species' original habitat (Roche *et al.*, 2014). Recent records from winter roosts suggest that the Lesser Horseshoe is increasing, but this should be viewed

FIG 258. Speckled Wood (*Pararge aegeria*), the commonest butterfly in and on the margins of Burren woodlands. Although it feeds on grass species as a larva, the adults seek honeydew from aphids in the tree tops. May. (Fiona Guinness)

as a recovery of numbers in past centuries, when more woodland habitat was available. The bat hunts in overgrown habitats and displays great manoeuvrability in flight. It employs higher-frequency calls for echolocation than other species and can follow insect wing-beats. A downside of this is that the calls do not detect distant objects, so the animals must follow features such as hedges and stone walls to reach their feeding grounds from the roost. One can imagine that they learn these routes over a long lifetime – the oldest animal so far recorded was aged 16 years (McAney *et al.*, 2013). Prey items collected in woodland include lacewings and moths, craneflies and winter gnats.

The woodland species Natterer's Bat (*Myotis natteri*) and Whiskered Bat (*M. mystacinus*) have not been recorded in the Burren, but both the Soprano Pipistrelle (*Pipistrellus pygmaeus*) and Common Pipistrelle (*P. pipistrellus*) do occur, as does Leisler's Bat (*Nyctalus leisleri*). Leisler's is the swallow of the night, commuting at high level to its feeding grounds, and there descending closer to lower levels to feed on dung flies and other insects.

The larger lichens and mosses form a small-scale community on tree bark. Rarely, they (or at least the algal part) are eaten by caterpillars such as the Common Footman (*Eilema lurideola*), Muslin Footman (*Nudaria mundana*) or a micro moth such as *Psyche casta*. This last species is fascinating for its caddis-like habit of sticking a case around the caterpillar. More often, the lichens provide camouflage for caterpillars such as Scalloped Hazel (*Odontopera bidentata*); this caterpillar has two forms, one brown and twig-like, and the other patterned with green and brown. Resting carpet moths (Larentiinae) and larger geometrids like the Peppered Moth (*Biston betularia*) are also found here, and in addition the epiphytes provide cover for unassuming denizens like barklice (Psocoptera) and tardigrades or water bears (Tardigrada).

At a few millimetres long, barklice are easily visible insects, especially as they are active, running around on bark or stones. They have incomplete metamorphosis, developing from an egg through various nymphal stages to an adult, when they develop wings for dispersal. They often appear short-bodied, and different ages may occur together. Their food is algae, lichens and fungi, and in this they resemble the minute tardigrades, which are usually only seen under the microscope when one is trying to identify something else. Tardigrades are media favourites because they can withstand extreme temperatures, pressures and levels of ionising radiation that no other life form can tolerate. They can also survive a complete lack of water and food, and can show no vital signs for at least 30 years, ending up with a body content of just 3 per cent water. They do not experience any such extremes in the Burren, but no doubt they will have survived some challenging times during their 530 million years or more of existence.

AFFINITIES BETWEEN BURREN AND OTHER WOODLANDS

If the Burren woodlands occurred in Britain, they might fit into an Ash–Rowan–Dog's Mercury (*Mercurialis perennis*) vegetation type, which is found down to sea-level on the north-west coast of Scotland. There and in Co. Clare, it is characterised by an abundance of ferns and bryophytes, including Male Fern (*Dryopteris filix-mas*) and, in Clare, Soft Shield-fern (*Polystichum setiferum*). The more frequent herbs include Barren Strawberry and Pignut (*Conopodium majus*), Common dog-violet, Wood Sorrel and Wood False-brome (*Brachypodium sylvaticum*). The obvious absence from the list is, of course, Dog's Mercury, although the species does occur in scrub in at least two places in the Burren, at Deelin Beg and near Corrofin. In both sites it covers a small area, perhaps 2 square kilometres in total, and there are both male and female plants. These may be its only native stations in the country, as in most other places it can be linked to estate planting of trees and would have been introduced on tree roots. The species is abundant in most of Britain, although it does peter out in north-west Scotland, probably owing to the high rainfall. Its two Burren stations are on pavement, which would have the greatest drainage possible in the region, and it may have had a wider native distribution in the past. No doubt it would have suffered during the wetter Atlantic climatic phase, when blanket peat started to grow in earnest, perhaps 5,000–6,000 years ago.

There is now so little woodland in the Burren that it may be splitting hairs to suggest that two types exist here. Nevertheless, a slight distinction can be drawn between the pure limestone woodlands surviving on Mullagh More, where Wood Melick (*Melica uniflora*), Goldilocks Buttercup (*Ranunculus auricomus*) and Woodruff

FIG 259. Limestone woodland, with Woodruff (*Galium odoratum*) and leaves of Wild Strawberry (*Fragaria vesca*) in a sun patch, May. (Fiona Guinness)

(*Galium odoratum*) are common, and the more extensive stands at Slievecarran and the Glen of Clab, with their traces of a calcifuge flora. Perhaps this is because there has been a longer period since the development of the woodland in the latter sites, where leaching occurs just as it does on the pavement, or perhaps because the original soil was deeper and contained more sandstone till.

BREEDING BIRDS OF SCRUB AND WOODLAND

The birds of scrub and woodland are relatively few in species; the Irish avifauna is distinctly poorer than that of Britain and the Burren avifauna similarly reduced by distance and lack of habitat (see p. 57). The scrub species are most frequent for obvious reasons, and Robin (*Erithacus rubecula*), Stonechat (*Saxicola torquata*), Blackbird (*Turdus merula*), Song Thrush (*T. philomelos*), Wren (*Troglodytes troglodytes*), Cuckoo, Whitethroat, Willow Warbler, Blue Tit (*Cynastes caeruleus*), Chaffinch (*Fringilla coelebs*), Goldfinch (*Carduelis carduelis*), Linnet (*Carduelis cannabina*) and Reed Bunting (*Emberiza schoeniclus*) are ubiquitous. The last species chooses taller cover than the Linnet, which replaces it on the Aran Islands.

FIG 260. Male Stonechat (*Saxicola torquata*). Hazel (*Corylus avellana*) scrub, Brambles (*Rubus fruticosus*) and Common Gorse (*Ulex europaeus*) form an ideal habitat for the bird, which usually reveals itself by its characteristic *chat* call. (Fiona Guinness)

FIG 261. Linnet (*Carduelis cannabina*), a widely distributed breeding bird that frequents scrub in the Burren, especially in the Aran Islands. It is less common higher in the hills. (Fiona Guinness)

Yellowhammers are still quite widespread in Ireland but becoming isolated by a range contraction towards the southeast. Their diet in the Burren would be interesting to determine as there are very few fields of grain here.

Birds of woodland are even more circumscribed. There are few, if any, Treecreeper (*Certhia familiaris*), Goldcrest (*Regulus regulus*) or Spotted Flycatcher (*Muscicapa striata*), and only small numbers of Long-tailed Tit (*Aegithalos caudatus*), Jay, Chiffchaff (*Phylloscopus collybita*), Mistle Thrush and Coal Tit (*Periparus ater*). The Woodpigeon, Great Tit (*Parus major*), Song Thrush, Blackbird, Hooded Crow and Magpie (*Pica pica*) are the few species that could be called numerous.

FIG 262. The adaptable Magpie (*Pica pica*) looking for parasites in a place other animals cannot reach. May. (Fiona Guinness)

Blackcaps are spreading each year (the Irish population increased by a factor of four in 1995–2010) and filling apparent gaps in the bird community. There are also relatively high numbers of Sparrowhawk (*Accipiter nisus*); the feathers of small birds plucked onto a mossy bough by the raptor is a typical sight in woodland. The Long-eared Owl (*Asio otus*), the only owl species to occur in the Burren, also has a wide distribution.

Many bird species have colonised Ireland from Britain in historical times. The Stock Dove (*Columba oenas*) has reached the Burren in the past but is now experiencing a contraction of range to the south-east. Magpie and Blackcap are the two current successes, but the Buzzard (*Buteo buteo*) is on the doorstep and the Great Spotted Woodpecker (*Dendrocopos major*) at the other side of the country. Subfossil evidence, however, shows that this last species was present in Co. Clare about 3,750 years ago, when there was much more forest (Scharff *et al.*, 1906).

CHAPTER 9

Turloughs

R AINFALL SOON SINKS INTO THE BURREN, finding its way through
cracks and joints to join the groundwater and augment it until it
flows to lower ground. The top of the groundwater is the water
table, which therefore rises and falls in response to the weather. Turloughs are

FIG 263. Turloughnagullaun in June, with a central island emerging as the water level
recedes. In winter, the water spreads well into the marginal trees. (Steve O'Reilly)

FIG 264. Turloughnagullaun at ground level, May. Algal sheets have been left high and dry after winter flooding and the central bushes are only just starting to leaf. (Fiona Guinness)

transient lakes that occur in hollows and appear when the water table breaks the surface and the water appears above ground. Where an ecologist may view them as seasonally dry lakes, a farmer may see them instead as seasonally flooded fields, as they often supply good grazing for animals. Turloughs are prevalent in the lowland, eastern Burren, where the water table is closer to the surface than it is in the uplands; the area around Mullagh More is particularly rich in sites – for example, Lough Gealáin in Gortlecka, Knockaunroe, Skaghard Lough and Pouleenacoona (Tulla). However, there are also several large turloughs in the high Burren, notably at Lough Aleenaun, Carran and Turloughnagullaun. In the Aran Islands there is an example on Inishmore, the most western island of the group.

Turloughs have a discrete groundwater connection – the water does not just seep into and out of the soil – and the drama of the first autumn rising has to be experienced in the field to be fully appreciated. Clear water trickles through the vegetation from one or several places in the rock, inundating the plant stems and leaves, which soon become covered in bubbles. The water collects in hollows or winding channels, before spreading imperceptibly to cover the floor over the course of a day. The clarity of the water is remarkable in a country like Ireland, where peat stains the majority of surface flows.

The openings from which the water flows, called estavelles or swallow holes, may occur at the lowest point of the basin and are guarded by rocks and stones, or may be found above it if the floor is sealed by glacial till. In some cases, the bottom of the depression may be covered by an accumulation of peat or marl that can be several metres thick. At Carran, the peat is so thick that it has risen above the margins to form a slight dome. Here, the openings are at the level of the semi-permanent marsh that fills the basin.

Some of the lowland turloughs lie on underground rivers, conduits through the rock that carry large volumes of water down the catchment. They flood only when the passing flow is too great to be accommodated below ground. In fact, all turloughs experience a flow-through of water; their levels fluctuate over the winter as more or less water passes, and they resemble floodplains in many ways. A series of high-water marks of plant debris are spread around the basin; they show how far the turlough water level rose during the winter and provide perfect contour lines for the temporary shore.

FIG 265. Carran turlough, filled with sedge vegetation in May, and grazed lightly by cattle and horses. (Fiona Guinness)

FIG 266. Remaining pond in Carran turlough in May, showing islands of peat, which are covered by water in winter but dry in summer. Bogbean (*Menyanthes trifoliata*) is growing in the foreground. (Roger Goodwillie)

FIG 267. Last year's vegetation petrified by an overwinter deposit of calcium carbonate. May. (Roger Goodwillie)

When the turlough dries out in spring, a white deposit of calcium carbonate is often left on the plant leaves, giving them a bleached look and petrifying old plant stems. This has precipitated out of the groundwater, mostly by physical means; sites that have been flooded for a longer period are generally whiter when they dry out. Some submerged plants use bicarbonate dissolved in the groundwater as a source of carbon dioxide and thereby precipitate additional carbonate. This process has the effect of returning to the surface some of the carbonate material that has been weathered previously from grikes and other cavities in the rock. Sites that dry later often support beds of stonewort algae (Charophyta), these dying off over the dry period. Their calcareous skeletons create a white sediment (marl) on the bottom of the basin, together with the shells of any snails or other animals that lived in the water.

A DYNAMIC ENVIRONMENT

In addition to their winter fluctuations, many turloughs show evidence of much longer changes, measured in hundreds, if not thousands, of years. Mushroom-shaped stones on the margins of some illustrate a past lake with a static water level that allowed wave erosion to act at the surface (Fig. 268). And if one looks at the sediments in the turlough floor, there may be discrete layers of marl and

FIG 268. Mushroom-shaped boulder in a turlough, showing wave erosion on its sides. It indicates an earlier permanent lake with a stable water level. (David Cabot)

peat, or even alternating layers. These show a sudden transition in the whole habitat, from a fluctuating waterbody to a permanent one or vice versa, and can be explained by the blocking or unblocking of flow lines in the bedrock below. That such changes are still happening can be seen by the occasional steep-sided collapse hollows that form in turlough floors, where a new conduit has opened up and is taking the sediment away from beneath the surface. This sediment will eventually reach the sea or fill up further cavities lower down the catchment, altering flows there. Given continuing erosion, the progression of lake to turlough must be more likely than the reverse – the reflooding of turlough basins to become permanent waterbodies (Proctor, 2010). Much of the water from the high Burren flows to the Fergus catchment (and the Shannon estuary), whereas water from the eastern Burren is released from intertidal springs at Kinvarra on Galway Bay.

The dynamic character of the turlough habitat is one of its fascinations. Every year is different and there is no set height to which inundation occurs. One does not know when a particular turlough will flood, how deep it will get and how wide the flood zone will be, how often it will dry out over the winter, or if this year it will become a permanent lake – something that may happen somewhere in Ireland once in a generation. Add in the vagaries of grazing animals, their trampling effect at a waterline in wet weather, their grazing of tall herbs and shrubs in summer, or their absolute numbers, and one has the elements of a complex habitat that is difficult – but not impossible – to understand. In recent years, some of the more intractable questions posed by the habitat have been answered by a major research project based in Trinity College Dublin. The functioning of what is a peculiarly Irish habitat has thereby been partly explained (Waldren, 2015).

Life in a turlough is challenging for both plants and animals. Plants are covered by water when they are still growing in autumn, and then have to start growing again underwater as the temperature rises in spring. Their roots suffer from a lack of oxygen as water displaces air in the soil, and there is an accumulation of sulphide that can be poisonous to finer roots. Aquatic animals find their habitat disappearing in spring, and become increasingly restricted to puddles that heat in the sun. Terrestrial species become concentrated around the margins in autumn as they retreat from the rising waters, and as a result may suffer undue predation. These factors select for temporary use of the waterbody – there are algal blooms in spring that disappear once drying occurs, and there are many tiny crustaceans or water fleas that feed on them and then form resting eggs that last all summer until the water returns. Vertebrates have a better time of it, as they can move elsewhere. Lapwing (*Vanellus vanellus*), Redshank (*Tringa*

FIG 269. Common Frog (*Rana temporia*), with charophyte algae. (Carl Wright)

totanus) and Snipe (*Gallinago gallinago*) nest in some turloughs, along with Mallard (*Anas platyrhynchos*) and possibly Teal (*A. crecca*) in the wetter ones, while Common Frog (*Rana temporia*) and Smooth Newt (*Lissotriton vulgaris*) are favoured by the lack of fish predators. Newt tadpoles take longer to develop than frog tadpoles and so favour sites that drain later. European Eels (*Anguilla anguilla*) were notable in some turloughs in the past, but with the decline in the Atlantic population, the only fish seen today are occasional Three-spined Sticklebacks (*Gasterosteus aculeatus*). Eels can travel over land in their young stage (as elvers) and therefore could reach turloughs, but the sticklebacks may persist over summer in the swallow holes if enough water remains.

FIG 270. Newly emerged stems of Water Mint (*Mentha aquatica*) are found in every turlough. The leaves become green later in the year. (Fiona Guinness)

A visit in spring to any turlough emphasises the aquatic nature of the habitat. The ground is soft, the air smells of a mixture of mud and hydrogen sulphide, and the leaves of plants are breaking through a skin of algal filaments that crunch on a sunny day as one walks over them. Many of the plants have started growth while underwater. There are festoons of *Fontinalis* moss left on fences or walls, weak stems of Amphibious Bistort (*Persicaria amphibia*) and Great Yellow-cress (*Rorippa amphibia*) lie stretched over the ground where they have collapsed after growth in the water, and streamers of Jointed Rush (*Juncus articulatus*) with hair-like reddish leaves add colour to the bleached vegetation.

Some plants have special underwater leaves that differ from those produced later in the year. Reed Canary-grass (*Phalaris arundinacea*) and Curled Dock (*Rumex crispus*) develop particularly narrow underwater leaves, while Creeping Buttercup (*Ranunculus repens*) has narrow, more aesthetically pleasing segments than normal. This last plant is a distinct variety, because the broadleaved Creeping Buttercup does not survive if transplanted into a turlough. The amount of leaf dissection varies with depth of flooding, even within the same turlough (Waldren *et al.*,

FIG 271. Dropping water levels in spring leave stranded stems of Amphibious Bistort (*Persicaria amphibia*) and allow the narrow leaves of the turlough form of Creeping Buttercup (*Ranunculus repens*) to break the surface. (Roger Goodwillie)

FIG 272. Plants growing through the algal mat. Turloughnagullaun, May. (Fiona Guinness)

2006). The plant seems to survive such submergence by maintaining a low level of photosynthesis underwater, including some possible use of the bicarbonate ion. This probably provides oxygen for the submerged tissue – it certainly enables considerable spring growth when the plant is underwater. Brian MacGowran (1985) reported 3 metre-long stems suspended in the water in a turlough near Gort. The buttercup is an example of a species ecotype that is developing a distinct genetic basis in response to selective pressure and may in time form a different species. There must also be others, waiting to be discovered. Creeping Bent (*Agrostis stolonifera*) is an obvious candidate for a turlough form, as it has been found to evolve rapidly in other habitats, changing both its physical structure and physiology to the extent that five separate varieties have been described (Cope & Grey, 2009).

Dried algal mats (or 'paper') spread in patches over the surface of stones or higher plants, or caught as curtains in barbed wire, are another feature of turloughs in spring. Small patches of algae are frequent, and usually comprise mixed filamentous species of *Cladophora*, *Mougeotia* and *Spirogyra*. Occasionally they may cover a hectare or more, forming large white sheets. This usually happens in the richer sites that have abundant phosphate, and the algal deposit on the surface provides a nutrient transfer to the soil. The most extensive sheets form in spring when there is a pause in the decline of water levels, giving time for maximum algal growth. Algae are consistently present in the turlough water; in the autumn, after only a few weeks of flooding, they build up to levels comparable with those found in permanent lakes (Allott *et al.*, 2015). Resting spores are held in the summer vegetation and filaments may also survive if protected sufficiently from desiccation. Algal growth is fast again in spring, as water temperatures rise and light levels increase.

Turloughs also support planktonic algae, which get stranded in the sheets of filaments. Diatoms are nearly always present, and sampling has shown them to be particularly common in winter (Cunha Pereira, 2011). The dominant phytoplankton group is the cryptomonads, single-celled algae with two flagella and an escape system based on projectile threads. These threads propel the organism away from predators or light in a zigzag motion. Other algal species are present in autumn and/or spring, among them *Chlamydomonas*, a textbook favourite.

PLANT SPECIES

The typical turlough appears in summer as a grassy hollow dotted with cattle or horses and crossed by walls that are blackened by the growth of the moss *Cinclidotus fontinaloides*. There are often boulders scattered on the floor, their size and location possibly suggesting that collapse and glaciation have both played a part in the formation of the basin. There may be shallow drains or pools beside the walls or elsewhere, indicating the lowest points of the basin, where water may lie for many months and fully aquatic plants occur. These may not be the points of contact with groundwater but, being low, they will fill first in autumn and dry out last in summer.

The presence of the blackish moss *Cinclidotus fontinaloides* gives a rough guide to the maximum height reached by the water in winter, although it does also grow in the spray zone of waves. At top levels it appears small and stubby, but lower down it flourishes indiscriminately on walls, rocks and tree branches – it is an odd sight growing as an epiphyte on Hawthorn (*Crataegus monogyna*) or Buckthorn (*Rhamnus cathartica*) twigs. At a certain height (or depth), *Cinclidotus* is replaced by the large-leafed Greater Water-moss (*Fontinalis antipyretica*), normally a species of lakes and rivers. This seems to be because Greater Water-moss shades out its smaller relative where the two occur together, but there may also be a physiological reason behind it. Greater Water-moss can persist in a vegetative form for many years and seldom produces capsules, whereas *Cinclidotus fontinaloides* spores regularly.

FIG 273. Boulders covered by the moss *Cinclidotus fontinaloides* at Turloughnagallaun, May. The scrub of Hawthorn (*Crataegus monogyna*) and Buckthorn (*Rhamnus cathartica*) behind indicates the winter water level. (David Cabot)

The changeover in mosses is one example of the zonation of vegetation that occurs in a turlough basin and is obvious as one walks towards the low point. In a perfect world, plant species would be arranged as rings around the basin, grading into each other to form distinct groupings or communities. Around the outside would be those species least tolerant of flooding, while at the centre would be those most tolerant. This rough pattern may be discerned at any site, but there is often the complicating factor of grazing, which reduces the overall height of the vegetation and selectively reduces some species. Animals also create breaks in the plant cover, and annual plants may spring up, regardless of the height in the basin.

As already noted, the flooding factor makes the turlough habitat difficult for many plants to survive, and the ones that do occur are often relatively common elsewhere – they tolerate a wide variety of environmental conditions. Compared to limestone grassland (Chapter 7), the number of species in a metre square is tiny and there are communities where five or 10 are all that can be found. The common ones such as Silverweed (*Potentilla anserina*) or Creeping Bent can tolerate damp or dry soils and occur throughout the basin. Creeping

FIG 274. Amphibious Bistort (*Persicaria amphibia*) in flower, growing with a thin stand of Shoreweed (*Littorella uniflora*). The leaves of the bistort float at the surface and the long stems indicate the earlier water level. (David Cabot)

Buttercup is similarly widely distributed in its specialised form. There is an obvious distinction between 'grassy' and 'sedgey' turloughs, which is based on their nutritional status. Grass-dominated basins such as Turloughnagullaun have a deeper soil that may include glacial till or some other source of nutrient, whereas the sedge-dominated ones like Knockaunroe have a bare scrape of peat or marl at the base and a smaller throughflow of water. Phosphorus is again a limiting factor, just as it is on the pavement soils (Chapter 6). However, even in the most oligotrophic sites there are patches of plants – Amphibious Bistort is a good example – that indicate some enrichment, perhaps along the lines of water flow or at the winter shoreline where debris accumulates. At Carran, a bed of the nutrient-demanding Reed Canary-grass and Bladder Sedge (*Carex vesicaria*) distinguishes one of the main swallow holes, the only place where these species occur in the basin.

Despite the prominence of 'common' plants, there are more specialised species that have a narrower tolerance and occur only at certain points on the transect down to the basin base. Around the margins of many sites Buckthorn is frequent, and the vicinity of a turlough is one of the best places to find it. Its toothed leaves, with their rather parallel veins, unroll in May and are immediately attractive to the Brimstone butterfly (*Gonepteryx rhamni*), one of the most iconic insects of the Burren in spring. This insect is a strong flyer, and is more commonly seen in the morning than in the evening (Harding, 2008). The male has primrose-yellow wings that end in a slight hook, while both sexes have wing venation that mimics the veins on the back of a leaf. Females lay eggs on the twigs or leaves of Buckthorn, and when they hatch, the caterpillars make little holes

FIG 275. Young Buckthorn (*Rhamnus cathartica*) foliage, characterised by the opposite leaves and parallel venation. (Fiona Guinness)

FIG 276. Female Brimstone butterfly (*Gonepteryx rhamni*), showing leaf-like camouflage. May. (Fiona Guinness)

in the leaves before graduating to eating entire ones. The larvae are fully grown after a month or so and pupate below the bushes, hatching mostly in August. The butterflies are again very visible then as they take nectar from various flowers for hibernation. There is a slow winding down in autumn, although they may come out on sunny days in October, before hibernating in nearby woodlands among Holly (*Ilex aquifolium*) or ivy (*Hedera* spp.).

Buckthorn is also the foodplant of a special moth, the Irish Annulet (*Odontognophos dumetata hibernica*), which is restricted to the Burren and seems to be endemic. It was discovered as recently as 1991 by Peter Forder while he was looking for the Burren Green moth (*Calamia tridens*), and has a related subspecies in Spain and France. It is a slightly

FIG 277. Irish Annulet
(*Odontognophos*
dumetata hibernica)
is restricted to the
Burren. (Dave Allen)

nondescript greyish geometrid whose caterpillar may be seen looping along the leaves or flowers of Buckthorn. The adults fly and lay eggs in August, so the larvae do not become active until the following year.

Mixed in with the Buckthorn is a prostrate form of Alder Buckthorn (*Frangula alnus*), which grows in several of the Burren turloughs where pavement is exposed, as does Shrubby Cinquefoil (*Potentilla fruticosa*). Although the Alder Buckthorn occurs in a wide variety of habitats in Ireland and Britain, it seems to need peat in the soil and is happiest in the company of Purple Moor-grass (*Molinia caerulea*) or Black Bog-rush (*Schoenus nigricans*). The Burren form is always prostrate and it is hard to imagine it as the 5 metre shrub that grows elsewhere. One must assume there is a genetic basis for the prostrate habit, and seeds collected in the Burren retained a prostrate form when grown on in Dublin. The population is probably genetically mixed since a prostrate plant transplanted by David Webb produced erect shoots in Dublin (Goodwillie, pers. comm.).

Shrubby Cinquefoil has a limited flowering period in the spring but is so well known from gardens that it can be recognised at any time. Looking into its buttercup-yellow flower, one can see both male and female parts, but only one of these is functional because the species is dioecious. It has a wide mid-latitude distribution in Europe, Asia and North America, and occurs with several chromosome numbers. Having originated in central Asia, the diploids seem to have spread along a southern route to the Mediterranean (Greece and the Pyrenees), while the tetraploids, perhaps with greater hardiness, followed a northern path, reaching the Baltic as well as Britain and Ireland in the postglacial period (Elkington & Woodell, 1963). As a native plant, Shrubby

FIG 278. Shrubby Cinquefoil (*Potentilla fruticosa*) on the shores of Lough Gealáin, Mullagh More, in May. (David Cabot)

Cinquefoil is much more frequent in Ireland than in Britain. It extends for more than 100 kilometres between Co. Mayo and Co. Clare, but is restricted to two small areas in northern England.

Midway down the gradient may be found the Fen Violet (*Viola persicifolia*), with its open flowers of the palest mauve. It grows at a lower level than Heath Dog-violet (*Viola canina*) and often hybridises with that species, the hybrid forming great drifts of colour. The flowers appear from May onwards, but later in the year the plant produces additional flowers that do not open and are self-pollinated. This cleistogamous habit is quite common in violets, and although it limits new genetic combinations it ensures the species survives even in unfavourable habitats when plants are on their last legs, as it requires less energy than making the petals and nectar that are needed to attract insects. The Fen Violet is usually distinguished by its narrow leaves, which are three or more times as long as wide, and its flowers have a pale green spur rather than the yellow of the Heath Dog-violet. It grows in both rich and poor sites, but it is a weak competitor and may be ousted from a richer site if the vegetation is not grazed enough to reduce the grasses,

FIG 279. The Fen Violet (*Viola persicifolia*), showing its narrow leaves and pale flowers. May. (Fiona Guinness)

FIG 280. A profusion of hybrid violets (*Viola riviniana* × *canina*) at Turloughnagullaun, May. (Fiona Guinness)

or if fertilisers are used. This may have been its fate along the Shannon valley, where it was recorded in former times (Preston *et al.*, 2002).

Deeper sections of the turlough that are flooded for longer may allow a temporary increase of Lesser Marshwort (*Apium inundatum*), Horned Pondweed (*Zannichellia palustris*) or Pond Water-crowfoot (*Ranunculus peltatus*). When they dry out, Needle Spike-rush (*Eleocharis acicularis*), Pink Water-speedwell (*Veronica catenata*) or Mudwort (*Limosella aquatica*) develop, diminutive plants that one may have to stoop to see.

VEGETATION ZONES

At the outset, a distinction should be made between nutrient-rich basins such as Turloughnagullaun, where there is a depth of soil, and nutrient-poor ones like Carran or Knockaunroe near Mullagh More. Both types of site may have boulders and rocks depending on the amount of clearance that has gone on in former years, and it is on the ground in between these where one can see a difference. The type of vegetation is intimately tied up with the levels of phosphate and nitrate available, which dictate the proportions of grass or sedge (Sharkey *et al.*, 2015).

At the richer sites, the dominant turlough grass is Creeping Bent, although at the upper levels Perennial Ryegrass (*Lolium perenne*) and Crested Dog's-tail (*Cynosurus cristatus*) may play a part, along with Daisy (*Bellis perennis*) where there is a deep enough soil. Meadowsweet (*Filipendula ulmaria*), White Clover (*Trifolium repens*) and Meadow Buttercup (*Ranunculus acris*) are ubiquitous also, although the last soon gives way to Creeping Buttercup at lower levels. Lady's Smock (*Cardamine pratensis*) and Hairy Sedge (*Carex hirta*) remind one of a connection to water, and there may be a few clumps of Soft Rush (*Juncus effusus*) or Yellow Flag (*Iris pseudacorus*), species that disappear lower down the profile as they cannot survive the particular flooding conditions. Grazing is common at these rich sites, but if it is not too severe Lesser Meadow-rue (*Thalictrum minus*) may be noticeable, adding interest to a generally mundane plant cover. Where the animals begin to have an impact, there are often annual species from farm surroundings, including Shepherd's Purse (*Capsella bursa-pastoris*), Chickweed (*Stellaria media*) and Pineappleweed (*Matricaria discoidea*). A waterline may sometimes be seen if the winter was wet and grazing cattle were limited to the shoreline, leading to an abundance of Great Plantain (*Plantago major*), Autumn Hawkbit (*Scorzoneroides autumnalis*) or Common Couch (*Elytrigia repens*), which flourish at these levels. Seldom seen in flower in turloughs, Common Couch is nevertheless identifiable (at least to most gardeners) by its scattered shoots with turned-over greyish leaves, and its often slightly hairy lower parts.

Turloughs that have a shallower, nutrient-poor soil support Purple Moor-grass as well as the small sedges (Common Sedge, *Carex nigra*, and Carnation Sedge, *C. panicea*), plants that are generally seen in abundance along a transect down the basin. In the upper parts, clumps of Tall Fescue (*Schedonorus arundinacea*) mark a shoreline of sorts, along with Heath-grass (*Danthonia decumbens*) or Quaking-grass (*Briza media*). Devil's-bit Scabious (*Succisa pratensis*), Grass of Parnassus (*Parnassia palustris*) and Tufted Vetch (*Vicia cracca*) are usually there, too, giving way to Tormentil (*Potentilla erecta*) and Silverweed lower down, along with the hook-tipped moss *Pseudocalliergon lycopodioides*. In most of the Burren turloughs the sedges rule supreme, whether the greyish types (Carnation Sedge and Glaucous Sedge, *C. flacca*), the yellowish Tawny Sedge (*C. hostiana*) or the greenish Common Sedge. The yellow flowers of the buttercups, Bird's-foot Trefoil (*Lotus corniculatus*) and Creeping Cinquefoil (*Potentilla reptans*) sprinkle the surface, but less common species appear too. The starched whiteness of Northern Bedstraw (*Galium boreale*) stands out in some sites, as does the off-white Sneezewort (*Achillea ptarmica*), a relation of Yarrow (*A. millefolium*). Violets (*Viola* spp.) will be at their most visible in May, but persisting later as leaves, and it will be worthwhile looking for the Adder's-tongue (*Ophioglossum vulgatum*) around the odd boulder.

Northern Bedstraw is true to its name in Britain and Ireland, being commonest north-west of a line from Limerick to Leeds. Classic research in Teesdale identified 21 different clones in that population, individuals or colonies with differing genetic make-up, shown mainly by visible leaf differences (Dale & Elkington, 1984). Such diversity endows the species as a whole with tolerance to many habitat factors, such as water content, pH and grazing, and illustrates its potential resistance to habitat change, however caused. The rarest plant species – relicts with one or two sites in any location – have a very limited genetic base and may not have the inbuilt capacity to survive climate change.

A particularly recognisable community in turloughs is the *Schoenus* fen, a stand of Black Bog-rush in which Meadow Thistle (*Cirsium dissectum*), Purple Moor-grass and Devil's-bit Scabious are the dominant species but allow in occasional orchids such as Early Marsh Orchid (*Dactylorhiza incarnata*) or Yellow Sedge (*Carex lepidocarpa*). The small Eyebright (*Euphrasia micrantha*) is also frequent and is notable in this species complex for the amount of purple on its white flowers. The community occurs on the upper slopes of the turlough and is found where there is a skin of peat on an infertile soil. Unexpected plants of drier soils may turn up on the tussocks of Black Bog-rush, including Creeping Willow (*Salix repens*), Heather (*Calluna vulgaris*) and Lady's Bedstraw (*Galium verum*). At Knockaunroe and Lough Gealáin, Shrubby Cinquefoil grows in this community, as well as in flooded pavement.

The plants that occur on the mid-slope and below include many of the sedges and grasses already mentioned, but also more strictly aquatic plants like Common Spike-rush (*Eleocharis palustris*), Water Forget-me-not (*Myosotis scorpioides*), Water Horsetail (*Equisetum fluviatile*) and Amphibious Bistort. Creeping Cinquefoil is also a constant presence, which may strike one as peculiar since it also likes the edges of dry paths and roadways. There may be patches where animals have damaged the turf, allowing a great spread of Curled Dock or annuals such as Northern Yellow-cress (*Rorippa islandica*) and Red Goosefoot (*Chenopodium rubrum*), which complete their life cycle between the turlough drying out and reflooding in autumn – sometimes a period of just a few weeks. Both flower when at a young stage, so that the final turlough Red Goosefoot may be only 4 cm high, compared to a farmyard plant at 70 cm.

Northern Yellow-cress was separated relatively recently, in 1968, from its commoner and more impressive relative, the Marsh Yellowcress (*Rorippa palustris*), which has twice the number of chromosomes and grows much taller. This species occurs by pools and other waterbodies, particularly those visited by wildfowl. It is adaptable and may germinate in August in a wet year when the water level stays up, flowering right through until October. Orange Foxtail (*Alopecurus*

FIG 281. a. The Meadow Thistle (*Cirsium dissectum*) is often an associate of Black Bog-rush (*Schoenus nigricans*) and Purple Moorgrass (*Molinia caerulea*). May. (Fiona Guinness) b. Bogbean (*Menyanthes trifoliata*) growing with Marsh Bedstraw (*Galium palustre*) at the base of a wet turlough. (Fiona Guinness) c. Marsh Marigold (*Caltha palustris*) grows near water at any level in the turlough. (David Cabot)

aequalis) is a similar opportunistic species, found on the fringes of the Burren at Termon Lough. It is by no means rare in Britain, so may easily have been missed elsewhere owing to its late flowering. In the spring, bare mud often reveals the narrow leaves of Thread-leaved Water-crowfoot (*Ranunculus trichophyllus*) or germinating bistorts such as Redshank (*Persicaria maculosa*) and Small Water-pepper (*P. minus*). It is also always worth looking for Mudwort, although the species is not common. It is a tiny plant related to figworts (*Scrophularia* spp.), unlikely though that may seem, as it has spoon-shaped leaves and single minute flowers. It forms stolons, so a single seed may give rise to a colony in a few weeks, making it much more visible. Like most of the 'mud' species, Mudwort's seeds are probably spread on birds' feet and remain dormant underwater until exposed by drawdown.

Towards the base of many turloughs, Bogbean (*Menyanthes trifoliata*) and Marsh Cinquefoil (*Comarum palustre*) may be found sticking through the grasses or sedges, indicating that water is just below the surface. But water is also exposed at many sites, certainly in the early part of the summer. Puddles remain, where water is held until it evaporates, and there are channels and drains that stay wet all year. Where there are enough nutrients, Amphibious Bistort and Reed Canary-grass dominate these surroundings, while elsewhere there may be Floating Sweet-grass (*Glyceria fluitans*) or Lesser Spearwort (*Ranunculus flammula*). Bulky plants of Marsh Marigold (*Caltha palustris*) or Fine-leaved Water-dropwort (*Oenanthe aquatica*) grow up through the strands of algae, while Broad-leaved Pondweed (*Potamogeton natans*) and Various-leaved Pondweed (*P. gramineus*) float quietly on open water, flowering and fruiting before they are left high and dry. The latter species is known as the Various-leaved because it can have distinct floating and submerged leaves. However, in turloughs the submerged leaves are seldom visible when the plant is accessible, and one is more likely to see shining green leaves sticking into the drying vegetation and supporting their fruiting spikes.

Lesser Water-plantain (*Baldellia ranunculoides*) and Shoreweed (*Littorella uniflora*) grow in marly, nutrient-poor sites, sometimes with Water Germander (*Teucrium scordium*), as in the south-east Burren. The fat brown shoots of the moss *Scorpidium scorpioides* are also seen, and sometimes the hair-like leaves of Bulbous Rush (*Juncus bulbosus*), reddened by their long submergence over the spring. It is worth looking out here for the marsh form of dandelion (*Taraxacum palustre*), which is commoner in the Burren than anywhere else in Ireland or Britain. It is probably the most distinctive of the 90 species of dandelion recorded in Ireland, and can be told by the undivided leaves, reddish stems and neat flowers (Dudman & Richards, 1997). A closely related form, also from the Burren, was named *Taraxacum webbii*; it seems to be endemic to Ireland, and also occurs in Co. Mayo

FIG 282. The base of an oligotrophic turlough, revealing the reddened leaves of long-submerged Bulbous Rush (*Juncus bulbosus*), May. (Roger Goodwillie)

FIG 283. A special
marsh dandelion
(no common name)
(*Taraxacum palustre*),
showing the untoothed
leaves. April.
(Jenny Seawright)

and Co. Sligo. Both of these species produce seeds asexually – they grow from
the ordinary cells of the parent plant without any meiosis taking place. Indeed,
they have both undergone mutations that prevent them from producing pollen,
and by saving this energy sink they may have benefitted for their survival in an
adverse environment.

Stranded plants of Floating Club-rush (*Eleogiton fluitans*) and even the Bog
Pondweed (*Potamogeton polygonifolius*) can be seen in nutrient-poor sites, left
lying on the surface after the retreat of the water. These and other species are
generally thought of as acid plants elsewhere, or at least plants of bogs. Their
presence in the Burren parallels terrestrial species like Heather and Goldenrod
(*Solidago virgaurea*), low-nutrient demanders that grow in both acidic and
alkaline habitats.

In turloughs with a central zone that scarcely ever dries, beds of Tufted Sedge
(*Carex elata*) and Great Fen-sedge (*Cladium mariscus*) may persist in nutritionally
poor sites, giving rise to an organic peaty soil. Carran turlough has such peat,
which appears to have been cut in the past. Common Club-rush (*Schoenoplectus
lacustris*) and Common Reed (*Phragmites australis*) are more characteristic of richer
sites. All these species survive six to seven months of water coverage, although
they seldom grow as tall as those in a more 'normal' lake or marsh.

Stoneworts, or charophytes, are found in many turloughs – Knockaunroe
has particularly fine beds. However, with the vagaries of flooding and the
seasonal nature of growth, the species found are mostly the opportunistic and
relatively common types. While no species is particularly associated with the
habitat, Common Stonewort (*Chara vulgaris*) and *C. hispida* seem to be the main

ones that occur. There is also a record of *C. curta* in one turlough, although this stonewort is generally found in permanent waters (Langangen, 2005). Aside from the distinctively charoid smell of these plants, the most noticeable feature on picking them up is the feel of the encrusting calcium carbonate that gives them their name and makes the segments sometimes snap off. *Nitella* is a related genus whose members do not become encrusted and appear brownish green rather than the greyish green of most *Chara*. Field botanists in the nineteenth and early twentieth centuries were keen to collect and study these algae, which were included in field guides and floras. They subsequently fell out of favour, only recently attracting renewed interest when the rarity of some species and their sensitivity to pollution became clear, aided by the easy availability of diving gear.

PLANT ECOLOGY

Research has shown that the factors that control where plants grow in the turlough basin are broadly the length of the submergence period and the level of phosphate found in the turlough water (Penck *et al.*, 2015). The level of soil phosphate might seem a more obvious measure, since plants absorb the mineral through their roots, but this was found to be so variable over most basins that it was much less easy to correlate with the flora. In terms of species, Shrubby Cinquefoil and Black Bog-rush grow with a short-duration inundation and low phosphate, while Meadowsweet and White Clover may be submerged for longer but require medium or high phosphate conditions, as does Amphibious Bistort (Sharkey *et al.*, 2015). Lesser Water-plantain and Shoreweed, as expected, grow with long submergence and low phosphate, whereas Fine-leaved Water-dropwort grows only with long submergence and high phosphate. Water Forget-me-not is one of few species that needs high phosphate but is unaffected by short or long immersion.

The period of submergence (hydroperiod) varies each year and also between sites. Figures of 4.5 months for Turloughmore, 5.3 months for Lough Aleenaun and 7.1 months for Knockaunroe were established in the years around 2008 (Naughton et al., 2015). These figures do not represent one complete fill that lasts the whole winter, however, it is more a reflection of the passing underground flow. Lough Aleenaun especially has periodic rises in water level corresponding to rainfall that decline rapidly, whereas Knockaunroe fluctuates much less often. The same story applies to the emptying phase, when the water table below the turlough declines sufficiently to allow water to drain out of the basin. In the absence of additional rain, Lough Aleenaun empties in about 10 days, while

Knockaunroe takes 50 days or longer. In exceptional years, the flooding period may be extended by several weeks until the end of April. This must interfere with the onset of spring growth for many species, which find themselves in an oxygen-poor soil just when they are adapted to making new roots.

When a turlough dries out and the soil becomes aerobic once more, there is a release of nutrients from decomposition and a consequent stimulus to plant growth. The plants get ahead of their animal grazers; repeated visits at this time illustrate the high growth rates possible – and the attraction of the vegetation to livestock and farmers.

TREES

One of the questions about turloughs that has not been answered in any detail is how deep into a basin woody plants would grow if there were no grazing. Shrubs and trees surround many turloughs – a fringe of Hawthorn, Buckthorn, Spindle (*Euonymus europaeus*) and Guelder-rose (*Viburnum opulus*) is the most common, perhaps with Creeping Willow towards the bottom. In nutrient-poor sites, Alder Buckthorn, Shrubby Cinquefoil and Eared Willow (*Salix aurita*) are often found. The absence of trees in turloughs has been commented on ever since Lloyd Praeger wrote his initial impressions of the habitat in 1934, but few people if any have experimented with trees at the lower levels, either in the field or in simulated conditions. It is a common experience that large Grey Willow (*S. cinerea*) trees occur towards the bottom of a few turlough basins.

FIG 284. The Guelder-rose (*Viburnum opulus*) is a distinctive shrub of wet places, with maple-like leaves, and heads of flowers in May and June. (Fiona Guinness)

Trees aged 50–60 years grow behind a protective wall on the southern side of Turloughnagullaun and it seems likely that they could grow lower down, although probably at progressively smaller sizes as they would be killed more often. Small bushes of Hawthorn and Guelder-rose are regularly found around boulders below the 'treeline', where they are protected from grazing to some extent and where pockets of air may be caught when the turlough is full of water.

Willows are particularly good at withstanding floods; if necessary, they can produce new roots at the soil surface, where oxygen conditions are better, and even their seedlings can withstand flooding for up to 12 weeks. Our native Grey Willow may not be as tolerant in this respect as the White Willow (*Salix alba*), which seems to be the champion European tree. It can tolerate being flooded for more than 90 per cent of the growing season on the floodplains of Continental Europe (Glenza *et al.*, 2006). Even the Grey Willow has enough resistance to grow halfway down most turloughs, if not to the base. A small turlough in woodland at Coole Park, just to the east of the Burren, is completely shaded by willows of several species.

Apart from willows, the Pedunculate Oak (*Quercus robur*) is tolerant of a good deal of flooding and grows on turlough edges in several places. In autumn,

FIG 285. Pedunculate Oak (*Quercus robur*) at the edge of Rockforest turlough, still in full leaf in October. (Roger Goodwillie)

the rising water infiltrates the soil while the tree's canopy is still leafy and laps around the trunk. The reduction of oxygen in the soil causes alcohol to be produced by the oak's roots, and this is carried to the leaves to be further metabolised into useful products (Copini *et al.*, 2016).

In spring every year, it is apparent that the woody plants growing lower in turlough basins produce leaves and flowers later than the upper ones. This is even seen on individual plants, giving a somewhat top-heavy appearance to bushes such as Hawthorns, when the top branches flower and the lower ones are still unrolling their leaves. It is particularly notable on different plants of Creeping Willow, when the catkins burst over an extended six-week period depending on where the plant is growing. Most likely this is a temperature effect as the water will be cooler than the air at this time, at least during the day, although it could also be the result of a lack of oxygen. But flooding eventually has debilitating effects on the plant; following record water levels in Co. Galway turloughs in the mid-1990s, dead trees were seen all around the basins. Dead Ash, Hawthorn and Blackthorn were a common sight in the most extreme areas. Higher up, the flooding killed Juniper (*Juniperus communis*), Holly and Common Gorse (*Ulex europaeus*). Some of the dead Blackthorns produced suckers from their roots after some weeks, re-establishing themselves that way, while the

FIG 286. Female Creeping Willow (*Salix repens*) plant shedding seeds in May. (Fiona Guinness)

Buckthorns resprouted from their trunk among a haze of dead branches. Such incidents are reflected in the wood structure of the trees: the underwater vessels collapse under the influence of flooding in some species, leaving a trace in the annual rings (Copini *et al.*, 2016). Another effect is the overgrowth of the breathing pores (lenticels) on the trunks of trees, particularly noticeable in Ash owing to its smooth bark.

Flooding prevents air reaching the soil, and any available oxygen it contains is used up within two to three weeks by microorganisms and plant roots. When plants are dormant in winter this scarcely matters, but in spring or summer, if the water is still there, the roots of most plants become stressed and eventually start to die. Amphibious plants produce trace levels of ethylene when stressed, which is one factor that causes stems to elongate; it may cause growth promotion itself or interacts with the normal hormones in the tissues to have this effect. It is also a stimulus for new roots to form at the base of the stem, where the soil surface has the highest levels of oxygen and life for roots is easier than at depth. Many of the herbaceous turlough species have stolons or rhizomes on the surface – White Clover, Silverweed, Creeping Cinquefoil and Creeping Bent are clear examples, and the mode of their spread also gives them a clear advantage in the exigencies of a flooded life when adjoining plants may die back and expose new soil. It may not be a coincidence that the lowest woody species in the modern turlough is often Creeping Willow – perhaps the vanguard of a woody army that would colonise the upper parts of many turloughs in the absence of grazing animals.

GRAZING PRESSURE

After flooding it is grazing pressure that defines the appearance of most turloughs, modifying the vegetation but also the habitat for invertebrates. The grassy turloughs are those most closely grazed and may appear like a lawn towards the end of the summer. But even here there may be areas behind walls that have not been grazed for some reason, offering a welcome relief to the nibbled vegetation elsewhere. Reed Canary-grass and Meadowsweet often dominate such ground, reducing the variety of other plant species, especially the violets, which depend on a decent light exposure. By way of contrast, the vegetation of a sedgey, nutrient-poor site is basically unpalatable, and less impact is therefore seen in the turloughs of the south-east Burren. Even here, however, there is some grazing, perhaps by the Irish Hare (*Lepus timidus hibernicus*), removing the flowers of the sedge species and frustrating attempts at their identification.

Grazing is a fact of life in almost every turlough, and the more nutrient-rich sites prove valuable for cattle, horses and, sometimes, sheep. This value has dictated the building of walls across many basins, dividing the best grassland between different owners and incidentally providing a resting site for the stones that are scattered on the turlough floor. The central part of the turlough is sometimes a commonage without walls, but other fields radiate outwards onto drier ground, with laneways for access. The number of grazing animals has seldom been counted, but research near Kinvarra measured it at two livestock units (adult cows) per hectare, a figure that must fluctuate with the market for beef animals and available subsidies (Ni Bhriain *et al.*, 2003). In this case, cows and their calves took 70 per cent of the herbage, and horses the remaining 30 per cent. Farm units are tending towards intensification and the number of cattle in Co. Clare is currently rising, so the pressure on turlough vegetation may yet increase. The behaviour of the animals in the turlough means that there is often a nutrient transfer from the central grassy basin to the surrounding scrub where animals take shelter, adding their dung to the flood debris that also accumulates there, and encouraging such plants as Chickweed (*Stellaria media*) and Common Nettles (*Urtica dioica*).

People have made other uses of the habitat in the past. Racecourses are marked in some basins on old Ordnance Survey maps – for example, just west of Gort – while Turloughmore is located beside a fair green. Sedge peat has been dug from several sites; the remains at Knockaunroe show that it was 1 metre thick in places (MacGowran, 1985).

FAUNA

The differing lengths of flooding in Ireland's turloughs reflect one aspect of the variation to which plants and invertebrate animals found here must adapt. One can imagine that they are constantly adjusting to the last flood event, extending down the basin in dry periods and retreating towards the margins when the water rises following a series of wet winters. Turloughs are very much boundary zones, and they suit animal species that are mobile enough to avoid the worst of the flooding, those that can multiply quickly to take advantage of the water rise, or those that can resist desiccation. Fully aquatic species add another level to turloughs that retain a marshy centre or a deep, lake-like hollow (e.g. Lough Gealáin). Here, the fauna is dominated by detritivores (animals that feed on dead plants), such as caddis flies, beetles and Crustacea. In contrast to the vegetation, the main influence on the invertebrate fauna seems to be hydroperiod rather than nutritional status – in other words, the availability of water rather than its quality (Porst & Irvine, 2009).

The turlough fauna was reviewed in 2016 by Julian Reynolds. It may be divided into permanent members that appear immediately the site is inundated, and ephemeral ones that have to find the site anew each year with their adult dispersal phase. This is probably the reason why different sites develop different faunas despite very similar environmental conditions. The animal community is inherently simple and unstable, just as the habitat is, and a newcomer may tip the balance one way or another. In some years, tolerant species survive through the summer and are ready to breed in the autumn, whereas in dry years they may have to reinvade.

The permanent fauna is truly adapted to the habitat. The animals have a resting or dormant stage in summer – usually eggs – and when the water returns, they hatch and take immediate advantage of the conditions. They are generally cold-water forms since the sites are wet only over the winter. Many become very numerous in the uncontested conditions – at least until predators arrive. Small crustaceans in the form of water fleas such as *Daphnia*, *Chydorus* and *Cyclops* occur in every site and are often concentrated in the puddles on the turlough floor, where they can clearly be seen, apparently in perpetual motion. More than 40 species of water fleas occur in turloughs in general and probably two-thirds of these are found in the Burren (Reynolds, 2016). *Eurycercus glacialis* is one of the more interesting; it is large for this group – up to 6 mm in length – and is found in fully drying turloughs that do not have a resident fish population. Like some of the plant species, it is a glacial relict, found today in Arctic and subarctic regions. So far, it has been recorded in the Burren from lowland sites, such as Roo in the east of the region.

The fairy shrimp *Tanymastix stagnalis* is one of the most mysterious members of the turlough fauna. First found in Rahasane Turlough to the east of the Burren, it was subsequently seen in a temporary pool near Castle Lough (and elsewhere) in 1976, but has not been reported since. It is colourless, long and narrow, and at more than 1 cm in length is also is the largest floating crustacean one would hope to see in turlough water. The animal has a number of bristle-edged appendages along its body with which it filters microplankton from the water (see Fig. 114). It has the endearing habit of swimming on its back, so that the rippling motion of the 'legs' is plain to see, as they propel food towards the mouth and also keep the animal moving slowly. It is usually considered a cold-water species, but in Ireland this means the eggs may hatch at any time outside winter, and the life cycle may be completed in two to four weeks. Many of the females recorded were carrying eggs, which are resistant to desiccation. They carry these until they die, so all are deposited in the lowest part of a turlough where the last of the water persists and where the eggs must easily be picked up by birds' feet. Snipe seem to be the prime suspect in this, because they favour small waterbodies and wet fields where new pools may form in wet weather. Once transferred to these new sites, the fairy shrimp may then hatch and develop without predators – a critical part of the complex equation that governs its lifestyle.

Larger Crustacea may include the freshwater shrimp *Gammarus lacustris*, or more commonly *G. duebeni*, which is widespread in fresh waters in Ireland in the absence of its relative *G. pulex*. The water louse (*Asellus* sp.) may also be found, especially in richer sites. Both the shrimps and louse are detritivores so have no direct impacts on other species. They seem to survive dry periods in swallow holes or deeper in the groundwater.

Some snails are numbered in the permanent turlough fauna. Most are common in other wetland habitats, including the Wandering Snail (*Radix balthica* (*peregra*)), but there is also the Marsh Snail (*R. palustris*) and the Smooth Ramshorn snail (*Gyraulus laevis*). On the smaller scale is the Common Bithynia (*Bithynia tentaculata*), one of the frequently occurring species and more of a specialist. It has an operculum and hence is able to protect itself from desiccation during the worst of droughts. One other hazard that snails suffer is attacks by the larvae of certain marsh flies (Sciomyzidae), which lay their eggs in water during the flooded phase. The larvae enter snails as parasites and may pupate there before emergence.

Aquatic species generally have good dispersal – one finds pond skaters (*Gerris* spp.) and water-boatmen (*Corixa* spp.) within a few days of filling a garden pond. So, these insects, along with mayflies, diving beetles and dragonflies, find new turloughs as they appear in autumn and lay eggs that will hatch and grow through the winter months. The beetle fauna is usually rich – 58 aquatic species have been recorded from the turlough habitat as a whole, with 14 captured in Knockaunroe alone (Bilton & Lott, 1991; Reynolds, 2016). Most of these are predatory and will eat anything that moves, from water fleas to tadpoles. The Lipped Diver (*Agabus labiatus*) is a rare species whose stronghold is turloughs because elsewhere it is easily taken by fish. It is one of few flightless water beetles, so dispersal is more of a problem than for most of the group. The semi-aquatic weevil *Bagous brevis* also occurs, with stations in Ireland restricted to the area around Mullagh More (Chapter 5).

One of the specialist damselflies associated with Ireland's turloughs is the Robust Spreadwing (*Lestes dryas*), although it occupies other marshes too. It seeks dense vegetation, laying its eggs in plant stems that will die and be flooded in winter (Nelson *et al.*, 2011). The eggs develop a little in autumn, but the larva does

FIG 287. Robust Spreadwing (*Lestes dryas*), a specialist of the turlough habitat, June. Note the parasites at the base of its legs. (Christian Fischer)

FIG 288. Female Common Darter (*Sympetrum striolatum*), July. The male has a redder body. (Carl Wright)

not hatch and escape into the water until the following spring. There it spends time swimming or hanging in the water, making it an easy prey for fish – and illustrating the species' evolutionary imperative for temporary waterbodies. The adults hatch in late June and may be seen until August. They are metallic green with a partially blue segment on the body behind the wings.

A close relative, the Emerald Damselfly (*Lestes sponsa*), is commoner overall in turloughs, and is distinguished by having a little more blue on the segments and a finer body. It shares the Scarce Emerald's weak flight and its habit of resting with wings half open.

The main dragonfly of turloughs is the Ruddy Darter (*Sympetrum sanguineum*), which appears roughly at the same time as these damselflies. The male has a blood-red body with a noticeable waist and is most frequent around turloughs that retain pools in summer. It also likes a woodland fringe to the turlough, as it brings its prey to a tree branch to eat. It differs from some larger dragonfly species by completing its life cycle in one year (rather than over several), so is well suited to the temporary habitat offered by a turlough.

Another adaptation to the habitat is shown by ground beetles that move in response to the flood waters and take refuge in surrounding mosses or other vegetation. Carabid and staphylinid beetles are usually present both during the flooded and dry stages of turloughs. They are largely nocturnal but can be found during the day under moss or organic litter. There is, in fact, a group of beetles known as the moss-edge community, which includes such interesting names as the Turlough Long-claw (*Dryops similaris*) and the Spotted Scavenger Beetle (*Berosus signaticollis*), both rare and recorded from the Burren (Foster *et al.*, 2009). This group is not well developed in Britain and is of high conservation value; it goes some way to rebalancing the value of the turlough habitat, as most people think of the region's flora first.

BIRDLIFE

Although the Burren turloughs are not full of birds, they do support characteristic species that one would expect to see in the region. As these waterbodies are transitory in nature, the bird species may change from day to day – the Black-headed Gulls (*Chroicocephalus ridibundus*) that may be present in spring while the water persists may be gone after a few days to more permanent lakes. Grey Herons (*Ardea cinerea*) are regular visitors as long as there are pools of water containing tadpoles or newts, while Pied Wagtails (*Motacilla alba yarrellii*) and Swallows (*Hirundo rustica*) feed on hatching flies, often chironomid midges. If

FIG 289. Pied Wagtail (*Motacilla alba yarrellii*) gathering food for its nestlings on the edge of Carran turlough in May. (Fiona Guinness)

some water remains in the basin, a few Snipe may nest, revealed perhaps by their footprints on the drying marl, but the quintessential wader of the turloughs is the Lapwing, even though it is becoming rare except on the Aran Islands. A pair or two of Mallards is usually a feature of the wetter turloughs, while Teal may be flushed occasionally from a dense bed of sedges.

Migratory waders spend little time on turloughs in their northward migration in spring, but in autumn, when there is a more leisurely movement, one could hope to see a few Golden Plovers (*Pluvialis apricaria*), Dunlins (*Calidris alpina*) and Whimbrels (*Numenius phaeopus*), and perhaps a Spotted Redshank (*Tringa erythropus*) or Ruff (*Calidris pugnax*) if the weather has been wet (Clare Birdwatching, 2018). Winter, however, brings in more birds, and the larger turloughs such as Carran or Castle Lough usually support some dabbling duck – Wigeon (*Anas penelope*), Teal, Mallard and, in small numbers, sometimes Shoveler (*A. clypeata*) and Gadwall (*A. strepera*). Swans are regular and there is a mobile population of Whooper Swans (*Cygnus cygnus*) in Co. Clare/Co. Galway that may supply 10–20 birds in any year.

Lapwing, Curlew (*Numenius arquata*) and Golden Plover are the frequent waders in winter. Redshanks advertise their presence when they see an observer, whereas Snipe lie hidden until approached. Both species are usually found in the richer sites (where there are more worms) when water levels allow access.

FIG 290. Whooper Swans (*Cygnus cygnus*) occur in small numbers in winter on the Burren turloughs. These individuals have iron staining on their necks from feeding on submerged vegetation. (Ken Kinsella)

The presence of birds in any number encourages scavenging by Red Foxes (*Vulpes vulpes*), which, with Irish Hares, are the commonest mammal species seen in turloughs. Sometimes, however, one may see a Pine Marten (*Martes martes*) on a rocky edge or, with luck, encounter the delightful sight of young Stoats (*Mustela erminea*) playing hide-and-seek in a mossy turlough wall.

CHAPTER 10

Lakes, Fens and Other Permanent Wetlands

T URLOUGHS MAY BE THE MOST unusual of the Burren wetlands (Chapter 9), but there is a suite of other wet habitats that may not have the same dramatic changes from winter to summer yet provide considerable interest nonetheless. Lakes are pre-eminent in the south-east of the area, their marly white shallows giving them a greenish or turquoise colour in aerial photographs. Some of them fluctuate considerably, flooding their margins for long periods, while others are more stable and have shores that are either rocky or covered by fens or reed beds. Smaller fens occur in seasonally wet hollows that have some groundwater influence, but do not experience the flooding that characterises a turlough. There are also flushes on hillsides where an impermeable band of rock brings water to the surface for a short or longer period. Finally, there is the Caher River, which flows along the surface of the Burren before discharging at Fanore on the west coast. The larger Fergus River, at the very southern point of our area, sinks and returns to the surface in several places near Corrofin.

LAKES

The lakes of the Burren are strongly alkaline, with a pH of 7–8, and are arranged in two series, separated by a drainage divide. The northern loughs – Bunny, Travaun and Skaghard – flow north through the karst to Kinvarra and have a relatively small catchment. The southern loughs, however, pick up some water from the lowlands near Lough Cutra and have a partially overland flow to the

FIG 291. Lough Bunny on a wild and stormy day. (David Cabot)

south-west, in some cases in artificial channels. In this system, the Castlelodge River first enters Muckanagh Lough, and then the water moves slowly through the Ballyeighter Loughs, Lough Cullaun and Lough Atedaun, before discharging into the Fergus River. This entire area is a network of shallow lakes and turloughs, flooded marginal fens and calcareous swamps, and is best viewed in winter from the summit of Mullagh More. Like the northern lakes, the water has a high alkalinity and there are extensive marl deposits, but it carries a greater range of minerals, derived from its sojourn on glacial till near the Slieve Aughty range. This is reflected by the populations of fish that succeed there. Formerly tenanted only by the native Brown Trout (*Salmo trutta*) and European Eel (*Anguilla anguilla*), the lake fauna has been augmented by introduced coarse fish, especially Perch (*Perca fluviatilis*), Tench (*Tinca tinca*), Rudd (*Scardinius erythrophthalmus*) and Pike (*Esox lucius*), which have relatively fast growth rates. Ballyeighter Lough holds an Irish record for Tench, whereas Perch and Rudd are widespread everywhere, and Pike is important in Lough Cullaun and Lough Atedaun.

Of all the loughs in the Burren, the northern ones are the purest examples of marl lakes, and of these, Lough Bunny is the most famous and well-researched example, not least because it sits beside a main road. Much of the lake is less than 3 metres deep, but two basins drop to 14 metres, separated by a saddle at a depth of about 5 metres (Lough Gealáin has a similar, 16 metre-deep basin). No streams

FIG 292. Lough Gealáin in Gortlecka, June. The deep central pool is surrounded by a flat shelf of marl with islands of fen. (Steve O'Reilly)

FIG 293. East end of Lough Bunny, with Cloondooan or Boston Castle in the distance. (Fiona Guinness)

flow into the lake, but submerged springs and drains do occur, much as in a turlough basin. Approaching the southern shore over limestone pavement that supports a full Burren flora, including Dropwort (*Filipendula vulgaris*), one notices first a marly skin of material covering submerged rocks and plant bases. This is krustenstein, a cyanobacterial crust formed as a hard calcareous mat surrounding the filaments of *Schizothrix fasciculate* (Roden & Murphy, 2013). Sometimes, sections of the crust break off and, rolling around on the lakeshore, develop into spherical pebbles. Other, ghostlier, spheres a few centimetres across may also be present in the water; these are the pale green jelly balls of *Ophrydium versatile*. This is a complex organism of ciliates and a green alga, a symbiotic arrangement also seen in sponges and jellyfish.

Pausing on the flat, pitted rocks of the shore, it is well to consider röhrenkarren, the vertical, closed tubes that develop by a natural dissolution of the underside of the limestone. They form at the top of air pockets trapped by the lake surface where condensing (pure) water vapour dissolves carbonate ions from the rock. Such erosion, which is also important in cave development, is much faster than that which is constantly reducing the general surface of the rocks and the width of grikes (Simms, 2003).

As one goes deeper into the lake, a number of stonewort (Charophyta) algae occur, their cell walls encrusted with the calcium carbonate that will add to the marl deposits when they die. *Chara curta* and *C. rudis* grow in waters of 1–3 metres, generally followed by *C. globularis* in deeper water. Below 7 metres there are very extensive beds of the similar but non-encrusted *Nitella flexilis*, which is able to grow at this depth thanks to the clarity of the water column above (Roden & Murphy, 2013). An earlier survey also found the rare *Nitella tenuissima* in the

FIG 294. The stonewort *Chara rudis* in 1.5 metres of water at Ballyeighter Lough. July. (Cilian Roden)

same lake (Langangen, 2005). All charophytes are things of beauty that should be observed under a lens or microscope. To give the best view, the encrusting lime should first be dissolved off some species, although there is not much one can do to eliminate the smell, variously compared to stale garlic or hydrogen sulphide. The algae have symmetry like a horsetail (*Equisetum* spp.), with long internodes separating whorls of branches that are decorated with spines. The internodes contain giant cells several centimetres in length, which are particularly useful in studying cell physiology – for example, where and how different substances are made in the cell, and how they are separated from each other and moved about. The similarity of this metabolism with that of higher plants suggests that charophytes are on a direct evolutionary route to higher land plants, despite previously being thought to be a side branch of evolution.

Charophytes occur in low-nutrient situations and many are sensitive to eutrophication. Their nitrogen economy is particularly important – it is reckoned that increased nitrate is 10 times more damaging to their presence than is an increase in phosphate (Lambert & Davy, 2011). This is partly because the phosphate is precipitated out in hard water, and partly due to the blanketing effect of other water plants when nitrate is applied. However, increased nitrate also has a toxic effect that disrupts the plant's cell metabolism. The absence of such pollution from Lough Bunny and its neighbours is a very valuable feature, as is the potential security endowed by its small, local catchment. In their 2013 survey, Cilian Roden and Paul Murphy considered that there had been no notable change in the lake's vegetation in at least the previous 20 years.

The branching patterns of charophyte algae provide a huge surface area on which other organisms can live. In some ways, they are a temperate equivalent of coral colonies, supporting smaller green algae such as diatoms as well as cyanobacteria, fungi, bacteria and protozoa. Consumers like water fleas and other crustaceans, beetles, mayfly and fish fry come into this community, which may be seen as the basis of aquatic life in such lakes. Charophytes also support wintering birdlife, as their reproductive oospores (to all intents like seeds) are eaten by diving waterbirds such as Coot (*Fulica atra*) and Pochard (*Aythya ferina*).

One of the interesting organisms that occurs independently of *Chara* is the tiny water beetle *Ochthebius nillsoni*, which is about 1.6 mm long (O'Callaghan *et al.*, 2009). It has been found in only two places in the world: the north of Sweden, and the Burren marl lakes such as Lough Gealáin and Coolorta Lough. It is thought to be a relict species that, like the water flea *Eurycercus glacialis* (which occurs in some turloughs; Chapter 9), invaded during the recession of the ice sheets and persisted in the absence of later-arriving predators (Chapter 5). The beetle is found in quite shallow water (50 cm or so), where it grazes on the biofilm

FIG 295. Charophyte beds in the foreground, encrusted with calcium carbonate and drying out with declining water levels in spring. Common Clubrush (*Schoenoplectus lacustris*) can be seen behind. (David Cabot)

of diatoms, cyanobacteria and true bacteria that accumulates on marl in the absence of plants.

The abundance of the charophyte flora in Lough Bunny may be the cause of a low planktonic population in the lake. During sampling in 1993, cryptophytes were the most frequently recorded organisms, as they are in turloughs (Roden & Murphy, 2013). Cyanobacteria were abundant too, especially *Anabaena*, as were dinoflagellates. Diatoms occurred more in winter than in summer, but only ever in small numbers (Pybus *et al.*, 2003). Much more significant are the bottom-dwelling diatoms, of which 96 different species have been identified from sediment cores (Foged, 1977).

In addition to the charophytes, the lakes support higher plants. In Lough Bunny, the Yellow Water-lily (*Nuphar lutea*) and Common Club-rush (*Schoenoplectus lacustris*) are scattered in the *Chara* beds, while Long-stalked Pondweed (*Potamogeton praelongus*) and Perfoliate Pondweed (*P. perfoliatus*) can grow down to a depth of 6 metres owing to the clarity of the water (Roden & Murphy, 2013). Around a submerged spring in this lake, Curled Pondweed (*P. crispus*) grows strongly, benefitting from the exposure to greater volumes of water from which to extract nutrients. It is normally a plant of eutrophic ditches and drains.

LAKE EDGES

The thin reed beds around the lake shores consist of Common Reed (*Phragmites australis*) or Common Club-rush, and there are often also colonies of Great Fen-sedge (*Cladium mariscus*). Peat records show the general succession as being Common Reed followed by Great Fen-sedge as the lakes fill up with organic material (Proctor, 2010). This is understandable, since the reed can grow strongly in 1 metre of water while Great Fen-sedge suffers from depths greater than 40 cm. The growth of Common Reed in today's alkaline lakes is seldom strong, but it is likely that after the retreat of the ice more glacial till was exposed and there were more nutrients in the lake water. It is an indication of Common Reed's versatility that it still occurs; it also grows on acid blanket bogs and in brackish saltmarshes. In fact, the species will grow in pH values of 2.5–9.8, a range with few parallels in the plant kingdom. It is probably the most widespread plant species in the world, found on all continents except Antarctica. As would be expected, there has been some genetic divergence in this range, so it is really a species complex. An unexpected fact that certainly has aided this cosmopolitan range is that Common Reed seeds are dispersed by wind, this aided by a tuft of long silky hairs at the base. Few other grasses have such an adaptation.

In many of the lakes in the south-east Burren, a shelf of marl extends from the shore before dropping rather rapidly into a deeper hole, which from the

FIG 296. Beds of Great Fen-sedge (*Cladium mariscus*) growing around the lough at the foot Mullagh More, May. (Fiona Guinness)

FIG 297. Seeding heads of the Common Reed (*Phragmites australis*), one of the most widespread plants in the world. (David Cabot)

bank often looks deep blue. Michael Proctor (2010) considered that this shelf represents the former bed of a marginal fen that was subsequently eroded by a mixture of peat cutting and wave erosion. Ballyeighter Lough and Lough Gealáin are good examples. Very little grows on this reworked marl – Shoreweed (*Littorella uniflora*) is often the only species found there, although there may be some Many-stalked Spike-rush (*Eleocharis multicaulis*) or starved shoots of sedges, these probably not able to flower. Islands of Black Bog-rush (*Schoenus nigricans*) are sometimes seen offshore, doggedly holding onto their own basal peat against the waves, and where enough peat remains on the shore, Autumn Hawkbit (*Scorzoneroides autumnalis*) and Common Yellow-sedge (*Carex demissa*) are usually found.

Shoreweed may appear a dull plant when first seen. It consists of a group of spiky, dark green leaves rising to 10 cm or so from an underground stem, and may have wispy stamens hanging at the top. It resembles its plantain relatives, except that it spreads rapidly by stolons to form a mat of growth. However, it conceals a peculiar way of life, in which the carbon dioxide needed by the leaves for photosynthesis can be absorbed from the soil as well as from water or the air (Boston & Adams, 1987). The roots are often larger than the tops of the plant, and all parts have longitudinal passages along which the gas can travel to the leaves. It is one of the isoetids, a group of plants named after the quillworts (*Isoetes* spp.), submerged fern relatives that share this habit of a rosette of leaves and live in oligotrophic waters that dry out. When submerged, Shoreweed obtains much of its carbon dioxide directly from the substrate, but when it is exposed it replaces the submerged leaves with an aerial type that is more efficient at extracting the gas from the air. The drying out of its habitat stimulates the plant to flower after three to four weeks, but it also makes life difficult for what is fundamentally an aquatic species. As a reaction to this selective pressure, Shoreweed has evolved a crassulacean acid metabolism, whereby it can accumulate carbon dioxide at night when potential water loss through the leaves is low. The gas may be absorbed from the roots or through the leaves, and it is then used during daylight for photosynthesis. Shoreweed is more adaptable than most of the isoetids in that it can make use of atmospheric carbon dioxide if the substrate becomes anoxic from eutrophication. One of the riddles of the western limestones is the occurrence of another such species, Water Lobelia (*Lobelia dortmanna*), which was once recorded in Lough Fingall near Kinvarra. This is normally a plant of acid lakes, but as we have seen, if the lake is sufficiently poor in nutrients it may not make much difference whether it is acid or alkaline.

The Common Reeds that grow on lake margins bear seeds that will germinate only on dry land, on the bare, moist soils revealed by spring drawdown. Their seedlings cannot tolerate water depths greater than 5 cm when young, but gradually grow out into the waterbody and in time form a floating swamp. They share this habit with various sedges, including Slender Sedge (*Carex lasiocarpa*) and Bottle Sedge (*C. rostrata*), which can grow in depths of 30 cm and 60 cm, respectively. The Tufted Sedge (*C. elata*) occurs more commonly on the exposed shores, where it maintains a tight peaty tussock and can reproduce from seeds more often. A long-lasting seed bank is important for this group of plants as not every year is suitable for germination. Many people have searched in vain for seedlings in the wild, so germination must be a rare phenomenon.

Such lake-edge fens are often almost pure stands of one species or another. Great Fen-sedge, however, is usually associated with the submerged Intermediate Bladderwort (*Utricularia intermedia*), whose large yellow flowers are a surprise when they appear from the invisible plant below. The plants flower sporadically for a short time, so in some years there will be flowers everywhere and in others very few. Bladderworts are closely related to butterworts (*Pinguicula* spp.) and are partly carnivorous, trapping water fleas in their bladders. They have rather fancy four-part (quadrifid) hairs in the bladders, which function as glands. These pump out water from the bladder, reducing the pressure inside so that an invertebrate bumping into the bladder's outer trapdoor is drawn inside, to be digested later with secreted enzymes. The hairs also occur in the spur of the flower, where they can more easily be seen.

Both Common Reed and Great Fen-sedge create a litter layer that few other species can penetrate. Reed stems contain a low level of nitrogen compared to carbon, so they decompose slowly – this is the reason why they are favoured

FIG 298. Common Cottongrass (*Eriophorum angustifolium*) grows happily in both acid and alkaline surroundings. (David Cabot)

for modern house thatch. Eventually, sufficient organic matter, or peat, is accumulated to allow the plants to rise above the water surface. Their rhizomes then lose some vigour and other species gain a foothold. One of the first may be the Blunt-flowered Rush (*Juncus subnodulosus*), with whitish flowers, but species with lower nutrient demands are often more prevalent, especially Black Bog-rush, Bog Myrtle (*Myrica gale*) and Purple Moor-grass (*Molinia caerulea*). These species form endless fens in the Ballyeighter area, along with such species as Meadow Thistle (*Cirsium dissectum*), Common Cottongrass (*Eriophorum angustifolium*) and Devil's-bit Scabious (*Succisa pratensis*).

In Britain, Great Fen-sedge swamps are often invaded by trees, with Grey Willow (*Salix cinerea*), Alder (*Alnus glutinosa*) and Alder Buckthorn (*Frangula alnus*) forming a likely partnership. In the Burren, however, open stands are more frequent; a few Grey Willow and Eared Willow (*Salix aurita*) may be scattered through, but the primary species is most often Black Bog-rush. This may be the result of burning in past centuries as the fens were extensively cut for fuel as well as grazed. Below Muckanagh Lough and around the Ballyeighter Loughs, peat cuttings are clearly visible in aerial photographs, if not so easily discernible on the ground. Harvesting of Black Bog-rush for thatching also had an impact on fens in the past – Martin Speight (1975) recorded extensive use of the plant in three-year cycles when the stems were almost 1 metre high. This thatch is likely to have been used extensively in the mid-nineteenth century when the local rural population was at its maximum.

Black Bog-rush is a wiry, rush-like plant that is named for its black flowers. It has rounded stems and inrolled leaves, these being about half the height of the flowering stems. The flowers are visible year-round, either the current crop or past flowers from the previous year or two. They appear first in autumn, and reach 4–5 cm long by the onset of cold weather. Growth recommences in March, with the flowers open in July and August. The stem, however, continues to grow taller, raising the seedheads as high as possible and contributing to the plant's distinctive appearance. The seed is held tightly until December, and some usually remains into the following spring, when it may germinate (Sparling, 1968). Like many plants, the Black Bog-rush has an insect species that is dependent on it, in this case a rather smart micro moth called *Glyphipterix schoenicolella*. This is a well-adapted consumer, found in 83 per cent of the flower heads in Sparling's study but still allowing every plant to produce some seeds.

FENS

As we have seen, fens are peat-forming habitats where water prevents or reduces the natural breakdown of the organic matter that builds up in a layer above

FIG 299. Northern Bedstraw (*Galium boreale*), with its broad, slightly bluish leaves, grows in fens and on lakeshores. May. (Fiona Guinness)

the ground. The water includes both groundwater and rainfall, so that the base content (usually calcium) is raised, and hence the plants found here are quite different from those of a rain-fed system or bog. At the start of such accumulation, the nutrient content is high and the plants are somewhat similar to those in marshes. Black Bog-rush may dominate, but there is room for a considerable variety of other species. Meadow Thistle, Grass-of-Parnassus (*Parnassia palustris*) and Sneezewort (*Achillea ptarmica*) are frequent, while multiple small sedges reinforce the fen atmosphere. In slightly wetter places, Bogbean (*Menyanthes trifoliata*), Marsh Lousewort (*Pedicularis palustris*) and Lesser Spearwort (*Ranunculus flammula*) will occur, and the vegetation may be peppered by the vertical shoots of Marsh Horsetail (*Equisetum palustre*). The Black Bog-rush tussocks are sometimes raised on pedestals, particularly if there is a fluctuating water level. This diversifies the habitat and allows a suite of aquatic mosses into the lower storeys. *Scorpidium scorpioides* or *S. revolvens* are commonly there along with the starry *Campylium stellatum*. Bog Pimpernel (*Anagallis tenella*), Marsh Pennywort (*Hydrocotyle vulgaris*) and sometimes Lesser Clubmoss (*Selaginella selaginoides*) crawl around where the larger plants allow this.

The accumulation of peat means that plants become further removed from groundwater with time and the nutrient status declines. Butterworts get around this by trapping and digesting insects on their sticky leaves, while Bog Myrtle has a bacterial symbiont that fixes nitrogen. Plants with lower nutrient demands take hold on the tops of tussocks; these include Tormentil (*Potentilla erecta*) and Heather (*Calluna vulgaris*). As the fen matures, *Sphagnum* mosses appear (there are a range of *Sphagnum* species adapted to different nutrient levels) and mediate the gradual transition of the habitat to a rain-fed, raised bog. This has seldom happened in the high Burren owing to fluctuating groundwater levels, but south and west of Lough Bunny it may be seen around the Ballyeighter Loughs. Much of the peat has now been removed, but at Rinroe the road runs on a causeway of peat 2–3 metres above the fen on the eastern side but in line with some remaining bog on the west. Bog Asphodel (*Narthecium ossifragum*), Cross-leaved Heath (*Erica tetralix*) and the reindeer lichen *Cladonia portentosa* occur here among tall Heather, and there are three or four *Sphagnum* species as well as Bogbean in a pool, and Royal Fern (*Osmunda regalis*) in a drain. Other patches of raised bog would have occurred on the periphery of the lakes wherever the alkaline floodwaters of winter failed to reach.

However, it may be for their orchids that fens are particularly appreciated. Common Spotted-orchid (*Dactylorhiza fuchsii*) and Heath Spotted-orchid (*D. maculata*) are frequently seen, as is the Fragrant Orchid (*Gymnadenia conopsea*). The species epithet of this last species, *conopsea* (meaning 'mosquito-like'), was given by Carl Linnaeus for the flower's long, pointed spur. The Early Marsh Orchid (*D. incarnata*) flowers first in May. It is quite variable in colour and in leaf spotting, and forms that were formerly regarded as subspecies such as *D. incarnata* spp. *cruenta* have now been shown to be genetically very close to the type. Later, the Irish Marsh-orchid (*D. kerryensis*) comes into form; it is generally shorter than the Early Marsh and has broader leaves. A relative is the Narrow-leaved Marsh Orchid (*D. traunsteinerioides*), which is slender as opposed to chunky, and pink rather than purple. It was formerly confused with other species and thought to be very rare, but has now turned up in a good number of fen habitats in both Britain and Ireland. A distinctive character is the long central lobe of the labellum (Curtis & Thompson, 2009).

Compared to these pink- and purple-flowered species, the Fly Orchid (*Ophrys insectifera*) is quite inconspicuous and may bring one up short when it is seen among Black Bog-rush tussocks. It is tall, reaching up to 50 cm, and narrow, with brown velvety flowers that are spread out on the stem. The upper flower segments

FIG 300. Irish Marsh-orchid (*Dactylorhiza kerryensis*), a small plant with broad leaves. May. (Fiona Guinness)

FIG 301. a. Fly Orchid (*Ophrys insectifera*) in a rocky habitat, May. The flowers mimic a solitary wasp and are pollinated by pseudo-copulation. One Irish insect, *Argogorytes fargei*, has antennae just like the two dark perianth segments at the top of the flower. (Fiona Guinness) b. Fly Orchid close-up, which to our eyes looks more like a monk than an insect. (David Cabot)

are yellow or green, but the labellum is coloured and shaped like an insect. It releases pheromones that attract solitary wasps – either *Argogorytes mystaceus* or *A. fargei* (Florian *et al.*, 2008). These insects attempt to copulate with the flowers, picking up the pollinia in the process. The wasps are black with yellow stripes and quite unlike the orchid labellum, so the pheromone released is critical in this sexual deception. Without it the Fly Orchid would not be visited and the labellum is largely redundant in this part of the plant's range. Even with the pheromone, seed-set success is quite low (less than 20 per cent and frequently down to 5 per cent). It seems that the chemicals involved (waxes and other hydrocarbons) were already part of floral protection and odour before being selected as a pollination attractant. A chance mutation would have provided the combination required to attract a particular insect species. A major result is that crossing with other orchid species, which is possible by artificial means, is prevented.

The last of the fen orchids, the Marsh Helleborine (*Epipactis palustris*), is showy when seen at close range but for some reason seldom stands out in the vegetation. However, once one is spotted there are usually others in the vicinity. It has a complicated labellum formed of an inner pinkish part and an outer white frill, with a hinge or lip in between. As in all helleborines, the flower does not have a spur and any nectar is produced on the lip, where there are two yellow honey guides. In contrast to the Fly Orchid, it is visited by a multitude of different insects – in the Netherlands, these were found to include wasps, honey bees, hoverflies and ants (Brantjes, 1981). The last three were all effective pollinators, but the honey bees removed some of the pollinia for their own use. Such insect attention means that all flowers are usually pollinated and that all develop the characteristic hanging capsules.

OTHER FEN AND MARSH HABITATS

Lake-edge fens cover by far the largest area in the Burren and include most of the land surrounding the south-eastern lakes that is flooded or at least infiltrated by water in winter. But there are other, smaller sites where a shallow basin is sealed from major groundwater flows either by its position at the head of a catchment or, more usually, by a layer of glacial till that was laid down during the Pleistocene and has persisted since. These habitats short circuit the reed-bed phase, although they may have scattered shoots of Common Reed growing through the more usual cover of small sedges, Purple Moor-grass and Black Bog-rush. Flushes may be encountered on the hills, most usually at the base of one of the steps in the profile where water issues from the limestone at its junction with a band of shale. The water seeps out of the rock, meanders through a fen habitat and disappears again into the pavement after a short distance.

One of the more unusual fens occurs on the northern slopes of Carnsefin above the coastline. Here, a spring line on the hillside just below the old road infiltrates the surface to such an extent that Black Bog-rush is the dominant plant on a 20–30-degree slope. The rock shows through in places, so there is a wide mixture of plants: Purple Moor-grass, Devil's-bit Scabious and Long-stalked Yellow-sedge (*Carex lepidocarpa*), with its characteristic downturned fruits, are mixed with Wild Madder (*Rubia peregrina*), Goldenrod (*Solidago virgaurea*) and Irish Eyebright (*Euphrasia salisburgensis*). There are occasional plants of Crowberry (*Empetrum nigrum*), Rowan (*Sorbus aucuparia*) and Bell Heather (*Erica cinerea*). Common Butterwort (*Pinguicula vulgaris*) and Pale Butterwort (*P. lusitanica*) grow on the mossy cushions of *Breutelia chrysocoma*, while Brittle Bladder-fern

(*Cystopteris fragilis*) and Maidenhair Fern (*Adiantum capillus-veneris*) appear in damp clefts in the limestone. The water disappears underground at the base of the slope, beside the current road.

Other small flushes above the road at Black Head show Black Bog-rush on relict peat that is almost 50 cm thick. This links in with the habitat on the tops of these hills, where the peat sustains Bearberry (*Arctostaphylos uva-ursi*) in a heath with Mountain Avens (*Dryas octopetala*) and Juniper (*Juniperus communis*). Nearby, on Gleninagh and Moneen Mountain, Deergrass (*Trichophorum germanicum*) persists on peat over clay and there is even *Leucobryum glaucum*, a hummock-forming whitish moss of rain-fed bogs.

The most famous flushes in the Burren occur both east and west of Ballyvaghan. Here, on Cappanawalla and Moneen Mountain, there are springs that discharge onto the rock around tufts of mosses and sedges – and the Large-flowered Butterwort (*Pinguicula grandiflora*). To see this in May is a sight worth every effort of the climb – many hundreds of plants shine out from the vegetation and leave no doubt as to the species' native status.

FIG 302. A flush where springs emanate from the south-east slope of Cappanawalla creates perfect conditions for the Large-flowered Butterwort (*Pinguicula grandiflora*). May. (Fiona Guinness)

FIG 303. Large-flowered Butterwort (*Pinguicula grandiflora*) in May. Note the flies trapped on the sticky leaves of the plants. (Fiona Guinness)

When Jack Heslop-Harrison first found Large-flowered Butterwort in this area in 1949 (Chapter 2), there were those who muttered that it could not grow on limestone and therefore must have been planted. The species was then known only on the acid rocks of south-western Munster. It is a Lusitanian species also found in north-west Spain, with the Burren being its most northerly station in the world. Here, it grows with *Palustriella falcata* and *P. commutata* mosses, which are usually orange in colour, as well as Devil's-bit Scabious and Carnation Sedge (*Carex panicea*) (Lyons, 2015). At Cappanawalla and Moneen, small amounts of calcium carbonate are precipitated from the water on its contact with the air and deposited as a skin on the limestone. The fact that the carbonate coats rocks and plants alike has given rise to the rather disturbing name of petrifying springs for these habitats. It is relatively minor in these cases but elsewhere a massive amount of precipitation can occur if the lime gets caught in mosses as tufa, which can build up 2 cm of spongy material in a year. Another spring at Black Head is similar and, although on a slope, supports the wine-red moss *Orthothecium rufescens*, which in Ireland otherwise occurs only in Co. Sligo and Co. Donegal. These depositing springs are listed as a priority habitat in the European Union Habitats Directive.

Red mosses always excite attention because they are unusual and often a rather 'unnatural' pinkish colour. The *Sphagnum* genus includes well-known examples, but red bryophytes are also scattered among other genera, including some liverworts – for example, the *Frullania* that grows widely on twigs all over the Burren. Why should some plants be red and not others? The proportion of red and yellow pigments is found to be highest in species growing in brightly lit places and lowest in deeply shaded woodland. Some are much redder than others, such as the *Orthothecium rufescens* at Black Head. The favourite explanation for this is that the red coloration is a form of photo-protection for the photosynthetic pigments, but this has proved difficult to demonstrate (Marschall & Proctor, 2004).

FIG 304. Marsh at Burren village, with beds of Bulrush (*Typha latifolia*) and Branched Bur-reed (*Sparganium erectum*). (Roger Goodwillie)

Ponds and marshes are fed by surface water flows and occur generally in the lowlands. The coast road west of Kinvarra passes a large reed bed at Burren village and approaches Lough Luirk and Lough Rask. These sites have a commoner type of vegetation, at least in terms of the rest of Ireland, with tall beds of Bulrush (*Typha latifolia*), Branched Bur-reed (*Sparganium erectum*), Water Horsetail (*Equisetum fluviatile*) and Bottle Sedge (*Carex rostrata*). An edging of Common Spike-rush (*Eleocharis palustris*), Purple Loosestrife (*Lythrum salicaria*) and Great Willowherb (*Epilobium hirsutum*) may also be prominent. Lough Rask appears slightly brackish and supports Grey Club-rush (*Schoenoplectus tabernaemontani*) and Slender Spike-rush (*Eleocharis uniglumis*), a situation mirrored in small wetlands on Inishmaan and Inishmore (Ivimey-Cook & Proctor, 1966). Brookweed (*Samolus valerandi*) and Mare's-tail (*Hippuris vulgaris*) feature at some of the more coastal sites.

When it comes to smaller wet habitats on wet soils in the Caher valley, some have similarities with the shale areas to the south of the Burren limestone. Sharp-flowered Rush (*Juncus acutiflorus*) is usually present in such places, often with Marsh Ragwort (*Senecio aquaticus*) or Meadowsweet (*Filipendula ulmaria*), and there is Silverweed (*Potentilla anserina*) and Meadow Buttercup (*Ranunculus acris*) too. These communities remind one what 'normal' Co. Clare fields look like in the fundamentally abnormal Burren.

FIG 305. Yellow Iris (*Iris pseudacorus*) is ubiquitous on wet Co. Clare soils and found sparingly in the Burren at marshes and turloughs. (David Cabot)

INSECTS

North-east of Lough Bunny, near Kilmacduagh, is an extensive fen basin
that includes beds of Great Fen-sedge, Common Reed and Black Bog-rush.
A few small lakes survive – Lough Loum is the largest – but, unlike the major
waterbodies, they have relatively stable shorelines that do not suffer wave
erosion and they lack the expanses of marl found elsewhere. Sampling in the
1990s showed that these areas support a very high diversity of ground beetles
(staphylinids and carabids), especially those that require stable natural habitats
and are therefore rather rare in our agricultural world (Good, 2004). The best
sites for beetles in the study were ecotones, the transition areas to dry land
with well-developed vegetation, and stable wetlands that aren't subject to the
catastrophic floods (for ground beetles) characteristic of turloughs. In terms of
numbers, 76 species of Staphylinidae and 35 species of Carabidae were recorded,
including several that are very local in Ireland and Britain. Among them was
the Crucifix Ground Beetle (*Panagaeus cruxmajor*), which has a black cross on
its orange wing-cases. It is declining in Britain but has been recorded at six or
eight sites in Ireland. *Pterostichus aterrimus* also deserves a mention as it may
have died out in Britain. It is a large and very shiny black beetle that looks wet,
in keeping with its habitat of choice.

FIG 306. Pond skaters (*Gerris* spp.) live
on the surface film of water and feed
on insects that land there or come up to
breathe. (Fiona Guinness)

Dragonflies and damselflies are ever
present in lakes and fens; they spend
one or two years as larvae but are seen
as flying insects from April to October.
Individuals of each species fly for several
weeks (up to two months in the case of
larger ones), and they have different flight
periods that give them the best chance
of finding a mate and laying eggs. In the
early part of the year, the blue damselflies
may seem to predominate: the Common
Blue Damselfly (*Enallagma cyathigerum*)
and the Blue-tailed Damselfly (*Ischnura
elegans*), and, if one is lucky, the Irish
Damselfly (*Coenagrion lunulatum*). This
last species was first recorded in Ireland
as recently as 1981 and has a range across
lowland areas in the north-west. It
resembles other blue damselflies, but the

FIG 307. Beautiful Demoiselle (*Calopteryx virgo*), characteristic of nutrient-rich sites. August. (Carl Wright)

FIG 308. Common Darter (*Sympetrum striolatum*), showing its hemispherical eyes, which give all-round vision. (Carl Wright)

background colour of the males is black rather than silver. The southern limit of its known distribution is in the Burren lakes (Lough Skeardeen), where it finds the combination of reed fringes and floating plants to its liking. It has a shorter flight period than many other species, flying in May and June, with a few in July (Chan *et al.*, 2014).

Apart from the Hairy Dragonfly (*Brachytron pratense*), which is an early flyer, dragonflies in general appear later, when the populations of their prey (flies) have built up. The Brown Hawker (*Aeshna grandis*) is quite noticeably brown, even on its wing veins, and its nymphal skins are some of the most likely to be found on waterside plants. It has been joined recently by the powder-blue Emperor Dragonfly (*Anax imperator*) in one of the most striking expansion of ranges of any insect. The Emperor first appeared in Ireland in 2000, and in the years since has colonised all but the north-western fringes of the island. Smaller or shorter-bodied species include the Black-tailed Skimmer (*Orthetrum cancellatum*), which again has a powder-blue body and is inclined to settle on bare ground. Darter dragonflies (*Sympetrum* spp.) finish off the season and are usually the last to be seen in October or November. The Common Darter (*S. striolatum*) and Ruddy Darter (*S. sanguineum*) are as similar to one another as the blue damselflies in the spring; the latter is the redder of the two.

The occurrence of similar but distinct species in these genera of insects gives rise to the question of why. One current theory based on North American damselflies is that there is significant sexual harassment of females by males during the breeding season (Takahashi *et al.*, 2014). The insects mate in the air, and the victims of harassment were found to produce fewer eggs. It is in the females' interests therefore to be slightly different in order to put off some of the more annoying males. Any small difference that arises in colour pattern or skeletal shape will be retained as long as some mating occurs. Fewer matings may therefore give rise to more offspring, which perhaps is counterintuitive but has a clear evolutionary advantage. Once the female has changed in some small way, males that are born with the necessary complementary adjustment will also win out. Twenty-three new odonate species have arisen in the last 250,000 years, which is rapid for a group of insects that appeared 290 mya (Turgeon *et al.*, 2005).

A group of insects that one cannot forget easily are the horseflies (Tabanidae). The Common Horsefly (*Haematopota pluvialis*) has a silent approach and appears moth-like when found biting one's skin. It has a pretty pattern of dots on the wings – unfortunately seldom appreciated by its victims. It is associated with all habitats in the Burren, including fens and bogs, but a specialist of these wetlands is the Twin-lobed Deerfly (*Chrysops relictus*). It has a much louder buzz, wonderful green eyes and blotched wings, and usually goes for the head. It may be worth remembering that the most efficient trap for a related species was found to be blue in colour, placed 1–2 metres above ground, and moving slower than 3 metres per second – features closely matched in a wandering naturalist (Russell *et al.*, 2002). Only the female flies bite, and they lay eggs on surfaces near water. The larvae drop into damp soil and feed on invertebrates there.

BIRDLIFE

A summer visit to the lakes in the south-east Burren will usually result in a sighting of grebes, either the Little (*Tachybaptus ruficollis*) or the Great Crested (*Podiceps cristatus*). Little Grebes are more frequent, and with their rippling calls enliven the nesting period in May. They have the habit of gathering in large groups after breeding; 94 were recorded on a turlough near Lough Bunny one July (Heery & Madden, 1997), while Liam Lysaght (2002) reported an assembly of 130 of these birds on Inchiquin Lough. Grebes are fairly conspicuous when nesting, but ducks of all sorts are discreet to the point of disappearance. Records exist of nesting Teal (*Anas crecca*) and Shoveler (*Anas clypeata*) in the eastern lakes, where there are many quiet waterbodies that are hard to reach. The Mallard (*Anas platyrhynchos*) is the one duck that will be seen, flying in pairs or leading ducklings out onto the water surface in April or May. It also nests in wetlands on the Aran Islands. Greylag Geese (*Anser anser*) are similarly conspicuous, and birds from introductions may sometimes be seen on Lough Bunny. This species is now following the example of the Mute Swan (*Cygnus olor*), an ancient introduction now accepted almost as a native. Individual pairs nest on many of the lakes and there is sometimes a gathering at moulting time in August at Lough Atedaun.

Moorhens (*Gallinula chloropus*) and Coots are both widespread. The former species is associated with wetlands of all sizes, even on the Aran Islands, and is sometimes seen in open fields and at other times hidden in drains. Its night-time calls may ring out over the rocks in unexpected places as it performs a territorial flight around a marsh that one did not know existed. Coots are restricted to lakes as they feed by diving rather than picking and they have difficulty taking off except by pattering along the water surface. They consume some vegetation, along with snails and other invertebrates. Both species may construct their floating nests in reed beds, but the Moorhen is much less prescriptive and can choose dense vegetation of any sort.

FIG 309. Common Sandpiper (*Actitis hypoleucos*) is scattered as a breeder in the south-east Burren. (John Fox)

Waders are seen in the breeding season, and while they do not occur in great numbers they are noticeable nevertheless. The Common Sandpiper (*Actitis hypoleucos*), Redshank (*Tringa totanus*) and Ringed Plover (*Charadrius hiaticula*) draw attention to themselves with their alarm calls, and Lapwings (*Vanellus vanellus*) by their aerial antics. These species are the most southerly breeders of our common waders, but all are declining, retreating northwards in Europe as a result of disturbance, predation and climate change (Balmer *et al.*, 2013).

The Common Sandpiper is an engaging bird, bobbing like a wagtail as it walks and flying with distinctive stiff wing-beats. Despite its association with lake shores (and watercourses), it eats much more terrestrial prey than aquatic prey; a study in Britain showed that three-quarters of its diet is made up of beetles, flies, ants, spiders and earthworms (Yalden, 1986). The remainder includes mayflies, stoneflies and caddis flies. The earthworms were taken primarily by adults in adjacent grassland when they first arrived in the breeding area in spring. The reason the Common Sandpiper favours lake shores is possibly that it relies on scavenging in an open habitat because its bill is not specialised enough to probe or filter. Both terrestrial and aquatic prey can be picked off a surface that wave action and winter flooding keep free of vegetation. Trapping in the same study showed that beetle numbers were relatively high on shingly areas. It also found that, when they hatched, the chicks first fed in short vegetation nearby but beyond an age of five days concentrated on the shingle habitat. This, in fact, may be the critical factor in the species' ecology and evolution. Only those chicks that spent time on the open shingle and rock surfaces survived predation in enough numbers to return to breed. The raising of young is often a time when the greatest selective pressures are in action.

A postscript to the Common Sandpiper story is that the birds leave very soon after breeding – no adults or young hang about during the summer, feeding on the lake shore and developing their life skills. Pitfall trapping showed that numbers of prey, particularly ground beetles, decline rapidly during July, so no food would be available for the birds anyway if they were to stay (Yalden, 1986).

Black-headed Gulls (*Chroicocephalus ridibundus*) and Common Gulls (*Larus canus*) are still seen in spring on these lakes. The Black-headed used to breed in numbers on Lough Bunny. The Cormorant (*Phalacrocorax cerbo*) also used to nest (in trees) at Lough Bunny. 20–25 pairs were seen in 1962 but deserted the next year (D. Cabot, pers. obs.). Gull Island is still marked on maps depicting the lake, and the noise of 500–600 birds was once the constant backdrop to a visit there. But in line with a general move of these birds to coastal colonies, most likely in response to American Mink (*Neovison vison*) predation, there has also been a huge reduction in the numbers in the Burren. The colony on Lough Bunny was first abandoned in 1994, but breeding continues sporadically on this or nearby waters. The colony was large enough to absorb a Common Tern (*Sterna hirundo*) or two in the 1990s, these birds taking advantage of the protection given by the gulls. The Black-headed Gull itself takes all sorts of aquatic insects – and terrestrial ones if it can catch them. It is a significant predator on hatching mayfly, particularly the large *Ephemera danica*, which hangs at the water surface as it moults into its aerial phase (Vernon 1972). The Cormorant (*Phalacrocorax carbo*) also used to nest (in trees) at Lough Bunny. In 1962, 20–25 pairs were seen but they deserted the next year (D. Cabot, pers. obs.).

In the fens, it may be a case of hearing the birds rather than seeing them. Water Rails (*Rallus aquaticus*) are present in the south-east Burren, although they prefer richer conditions than those offered by the marl lakes. The Snipe (*Gallinago gallinago*) is also distinctive both day and night in its drumming and display calls. It is not yet uncommon, but having been a feature of fens and rushy fields everywhere, it is showing the same decline as other waders, although at a slower rate (Balmer *et al.*, 2013). The contraction in its range is to the north and west; for comparison, the nesting population in Scotland has increased by 30 per cent but in England it has declined by 13 per cent. No such decline is apparent in the smaller wetland birds, and it is the urgent song of the Sedge Warbler (*Acrocephalus schoenobaenus*), or the more relaxed songs of the Grasshopper Warbler (*Locustella naevia*) and Reed Bunting (*Emberiza schoeniclus*), that one hears here the most.

THE CAHER RIVER

The Caher River rises near Caherbullog, a circular stone fort east of Slieve Elva, and flows on a layer of glacial till through an open valley that is edged by Hazel scrub and limestone pavement. For most of its course, the gradient is low and the river may flood into its marginal scrubland, but in the final 2 kilometres towards Fanore it plunges down a series of rapids, losing 60 metres in height. In this stretch it has excavated a wonderful cross section through a ridge of glacial till, one of several that extend from north-east to south-west in the lee of Gleninagh Mountain. The till is fine-grained and pale grey, and shows the type of parent material that gives rise to some of the soils in the Burren. It is in two layers, representing two ice advances.

The river flows for most of the year, but in prolonged droughts it runs underground in part of its descending section. The whole catchment has very little agriculture; only 12 per cent of the area is managed grassland, while the rest is natural grassland, scrub and pavement (Kelly-Quinn *et al.*, 2003). The water is very clear, and is loaded with calcium carbonate but rather low in other nutrients like magnesium and potassium. The carbonate is deposited on the pebbles and rocks, giving the whole channel a whitish look, particularly

FIG 310. Caher River in extreme low flow, with mounds of river mosses and dripping cones of algae. (Fiona Guinness)

FIG 311. Above: the descending section of the Caher River, cutting into glacial till. Left: the till deposit of mixed clay and stones fringed by Juniper (*Juniperus communis*) and Mountain Avens (*Dryas octopetala*). (both photographs by David Cabot)

FIG 312. The brownish leaves of the hybrid pondweed *Potamogeton* × *lanceolatus* in the river just above Caher Bridge, May. (Fiona Guinness)

where the flow is greatest. The lime precipitates onto plants and the stream bed, cementing the pebbles there and limiting the animal species that would normally live in these spaces. It is also deposited on the *Nostoc* cyanobacterium (Chapter 6) that characterise the upper part of the river. The selection of plants that occur is not large but includes a hybrid pondweed, *Potamogeton* × *lanceolatus*, which grows in little reddish tufts at the edges of the stream (Chapter 5). Often it has only submerged leaves, but in quieter reaches it may develop a few floating leaves. It is a cross between two common species, the Fen Pondweed (*P. coloratus*), which also occurs in the Caher, and the Small Pondweed (*P. berchtoldii*), but has arisen very seldom in nature. It was described from Anglesey in 1806, where it seems to have died out, but remains common in the Caher River as well as in a few places in east Co. Galway and Co. Mayo (Preston, 1995). Once it is in a river it may spread vegetatively, but as it does not produce fertile seeds it remains quite uniform. It occurs without one of its parents in the Caher River – the Small Pondweed does not have a rhizome and cannot persist in flowing water – a fact that suggests the hybrid arose in a pond somewhere beside the channel or in a preceding lake before the river cut down through its limiting band of till.

Other plants in the river are relatively common species, including Fool's-water-cress (*Apium nodiflorum*), Lesser Spearwort and Jointed Rush (*Juncus articulatus*). Purple Loosestrife and Marsh Marigold (*Caltha palustris*) brighten the banks in many places. Shoreweed does occur sparingly, but it is perhaps among the algae that one should look for rarities. The freshwater red alga *Bangia atropurpurea* has been found here; it also occurs in the Corrib and Shannon.

The invertebrate life is unexceptional according to Mary Kelly-Quinn *et al.* (2003), and resembles that of other clean-water rivers in limestone areas. Stoneflies (Plecoptera) and mayflies (Ephemeroptera) are major groups; the most frequent stonefly recorded was *Dinocras cephalotes*, a generally uncommon species from mountainous areas. The caddis flies (Trichoptera) were limited in numbers, as were water beetles, except for the 2 mm riffle beetle *Elmis aenea*. This was the dominant form and the most numerous invertebrate sampled. It is one of the group of 'clingers and crawlers' that feed on algal films and detritus in fast-flowing streams. Such conditions also suit the Brown Trout, which is the major fish species found and often seen from Caher Bridge. In addition, electro-fishing also has brought small numbers of European Eels to light.

The river has enough invertebrate food to support a few pairs of Dipper (*Cinclus cinclus*) and more of Grey Wagtail (*Motacilla cinerea*). Dippers in Ireland have a rusty transitional band on their breasts between the white and brown patches, and are classified as the subspecies *hibernicus*, a race that also occurs on the west coast of Scotland. Genetic studies show that it has more similarities with the central and southern races of the species than with the British one, so it may

FIG 313. Dipper (*Cinclus cinclus*) with food for its nestlings. Caher Bridge, May. (Fiona Guinness)

FIG 314. Grey Wagtail (*Motacilla cinerea*) on the mossy boulders of the Caher River. (Fiona Guinness)

have invaded after the ice age from a refuge in the Pyrenees (Lauga *et al.*, 2005). Either because of this link with nationality or the general attractiveness of the species, it has been the object of considerable study. Irish Dippers start breeding a week or two earlier than others – often in March – because their prey is also more abundant then (Smiddy *et al.*, 1995). In average river flows they concentrate on caddis fly larvae as food, whereas in spates they take more mayflies dislodged off surfaces. They have been found to feed their young on the soft-bodied caddis flies *Hydropsyche* rather than those with stony cases. Survival in Ireland must be better than elsewhere in the range, as the Irish birds lay fewer eggs and seldom have a second brood.

The same breeding pattern occurs in Irish Grey Wagtails: early nesting, fewer eggs and therefore fewer young than in Britain (Smiddy & O'Halloran, 1998). They occupy similar habitats to Dippers, but feed mostly on the aerial stages of aquatic insects, rather than the larvae.

CHAPTER 11

Maritime Habitats

T HE LIMESTONE COASTS OF THE BURREN vary in their exposure to
Atlantic waves. The most extreme areas are the south-west coasts of
the Aran Islands, where the ocean waves crashing into the cliffs have
been generated over a 6,000 kilometre fetch. These islands offer some slight
protection for the mainland shore from Poulsallagh to Black Head, but it is
along the northern coast that there is actual shelter, in the bays and inlets around

FIG 315. Coastline south of Fanore looking north, with the Fanore dunes in the distance.
(David Cabot)

Ballyvaghan, Bell Harbour and Aughinish Bay, where the tide infiltrates the land. What the west coast lacks in subtlety, it makes up for in salt spray, which draws the maritime effect far inland. The concentration of salt in the atmosphere and rainfall is at least three times higher in the Burren than in inland areas, adding to the scorching and dieback of foliage during summer storms.

EXPOSED SHORES

The coast road at Poulsallagh runs along limestone pavement about 20 metres above the sea and at the base of the first terrace in from the cliff edge. Waves breaking on the cliffs create fountains of spray that are swept horizontally across the rocks. The immediate strip of maritime communities occupies the first 40–50 metres in from the coast, but individual specialists continue further inland, mixed into the turf of the limestone pavement. Buck's-horn Plantain (*Plantago coronopus*) and Sea Spleenwort (*Asplenium marinum*) can be found on the landward side of the road, while Rock Samphire (*Crithmum maritimum*) occurs fully 200 metres from the sea on further inland terraces. Spray dripping down from the roadside cliff creates a zone of saltmarsh, full of Distant Sedge (*Carex distans*).

FIG 316. Rock Samphire (*Crithmum maritimum*) is frequent on cliffs and rocks by the sea, and sometimes inland in sites that are influenced by sea spray. (Fiona Guinness)

FIG 317. Exposed shoreline from Black Head to Poulsallagh. Note the broken blocks of limestone at the foot of the cliffs. (Steve O'Reilly)

The sea-lavender *Limonium recurvum* ssp. *pseudotranswallianum* is one of the more famous species to grow here, its small tufts visible in chinks in the rock and on pavement nearer the sea (Chapter 5). It also occurs on Inishmore in the Aran Islands. *L. recurvum* is an apomictic species, producing seeds without fertilisation (and without meiosis). Dandelions (*Taraxacum* agg.) and hawkweeds (*Hieracium* agg.) do the same, and all give rise to micro-species. Every offspring is identical to the parent but mutations are still possible during the development of the seed, although at a much slower rate than with 'normal' plants. As a group, the sea-lavenders of rocky habitats have accumulated mutations of their own, and the further away colonies are from each other, the more numerous the differences between them (Ingrouille & Stace, 1985). The Burren plant is one of the smaller versions, 10–12 cm high when in flower. It is endemic to Ireland, although other subspecies of *L. recurvum* are scattered more widely.

The seaside pavement supports a diverse maritime flora. Thrift (*Armeria maritima*), Sea Campion (*Silene uniflora*) and Common Scurvygrass (*Cochlearia officinalis*) grow among Bird's-foot Trefoil (*Lotus corniculatus*) and Irish Saxifrage (*Saxifraga rosacea*), in many places producing a photogenic mix. Saltmarsh

FIG 318. Shoreline beach in the Poulsallagh area, June. The bank on the left is one of the locations where Pyramidal Bugle (*Ajuga pyramidalis*) is found. Other conspicuous flowers here are Thrift (*Armeria maritima*) and Primrose (*Primula vulgaris*). (David Cabot)

plants such as Sea Plantain (*Plantago maritima*) and Sea Aster (*Aster tripolium*) occur here too. There are solution hollows with Saltmarsh Rush (*Juncus gerardii*) and Sea Arrowgrass (*Triglochin maritima*), and on rock where water lies or flows with spray, a sward of Red Fescue (*Festuca rubra*), Common Saltmarsh-grass (*Puccinellia maritima*) and Thrift occurs, along with the saltmarsh specialists Distant Sedge (*Carex distans*), Long-bracted Sedge (*Carex extensa*) and Dotted Sedge (*Carex punctata*). In other situations these plants are covered by high spring tides, which here are simulated by sea spray. Hemp-agrimony (*Eupatorium cannabinum*) is conspicuous quite close to the sea, if somewhat unexpected. The neat fronds of Sea Spleenwort appear in many of the grikes. This is a fern with a wide Atlantic distribution, from the Azores to Norway. It is very tolerant of salt, although it does not require it, and in a few places (e.g. Co. Fermanagh and Co. Donegal) it grows inland. Unlike the nearby Maidenhair Fern (*Adiantum capillus-veneris*), which suffers in any salt spray, the Sea Spleenwort fronds seem to shine with good health. This is thanks to their thick cuticle and slight succulence, which provide resistance to the salt. The plant's reproductive stage, or prothallus, is much more sensitive, producing only male gametes if the salt content is too high (Pangua *et al.*, 2009). The spores themselves will not germinate in full sea water, although they can recover if given pure water afterwards. The plant therefore depends on frequent rainfall to dilute the worst excesses of the sea.

Sea Spleenwort is part of a tough group of plants that occur in the splash zone of the waves, but it does need some shelter. In the more exposed places Rock Samphire is ubiquitous, clinging onto the crests of exposed cliffs; it is very frequent on the Aran Islands too. Lichens alone grow in the most exposed places and occur in noticeable bands that were first described in Ireland by Matilda Knowles in 1913. At the base of the shore is usually a black or *Verrucaria* zone, named after the lichen *Verrucaria maura*, which looks like tar on the rocks. It is punctuated by the orange scales of *Caloplaca microthallina*. This is followed upwards by the orange zone, which receives a lot of salt spray, and finally the terrestrial grey zone. Conspicuous orange lichens include other *Caloplaca* species such as *C. marina*, as well as *Xanthoria aureola*. The grey or white lichens are primarily terrestrial species in the Burren, and the well-known strap-like

FIG 319. Thrift (*Armeria maritima*) and Bird's-foot Trefoil (*Lotus corniculatus*) at Black Head, May. The black lichen zone on the lowest rocks is conspicuous behind. (Roger Goodwillie)

FIG 320. a. Irish Saxifrage (*Saxifraga rosacea*) is prominent along the Burren coast and in the Aran Islands. May. b. Sea Spleenwort (*Asplenium marinum*) grows in grikes and rocks along the coast, where it is often sprayed by the sea. Its fronds are shiny and rather leathery. The plant's common name comes from its supposed resemblance to the human spleen, as noted by the Greek physician Dioscorides about 2,000 years ago. c. Sea Campion (*Silene uniflora*) is found on the rocky coast and occasionally on mountain tops. In coastal locations it is the foodplant of the Grey moth (*Hadena caesia mananii*). May. (all photographs by Fiona Guinness)

Ramalina siliquosa is commoner on other coasts where siliceous rocks prevail. However, it does grow in the Burren on granite erratics close to the sea.

The seashore vegetation provides a home for many invertebrates. The large bristletail *Petrobius maritimus* frequently takes refuge in the tufts of Sea Spleenwort, and is often seen on the surface of rocks or in clefts when a stone is moved; it is somewhat darker than the inland species *P. brevistylus* (Chapter 6). These animals are primitive wingless insects up to 2.5 cm long and jump when threatened. They feed by scraping lichens and unicellular algae off the rocks throughout the year. They can live for at least 16 months, so some individuals – especially females – overwinter here (Joose, 1976). They also have overwintering eggs that are laid in autumn in clusters of up to 100 by several females. As in all insects, the bristletails have to moult to allow growth – laboratory specimens have moulted 50 times over their lifetime. This is an advantage for their habitat of hard rocks and waves, in that limbs or antennae lost in the extreme conditions can be replaced at the next moult.

Several unusual moths occur in the coastal habitat. A form of the Grey (*Hadena caesia mananii*) is exclusive to western Irish rocky coasts where Sea Campion grows, while the White-line Dart (*Euxoa tritici*) and Northern Rustic (*Standfussiana lucernea*) are similarly rare.

OFFSHORE HABITATS

The Poulsallagh cliffs are difficult to access in many places, but at Doolin to the south or Fanore to the north, a similar limestone shore consists of several terraces at progressively lower levels. The rock is pockmarked with tiny pools filled with marine life. Covered by acorn barnacles and sea anemones, and full of small seaweeds, their walls overhang the pools as the plants and animals produce slightly acid conditions that erode the rock. Small Common Mussels (*Mytilus edulis*) crowd each other on some of the intertidal surfaces like a felt, and Limpets (*Patella vulgata*) are ubiquitous. The black lichen zone extends downwards to the mean high-water mark of spring tides, so is wetted by waves for several hours most days of the year, whereas the orange lichens stop some 1–2 metres above this and remain dry except during storms (Ryland & Nelson-Smith, 1975). One of the black lichens, *Lichina pygmaea*, extends even further, growing down among the tufts of Chanelled Wrack (*Pelvetia canaliculata*) and the topmost fucoid species, an exposed form of Bladder Wrack (*Fucus vesiculosus*) that lacks air bladders. The other brown seaweeds are usually lower down the profile, with the lowest, Serrated Wrack (*F. serratus*), seldom growing above the low-water

FIG 321. Zonation on the shore at Fanore: in the distance are permanent rock pools and acorn barnacles; there is then a zone of Common Mussels (*Mytilus edulis*), out of reach of most of the species' predators; and in the foreground is the upper tidal zone, exposed to breaking waves and physical erosion but supporting little marine life. (Roger Goodwillie)

FIG 322. Typical biokarst shoreline south of Black Head, May. There are two levels of the shoreline: the lower is covered by every high tide and festooned by Bladder Wrack, and the upper zone (to the right) is black with lichen growth. (David Cabot)

FIG 323. Bladder Wrack (*Fucus vesiculosus*) was the original source of iodine, discovered by the French chemist Bernard Courtois in 1811, and was used extensively to treat goitre, a disorder of the thyroid gland. (David Cabot)

FIG 324. The brown alga *Bifurcaria bifurcata* growing in a rock pool near Fanore, along with Purple Sea Urchins (*Paracentrotus lividus*) and sea anemones, and surrounded by acorn barnacles. (Roger Goodwillie)

level of ordinary tides. Together with the larger kelps, it is seen floating in the swell at the base of the cliffs.

During spring tides, when low water is below its normal level, the kelps become more exposed and the *Laminaria* and *Saccorhiza* species may be picked out. The largest of these, *Laminaria digitata* and *L. hyperborea*, grow on stalks, these more flexible in *L. digitata* and likely to stand out of the water in *L. hyperborea*. The latter species provides the sea rods that were once widely collected for the alginate industry. It is likely that another kelp species, *L. ochroleuca*, may also occur here in the future. It is a warm-water seaweed, first seen in Britain in 1940 and slowly colonising the southern coast there. It is a good example of a Lusitanian species, also occurring in Spain and Portugal, but with a distribution that extends northward along the seaboards of western France and south-west Britain. Indeed, the sea provides a host of other organisms whose immigration is so much simpler than for terrestrial organisms. The Purple Sea Urchin (*Paracentrotus lividus*) may be the best-known animal; its range extends from the Mediterranean to a few sites in north-west Scotland, but it is nowhere commoner than in the Burren. It uses its teeth and spines to bore into the limestone and expands its burrow as it grows. Where it is numerous, the rock may be honeycombed with burrows; indeed, these animals may initiate many of the intertidal pools seen along the coast here. The cavity gives the animal protection from the waves and reduces the risk of it drying out at low tide, although many of the urchins live sub-tidally, where they grow rather larger – up to 7 cm across excluding spines. The sea urchin is herbivorous, grazing on algae of many sorts but preferentially on young specimens.

The brown alga *Bifurcaria bifurcata*, with cylindrical, dichotomous branching, is one of the more distinctive Lusitanian seaweeds and grows in many of the rock pools. There are also red algae with this distribution pattern, reaching their northern limits in western Ireland and extending south to Iberia. In general, red seaweeds outnumber the brown and green total by two to one, and contain some of the most beautiful, but smaller, species. Many are finely divided and flimsy looking, so can only grow sub-tidally, where wave damage is less likely. When cast up on beaches, they appear as tangled tufts; their beauty is appreciated only if they are refloated in water. Some red seaweeds, such as the oak leaf-like *Phycodrys rubens* or the feathery fans of *Plocamium cartilagineum*, choose to grow on the stipes of kelp. Others coat rocks; the baby pink of *Lithophyllum incrustans* is characteristic of sea urchin beds, as it is too small or unappetising for them to eat. Other encrusting red algae are also easy to see in rock pools. *Corallina officinalis* produces tough feather-like fronds, while the eccentric *Mesophyllum lichenoides*, which looks like a fungal fruit body, grows epiphytically on *Corallina*. A number of crustose, calcareous red algae (Corallinaceae) also grow detached in shallow waters and accumulate to form large beds of stone-like sediment called maerl. A mix of *Phymatolithon calcareum* and *Lithothamnion corallioides* form most of the maerl that is occasionally thrown up to form 'coral' beaches in Co. Galway.

The green algae *Ulva* (*Enteromorpha*) *intestinalis* is widespread on the shore but is particularly noticeable where fresh water seeps from the land or falls from cliffs onto the rocks below. In this habitat, it displaces other species because it has such a broad tolerance to salt concentration. It is able to survive for a few days in pure fresh water, but is also able to withstand four times the usual concentration of salt (Kame & Fong, 2000). It seems that its growth increases as the salt concentration becomes closer to the average for sea water, and that individual populations show a genetic adaptation to the variations of a particular site (Reed & Russell, 1979). Sporelings grow best in the conditions their parents experienced.

Seaweeds spread freely once conditions are right, none more so than Wireweed (*Sargassum muticum*). This is an invasive brown alga that was first seen in Ireland in 1995 and spread all around the coasts over the following 10 years. In its native home in Japan, it is a relatively small plant of 1–2 metres, but here it is larger, presumably because of a lack of pests; in mainland Europe, plants that are 16 metres long have been found. Wireweed is a perfect weed: it grows fast, up to 4 cm a day; it lives for three to four years, reproducing in each of them; it dies back in winter and so withstands wave action; and it tolerates a wide range of conditions in life, such as full sunlight, drying out, and variations in salinity and temperature (–1 to 25 °C) (Guiry, 2018). It can therefore occupy habitats ranging from intertidal rock pools to sub-tidal shores, and surfaces from exposed rock to stabilised sand, as in eelgrass (*Zostera* spp.) beds. Its main impact is to shade out other species and bring about a simplification of the community, although it can occasionally add habitat to places that lack algal growth. Generally, it competes with the native seaweed species, noticeably on the Burren coast with *Bifurcaria bifurcata*, and probably with some colonies of the Purple Sea Urchin.

Wireweed is not the only introduction – the oceans are beset by alien species as much, if not more so, than the land. *Asparagopsis armata* is a small tree-shaped red alga widely found on the north coast of Co. Clare that was introduced in the 1920s by shipping from the southern hemisphere. The lacy *Dasysiphonia japonica* has a Pacific origin, and has so far been found at New Quay and Finavarra. The acorn barnacle *Elminius modestus* from Australasia has spread all round the Irish coast in sheltered places.

SHELTERED SHORES

One of the features of a karstic shoreline is that the sea may penetrate some distance through the grikes, and consequently seaweed can grow quite far inland. At Finavarra, for example, behind the Flaggy Shore, the coast road runs on a ridge of glacial till overlying limestone pavement. The pavement reappears inside the road in the middle of a field and there is a tidal pool about

ABOVE: **FIG 325.** Sheltered inner shore of Galway Bay, north of Abbey Hill, with extensive beds of wrack (*Fucus* spp). (David Cabot)

FIG 326. Limestone pavement in the middle of a field at Finavarra, with *Ulva* and *Fucus* seaweeds. (Roger Goodwillie)

250 metres from the shore. The rocks here support Bladder Wrack, Channelled Wrack and Spiral Wrack (*Fucus spiralis*), while between them large glasswort (*Salicornia* spp.) plants spread up the muddy grikes through drapes of green *Ulva* algae.

There are further intertidal habitats on the low shores of the north coast, including a number of small saltmarshes and some, more impressive, lagoons. Set in the sandy or rocky ground, both of these habitats are situated at a height within the tidal range, so are affected at least by spring tides and brackish conditions. The lower ground of saltmarshes is flooded by practically every high tide and so is bathed by sea water regularly, with little evaporation taking place. The upper levels, however, are inundated only by spring tides and in between can dry out significantly. The result is that they experience very high salt concentrations in the soil at times, and both flora and fauna must be adapted to deal with these challenging conditions.

The sheltered part of Ballyvaghan Bay behind the Rine Peninsula has a considerable fringe of saltmarsh dominated by grasses and Thrift. This runs smoothly outwards from the rocky grassland quite without a zone of Sea Rush

FIG 327. The remains of the Rine Peninsula at the mouth of Ballyvaghan Bay, soon to be a victim of rising sea-levels. (Roger Goodwillie)

FIG 328. Erosion of sub-sea glacial till on the Rine Peninsula. (Roger Goodwillie)

(*Juncus maritimus*), which might be expected. The substrate here is not typical of saltmarshes and appears windborne (loess) rather than waterborne (Sheehy-Skeffington & Wymer, 1991). Postglacial sea-level rise may have flooded the shore and allowed saltmarsh to develop on a formerly dry habitat, a parallel development to the saltmarshes on peat in the west of Ireland. Certainly, the fretted islands and spits of this area suggest a drowning process. The Rine Peninsula itself hangs onto land by a thread, much narrower now than when mapped in the 1840s. Recent wave attack has exposed great lumps of till on the outer shore, where they are winnowed further and their finer sediments washed offshore.

Ballyvaghan Bay is full of seaweed, and bands of wrack – largely Spiral Wrack – are spread on the contour corresponding to the current high-tide mark, defining the outlines of the reefs and included islets. Sea Aster, Lax-flowered Sea Lavender (*Limonium humile*) and Greater Sea-spurrey (*Spergularia media*) are widespread in the patches of saltmarsh. Towards the base of the bay, Common Glasswort (*Salicornia europaea*) and Annual Sea-blite (*Suaeda maritima*) come into the picture, and there are a few places out to sea where Eelgrass (*Zostera marina*) grows, especially on the sand at Bishop's Quarter to the east. Close to the laneway down to the saltmarsh, Sea Wormwood (*Artemisia maritima*) occurs, its white shoots standing out against the prevailing greenness and bringing to mind some garden plant that has escaped. It is a peculiarly local species in Ireland as compared to Britain, growing in scattered colonies on both east and west coasts, particularly in the Shannon estuary. It also grows on Inishmaan.

Further areas of saltmarsh occur at Finavarra around Scanlan's Island, in Muckinish Bay and Aughinish. Like many on the west coast, they include rather little of the shrubby Sea-purslane (*Atriplex portulacoides*) – which suffers from grazing – and so appear grassy. They also support outlying populations of the glaucous Sea Couch (*Elytrigia atherica*), which is such a feature of south-east coasts in Ireland. Frequently one sees turf fucoids, terrestrial forms of wrack quite without bladders and clustered like mosses among the Thrift and Sea Plantain. This entity has had a variety of names; *Fucus vesiculosus* var. *muscoides* is a good descriptor, but *F. cottonii* is the currently accepted name. Recent DNA research in the west of Ireland has shown that in some places the plant seems to be pure Bladder Wrack but in others it has hybridised with Spiral Wrack and has developed a number of different polyploidy levels (Sjøtun *et al.*, 2017). The fact that the turf fucoids are common in west coast saltmarshes but absent from those in the east is put down to the wetter climate here, which prevents the soil drying out and developing damaging levels of salt (Sheehy-Skeffington, 2000). A dieback of the fucoids was noted in August and September at one site during a period of low rainfall.

Most of the saltmarshes around the coast are grazed by cattle and the lawn-like quality of the vegetation is remarkable. There is much more herbage in areas domestic animals cannot reach, including Sea-purslane, Sea Couch and sea-lavender plants jostling for space. Occasional Sea Beet (*Beta vulgaris* ssp. *maritima*) and Spear-leaved Orache (*Atriplex prostrata*) scramble through the other species, and there is an abundance of vegetation and numerous grazing invertebrates to eat it. In a cattle-grazed situation the invertebrates are very different, many of them feeding on detritus left by the tides (Andersen *et al.*, 1990).

LAGOONS

A lagoon is a waterbody separated from the sea but connected to it through sand or rock. It is set at a height between high and low tides, and is consequently exposed to both seawater and freshwater influences. The water in the lagoon is brackish and subject to rapid changes in salinity in response to high tides or high rainfall, creating a relatively difficult habitat that few plants and animals are able to adapt to successfully. The difference in osmotic pressure between an organism and the water means that its tissues tend to lose water when salinity goes up and gain it when it goes down – both of which can be lethal. The organisms living in lagoons also have to tolerate quite rapid changes in temperature, since the water is generally shallow and heats up quickly. These factors result in a general paucity of lagoon species, the same organisms turning up over a wide range in Europe. As a consequence, the few species create a simplified ecosystem where there are limited feedbacks or controls; individual species may increase without a corresponding increase in their predators. This in turn leads to blooms of plant or animal species. Any change in the environment leads to a huge die-off, which in turn results in a lack of oxygen for the organisms that remain.

Natural lagoons are relatively rare on a European scale because so many have been modified for boating or for aquaculture. They typically form behind a beach ridge where fresh water accumulates but where the sea seeps through the barrier when the tide is high. Only one of the Burren lagoons has such a barrier (Aughinish), but there are two on the northern shore of Inishmore in the Aran Islands, especially the prominent Loch Phort Chorruch, with its man-modified storm beach. Most other lagoons in our area, such as Lough Murree, have influxes of sea water through underground rock fissures.

Lough Murree received early attention and there is a university field station close by that specialises in the marine environment. The lake is based on rock and is shallow (mostly less than 2 metres deep) but quite large at 14 hectares. The salinity fluctuates over the year depending on rainfall and tidal heights, but is generally 30–60 per cent that of the sea. Even across the lake the salinity varies, being highest beside the shore barrier, where there are a few tidal 'springs'

FIG 329. A tidal spring runs into Lough Murree when the tide in the bay outside is high enough. (Roger Goodwillie)

– features that come alive when the tides are high enough. The lake contains many of the classic brackish-water species – for instance, Beaked Tasselweed (*Ruppia maritima*), Spiral Tasselweed (*R. cirrhosa*), Fennel Pondweed (*Potamogeton pectinatus*) and the unattached green alga *Chaetomorpha linum* (Healy & Oliver, 1996). In addition, two of the rarest charophytes grow there – *Chara canescens* and the so-called Foxtail Stonewort (*Lamprothamnium papulosum*). The former is the less frequent of the two, found in eight or nine sites in Ireland (and four in Britain), whereas the Foxtail Stonewort occurs in 14 different lagoons, clustered in Connemara and Co. Wexford. In Britain, it is restricted to the Hebrides, Anglesey and the south coast of England.

Lough Murree is highly eutrophic (a state partly caused by local cattle) and subject to algal blooms, when green algal species create shade for the bottom flora and oxygen stress among the fauna. This reduces the area inhabited by stoneworts and also limits the diversity of invertebrates. One of the species that does flourish is the Common Ditch Shrimp (*Palaemonetes varians*), a colourless crustacean that is a lagoonal specialist in Ireland, probably needing such an environment for reproduction (Oliver, 2007). It also tolerates lower oxygen levels in the water than its relatives and was found in 70 per cent of sites during the national study of Irish lagoons organised by Brenda Healy and others in the 1990s (Healy & Oliver, 1996). *Hydrobia* snails are also common and include *H. (Ecrobia) ventrosa*, a brackish-water species in Ireland and the rest of Europe. The commonest fish is the Three-spined Stickleback (*Gasterosteus aculeatus*), which despite its frequency in stream and rivers, is basically a sea fish with some colonisation of brackish or freshwater conditions.

Lough Murree certainly has had saline fluctuations in the past, as evidenced by marine and freshwater species that are unknown today but found deposited at certain depths in the sediment (Cassina *et al.*, 2013). It supports a widespread species of freshwater shrimp, *Gammarus duebeni*, but Aughinish lagoon has the distinction of harbouring *G. chevreuxi*, which is much less common. It was recognised only recently as an Irish species (from about four sites) and is similarly rare as a lagoonal species in Britain (16 sites).

Aughinish (translated as Horse Island) is an island connected to the mainland by a modern causeway. It is located in the north-eastern corner of our area, in line with the first rise of the bare limestone of Abbey Hill. The lagoon on its north-eastern coast is currently almost fully saline, although this may change over time. It is sandy with occasional flat outcrops of rock on the floor and is linked to the sea through a shingle bar. *Ulva (Enteromorpha) intestinalis* is the dominant algal species when the lagoon is flooded, but as the water levels drop it becomes stranded on the surface as large decaying flakes. In between, the ground is liberally scattered with the tiny twisted Frog Rush (*Juncus ranarius*), as well as Common Glasswort and Annual Sea-blite. Spear-leaved Orache occurs on the sand, in this case conforming to its Latin species epithet, *prostrata*.

One of the main tidal springs feeds the north-west corner of the lagoon, and has initiated a nice sequence of particle size from the stones of the beach to increasingly finer sediments stretching into the shallow waters behind. A recent storm has brought to light seeds of the Yellow Horned-poppy (*Glaucium flavum*) in one of its only west coast sites.

In any survey of lagoons on the Aran Islands, the rock-girt Loch Mór on Inisheer must be mentioned, if only for its setting. It occurs on a fault line in the rocks and is 25 metres deep, although just over 6 hectares in size. For such a sizeable body of water it has few species, and its chief claim to fame is the palaeoecological work carried out by Michael O'Connell and his team on cores extracted from its bed and quoted elsewhere in this book (Chapter 4) (Molloy & O'Connell, 2014). A more important lagoon on these islands, ecologically speaking, is the linear waterbody a kilometre north of Kilronan on Inishmore. It contains a high number of what are classified as lagoonal specialists, including the water-boatman *Sigara selecta*, known in only one other site in Ireland (National Parks & Wildlife Service, 2015). It also holds supports both Beaked and Spiral tasselweeds.

FIG 330. The northern end of Loch Mór on Inisheer, close to the intertidal connection with the sea (on the right). (Roger Goodwillie)

STORM BEACHES

Storm beaches are characteristic of exposed coasts on the western seaboard that have a suitable shore profile. Aughinish lagoon is held back by such a beach but it is relatively low owing to the shelter provided by Galway Bay. There is an impressive storm beach south of Poulsallagh, but the largest are on the Aran Islands, near Gort na gCapall on Inishmore and at Trácht Each on Inishmaan. They reflect the ongoing power of winter storms, whereby rounded cobbles and stones are piled in a tall ridge far above the normal high-tide mark. The true power of the waves is seen in the huge slabs of rock broken off the Aran cliffs and deposited on the adjoining pavement. Measurements suggest that slabs weighing 250 tonnes (or about 10 × 5 × 2 metres) have been moved at sea-level, slabs of 117 tonnes have been deposited at 12 metres ordnance datum (OD) and smaller pieces of 2.9 tonnes at 50 m OD (Williams & Hall, 2004). Indeed, there were observations of 40–80 tonne blocks being moved by a storm in 1991. Comparing early Ordnance Survey maps to current ones reveals that individual ridges have moved tens of metres inland, overrunning former fields and field walls (Cox *et al.*, 2012). This has occurred during 'ordinary' storm activity – no tsunami was responsible, which was one of the previous theories. It illustrates the awesome

FIG 331. Cliff collapse behind Trácht Each, Inishmaan, seen from the storm beach. (Roger Goodwillie)

FIG 332. Thrift
(*Armeria maritima*) in
the saltmarsh behind
the storm beach at
Poulsallagh, along with
Sea Plantain (*Plantago
maritima*). May.
(Fiona Guinness)

power of storms such as the 'Night of the Big Wind' (1839), when herrings
were blown almost 10 kilometres inland on the west coast of Ireland and trees
75 kilometres from the coast were salty to the taste (Shields & Fitzgerald, 1989).
Such a storm can hugely alter the coastline; the jumbled collapse of cliffs above
the saltmarsh behind Trácht Each on Inishmaan, for example, looks as if such a
storm has only just left it.

The ecology of such wave-based deposits is primarily terrestrial. Once the
salt has been washed off by rain, organisms are free to colonise the surfaces and
lichens are quick to arrive. The lower reaches of shingle that remain wet enough
develop an open vegetation, often in combination with species that also occur
as weeds in tilled fields. Field Sow-thistle (*Sonchus arvensis*), Common Couch
(*Elytrigia repens*), Goosegrass (*Galium aparine*) and Curled Dock (*Rumex crispus*)
are frequent. Sea Radish (*Raphanus raphanistrum* ssp. *maritimum*) is common on
the northern Burren coast and on the Aran Islands, where it grows on stones
and sand. It has a large seed, giving it a good start even on this sub-optimal
substrate. Sea Mayweed (*Tripleurospermum maritimum*) is ubiquitous and often
resembles its close relation, the annual Scentless Mayweed (*T. inodorum*), which
colonises farmland. Only a few beaches support special plants, like Tree-mallow
(*Malva arborea*) at Bell Harbour and Sea-kale (*Crambe maritima*) on Inishmaan
and Inishmore. The Rine Point spit offers additional potential habitat, as yet
colonised only by prostrate Blackthorn (*Prunus spinosa*) and Common Stork's-bill
(*Erodium cicutarium*).

FIG 333. Sea Radish (*Raphanus raphanistrum* ssp. *maritimum*) at New Quay, May. The plant is frequent along the coast from Ballyvaghan to New Quay and scattered elsewhere. (Fiona Guinness)

FIG 334. Tree-mallow (*Malva arborea*), planted here at Muckinish Bay, is native on Inishmore. (Fiona Guinness)

DUNES

Sand dunes are occasional in our area and only Fanore offers an extensive stretch on the mainland. There are other fragments on the northern side of Aughinish Island and at Bishop's Quarter in Ballyvaghan Bay, although this inlet is rocky in general. One has to go to the Aran Islands to see the habitat again, where dunes occur on the relatively sheltered northern side of each island. In fact, each island's airport is located on 'back dunes', the stable part of the system behind the mobile beach where shell sand has been blown inland to cover the rocky basement. Inisheer has a well-used beach near the harbour, where frequently used paths (and erosion) cut through the dunes. Inishmaan has a more natural set-up at its north-east point, although this area is becoming smaller by the year. Inishmore has the added features of a tombolo and spit at its eastern end, cutting off a backstrand (An Trá Mhór). Offshore, Straw Island is also a dune system on a rocky base and was formerly linked to the 'mainland' by another tombolo. It was an old station for the now extinct Sea Stock (*Matthiola sinuata*), which also grew at Bishop's Quarter at one time (in 1912, as recorded by Praeger (1934)).

Foredunes – the initial small dunes on the back of a beach – are in good shape at Bishop's Quarter and on the Aran Islands. They are more limited at Fanore owing to a general retreat of the coastline but they do occur at the mouth of the Caher River. Sand Couch (*Elytrigia juncea*), Sea Rocket (*Cakile maritima*) or Frosted Orache (*Atriplex laciniata*) trap an initial pile of sand, replaced by Prickly Saltwort (*Salsola kali*) at the eastern end of Inishmore. Marram (*Ammophila arenaria*) grows at higher levels as it is more sensitive to salt. The loose sand is soon enlivened by the exotic blooms of Sea Bindweed (*Calystegia soldanella*), the regimented shoots of Sea Sandwort (*Honckenya peploides*) or, on the dunes above, by Sea-holly (*Eryngium maritimum*), which takes advantage of bare sand anywhere.

Sea Spurge (*Euphorbia paralias*) often shares the foredune habitat with Sea Bindweed and is prominent at Fanore. The tolerance of these plants to salt spray, moving sand, high temperature and low-nutrient status is noteworthy. Sharp sand grains whip along the surface, tiny shells are fragmented by impact with each other, and the grass leaves draw arcs on the surface, but despite all this disturbance the plants look fresh and resilient, no matter what the conditions. Part of their secret is that they contain latex, a white milky fluid that is acrid to the taste and seems to protect them against grazing. So, the snails one meets on sand dunes everywhere in wet weather pay little attention to them. However, they do also have other adaptations. In the bindweed, salt glands are visibly scattered

FIG 335. a. The serried ranks of Sea Sandwort (*Honckenya peploides*) invading from the back of the beach. Inishmaan, August. b. Sea Bindweed (*Calystegia soldanella*) growing on foredunes with colonising Marram (*Ammophila arenaria*), July. c. Sea Spurge (*Euphorbia paralias*) growing among Marram on Fanore dunes, May. d. The complex flowers of Sea Spurge, with a single female flower and three male flowers in the cup made by two bracts. (a. and b. Roger Goodwillie; c. and d. Fiona Guinness)

over the leaf surface, these enabling it to excrete salt and grow lower on the shore than it otherwise would. Its leaves also lie close to the sand surface, where evaporation is slightly reduced. By contrast, the spurge grows vertically into the wind, but its overlapping leaves create their own cocoon of still air. It can further conserve water by using the crassulacean acid metabolism for photosynthesis (El Haak *et al.*, 1997). This involves taking in carbon dioxide at night, when there is no danger of evaporation, and storing it for use during daylight hours when there is adequate light for photosynthesis.

Fanore is located in a shallow bay where glacial debris was dumped offshore and has subsequently been sorted and blown on land by the prevailing (south-westerly to westerly) winds. The Caher River discharges through the southern end of the beach but does not supply much additional sediment. The dune system is somewhat overrun by caravans and day visitors, but there are sections without such impact where only Rabbits (*Oryctolagus cuniculus*) and cattle have an effect. It appears to be in gradual retreat; there are no embryonic dunes on the main beach, which instead is backed by a sheer slope cut into existing dune grassland and bearing clods of vegetation broken off from the crest. Blow-outs and paths create erosional pockets, so that there is actively growing dune habitat but only on sporadic piles of sand. Because of the conditions, weedy species

FIG 336. Mobile homes at Fanore dunes. (Fiona Guinness)

FIG 337. Limestone erratics on pavement hidden in a former blow-out in the Fanore dunes. (Roger Goodwillie)

are more frequent than might be expected and it is not uncommon among the Marram to meet Groundsel (*Senecio vulgaris*), Scarlet Pimpernel (*Anagallis arvensis*) and dandelions. Away from the sea, the Marram persists but is joined by other species as organic matter accumulates in the soil. There is much False Oat-grass (*Arrhenatherum elatius*), along with occasional Meadowsweet (*Filipendula ulmaria*) and Common Valerian (*Valeriana officinalis*), reflecting the local rainfall. Wild Angelica (*Angelica sylvestris*) and Hemp-agrimony grow in similar places in the dunes on Inishmore.

Fanore specialises in stabilised dune grassland, of which there is about 60 hectares in untrampled places. Some of this is based on Red Fescue, with Squinancywort (*Asperula cynanchica*), Bird's-foot Trefoil and Wild Thyme (*Thymus polytrichus*) in the shorter turf. Annuals are scattered wherever open sand is exposed, and include Common Centaury (*Centaurium erythraea*), Yellow-wort (*Blackstonia perfoliata*), Black Medick (*Medicago lupulina*) and Fairy Flax (*Linum catharticum*). In places, Devil's-bit Scabious (*Succisa pratensis*) turns the surface blue in August since it is unpalatable to Rabbits; Biting Stonecrop (*Sedum acre*) is also noticeable, both in flower and fruit. On the eastern side of the dunes, colonies of Blue Moor-grass (*Sesleria caerulea*) are expanding slowly, like islands in a sea of turf.

FIG 338. Back-dune grassland on Inishmaan, now enclosed as a field. A blue haze of Harebell (*Campanula rotundifolia*) is joined by Yellow-wort (*Blackstonia perfoliata*) and Squinancywort (*Asperula cynanchica*) in the foreground. July. (Roger Goodwillie)

There are more interesting plants at Fanore too, and the sight of Spring Gentian (*Gentiana verna*) or the Irish Eyebright (*Euphrasia salisburgensis*) here remind one of the proximity of the limestone Burren. Dodder (*Cuscuta epithymum*) is also widespread, as it is on the dune grassland on Inishmaan and the eastern end of Inishmore. Pyramidal Orchid (*Anacamptis pyramidalis*) is everywhere, but one should look out also for Thyme Broomrape (*Orobanche alba*) in the spring and Autumn Lady's-tresses (*Spiranthes spiralis*) in August.

The sand dune flora of the west of Ireland has long been noted for its lack of diversity (Praeger, 1934). There are very few annual clovers, no Wild Asparagus (*Asparagus prostratus*), and no Hound's Tongue (*Cynoglossum officinale*) or Viper's-bugloss (*Echium vulgare*), which do occur on the eastern side. Where the Burren

FIG 339. Irish Eyebright (*Euphrasia salisburgensis*), one of the specialities of the Burren and a hemi-parasite of Wild Thyme (*Thymus polytrichus*). The plant can grow up to 10 cm in height. Fanore, May. (Fiona Guinness)

FIG 340. Dodder (*Cuscuta epithymum*) growing in sandy grassland by the Inishmaan shore, July. (Roger Goodwillie)

(or, rather, the Aran Islands) confounds this pattern is with Wild Clary (*Salvia verbenaca*) and Purple Milk-vetch (*Astragalus danicus*), the latter with the flowers of a large clover and the leaves of a vetch. As noted in Chapter 5, the milk-vetch grows on the dune grassland and some cliff tops on Inishmore and Inishmaan, but nowhere else in the country. In Britain, it is widespread along the northeastern coast and extends inland on chalk from East Anglia to Bristol. However, this eastern bias is matched by sporadic occurrences on the west coast, on the Isle of Man, on the Galloway coast and on Tiree in the Inner Hebrides. This suggests a coastal immigration during the late-glacial period and a subsequent restriction to a substrate whose pH is high enough to withstand the leaching effect of rain; grey limestone sand and shell fragments are a feature of the Aran beaches. However, Purple Milk-vetch's unusual distribution could just be the legacy of a seafaring Celtic tribe who regarded it as a lucky plant to be brought to their ports of call. It has the reputation of increasing the milk supply from goats. Christopher Tolan-Smith (2008) postulates that the third phase of human expansion in Britain and the first in Ireland (6000 BCE) were seafaring movements, bringing people to the Irish and Scottish coasts, including the islands. DNA studies of the remains of these people show that they had an exclusively maritime diet, and genetic studies of the disparate populations of the milk-vetch might also be revealing. Many other plants formerly thought to be native have been proven or suggested as introductions to Ireland, for example Slender Rush (*Juncus tenuis*), Kerry Lily (*Simethis mattiazzii*), Irish St John's-wort (*Hypericum canadense*) and Mackay's Heath (*Erica mackayana*).

Wherever it grows, Purple Milk-vetch requires good drainage and poor soil. Some of its associates are Red Fescue and Crested Hair-grass (*Koeleria macrantha*), as well as Wild Thyme and Biting Stonecrop. Mosses and lichens are also prominent in places – clumps of *Hypnum cupressiforme*, *Hylocomium splendens* or *Homalothecium lutescens* look unbeatable after wet weather but shrink to yellowish tufts in drought, when only *Rhytidiadelphus triquetrus* stands proud. *Cladonia* and *Peltigera* lichens are present too, showing a certain amount of stability in the vegetation.

The Bishop's Quarter dunes have a particularly white sand, borne of the shellfish living in the shelter of Ballyvaghan Bay rather than the grey limestone sand found elsewhere. This, and a lack of Rabbits, gives it a much richer flora than Fanore, and the amount of Harebell (*Campanula rotundifolia*), Greater Knapweed (*Centaurea scabiosa*) and Autumn Gentian (*Gentianella amarella*) that results is a delight.

FIG 341. Rabbits (*Oryctolagus cuniculus*) exert a large effect on sand dune vegetation. Here at Fanore in June, mosses can be seen in the foreground but there is a complete absence of flowers on the herbaceous plants. (Roger Goodwillie). Inset: Rabbit, Fanore. (Fiona Guinness)

FIG 342. Lady's Bedstraw (*Galium verum*) manages to flower when far enough away from Rabbit (*Oryctolagus cuniculus*) burrows. May. (Fiona Guinness)

Mosses are particularly prominent where there are Rabbits. A large warren occurs at the northern end of the Fanore dunes, where burrows pepper the sides of an old blow-out, confining much of the flora to non-flowering shoots. Ribwort Plantain (*Plantago lanceolata*), Lady's Bedstraw (*Galium verum*) and Common Violet (*Viola riviniana*) survive this treatment in non-flowering form and there are centimetre-high plants of eyebright (*Euphrasia* spp.) and Black Medick. Scarlet Pimpernel is not grazed, so is common, while Spear Thistle (*Cirsium vulgare*) suffers pruning but does not die out. The feeding habits of the Rabbit are quite specific, as might be expected for any grazing animal. As well as having a notable distaste for Common Ragwort (*Senecio jacobaea*) and mosses, Rabbits in Wales have been shown through faecal analysis to choose particular grasses in a mixed sward, taking them just as the flowers are being produced and again after any seeds have formed (Soane, 1980). On Inishmore, exclosures installed by the AranLIFE project illustrate that Rabbits there prefer grasses over Sand Sedge (*Carex arenaria*). When grazing is prevented, Red Fescue and other species grow tall where previously they seemed almost absent.

Around the burrows there are obvious latrines where nutrient enrichment occurs and the grass consequently has a considerably deeper colour. These areas continue to be grazed to some extent as well as containing daytime faeces (night-time pellets are produced underground and eaten by the Rabbits for their vitamin B content). Most herbivores avoid their dunging grounds, which has the consequent effect of reducing parasite infestation. The fact that Rabbits can cope with this reflects an adaptation to living in large groups, tolerating a reasonable parasite load and developing an immunity to some of the major organisms, particularly the protozoan *Eimeria steidae*, which causes coccidiosis. Research has found that young Rabbits develop greater resistance to parasites if they have a happy, unstressed upbringing (Rödel & Starkloff, 2014).

Dunes in several places are notable for having depressions that reach down to the basement rock. At both Fanore and on Inishmore, hidden among the

FIG 343. Top: Rock Samphire (*Crithmum maritimum*) and Marram (*Ammophila arenaria*) trapping sand on the shore, September. Above: exposed limestone outcrops in the machair on Inishmore, colonised by *Cladonia foliacea* and other lichens. September. (both photographs by Roger Goodwillie)

dunes that accumulated subsequently, is pavement with large erratics. At Fanore the erratics bear Sea Spleenwort and the pavement has an abundance of the introduced White Stonecrop (*Sedum album*), which turns red in autumn, fringing each rock. Recent erosion along the coast has revealed other pavement, and Rock Samphire has spread to dominate the grikes and accumulate its own small linear dunes.

At Fanore one should see the excellent exposure of glacial till just to the north of the beach, where a linear drumlin-like deposit is being eroded by the sea (Croot & Sims, 1996). Limestone blocks at the base were dragged along and partially crushed by the moving ice, and above them is a deposit that was transported from further away. Angular gravel and stones are included in a lower layer, showing more crushing, whereas much finer silt is prominent in the upper layer, along with occasional rounded stones. Here it is thought that the particles were broken up by abrasion, rubbing together as the till was transported by a later ice advance.

Dune-slacks are quite rare in the Burren owing to the prevalence of free-draining rock below the sand. Bishop's Quarter has a shallow one where a blow-out has reached the water table and supports marsh plants, in this case Distant Sedge and Marsh Helleborine (*Epipactis palustris*). A drier site at Inishmaan maintains Silverweed (*Potentilla anserina*) and Lesser Hawkbit (*Leontodon saxatilis*) but still has Sand Sedge. Clumps of Creeping Willow (*Salix repens*) occur there and on Inishmore, reminding one of many other dune willow habitats along the west coast, although none occurs in the Burren.

MACHAIR

Machair is a special landscape found in the most exposed places on the west coast where shell sand is swept across a coastal plain that sits close to the water table. The strong winds prevent any build-up of dunes, but some sand lodges where there is enough dampness or in hollows between the rocks. A mosaic of habitats develops, comprising dry and damp grassland, small pools and marshes. There is no machair on the Burren mainland, but on Inishmore in the Aran Islands there is about 30 hectares or more at the eastern end. It is not the most diverse example of its type – in fact, it is the most southerly machair and not as varied as sites in Co. Mayo and Co. Donegal, containing very few damp habitats. One section of the machair has been taken over by the airport and a sports pitch, but flights are infrequent enough to allow some birdlife to persist, if not the original vegetation.

The main site extends up the eastern peninsula from Iarárine as shallow sand over limestone bedrock, which becomes more exposed the further north one goes. There are Marram dunes around the margins and occasional wisps of the grass internally if sand accumulates between the rocks. Flat sections of pavement occur, but more often there are low, whale-backed ridges covered by lichens such as *Aspicilia calcarea*, *Collema multipartitum* and an abundance of *Cladonia foliacea*. Granite erratics are always in view, both large and small.

Tiny plants of Wild Thyme or Lesser Hawkbit cling onto any cracks in the rock, while the sand in the hollows allows Field Wood-rush (*Luzula campestris*), Red Fescue and Fairy Flax to grow. Rabbits and their attempted burrows are everywhere, and large clumps of the lichen *Cladonia rangiformis* (which they do not eat) cover square metres wherever the surface is stable enough. The Wild Pansy (*Viola tricolor*) is shown up as the most distasteful species here, growing unmolested at the mouths of burrows and through the turf. It is uniformly yellow-flowered but resembles the dune subspecies *V. tricolor curtisii* in most ways. Recent work on Burren specimens reinforces the view that this plant – which has some of the characteristics of the Mountain Pansy (*Viola lutea*) – has an uncertain pedigree (Porter & Foley, 2017). Where the soil is thicker, a more typical dune flora based on grasses occurs, with species such as the Daisy (*Bellis perennis*), Cat's-ear (*Hypochaeris radicata*), Self-heal (*Prunella vulgaris*) and Sand Sedge. Squinancywort and Common Milkwort (*Polygala vulgaris*) add a bit of colour, while the flowers of Purple Milk-vetch may be picked out in a few places. Slightly wetter ground in the machair brings in the tree-like moss *Climacium dendroides* and also higher plants such as Bog Pimpernel (*Anagallis tenella*) and Marsh Pennywort (*Hydrocotyle vulgaris*).

Little work has been carried out on the invertebrates of the Burren dunes. A classic study on leaf-hoppers found 128 species in the Burren as a whole, but only eight at Fanore (Morris, 1974). This is not altogether surprising, because Fanore is an isolated area on the west coast that is not easy for sand dune specialists to reach. A species that has colonised the dunes, however, both here and on Inishmore, is the tiny Narrow-mouthed Whorl Snail (*Vertigo angustior*), which

FIG 344. The Mountain Pansy (*Viola lutea*) occurs in south Co. Clare and may have influenced the seaside populations of Wild Pansy (*V. tricolor*) in the Burren and Aran Islands. June. (David Cabot)

lives in dune-slacks and wet places in the machair. It is one of three rare *Vertigo* species that are specially protected in Europe, partly because they are indicators of particular kinds of habitat that might otherwise be missed in a conservation programme. The Narrow-mouthed requires a fine degree of wetness in the habitat, and may occur in moss, grass tussocks or under litter depending on the humidity (Moorkens & Killeen, 2011). If the water level rises, it will climb higher into the vegetation, and if it falls, it will bury itself. It is regarded as a relict species because its small populations are spread widely in Ireland's western coastal counties and the inland sites are isolated in the east Midlands. It seems to have been more frequent in the early part of the postglacial (8,000 years ago), when the climate was warmer than at present.

There are more conspicuous organisms at Fanore, especially the beetles one meets in the Rabbit-filled areas. The chunky bluish dung beetle *Trypocopris vernalis* and the brown scarab beetle *Onthophagus fracticornis* fly around in sunny weather, while the click beetle *Agrypnus murinus* may land anywhere unannounced. At first sight, it seems to have sand grains stuck onto its wing-cases, but these are small hairs in what is a case of effective camouflage. In the weevil group, Michael Morris (1967) found 13 species at Fanore as compared to 63 in the whole Burren. Three of these were restricted to the dunes but they were not of noteworthy species.

Bumblebees (*Bombus* spp.) fly relatively well and have a greater likelihood of reaching an isolated dune area than most other insects. In addition, they are generalists and able to feed from many different flowers. Good numbers have been found at Fanore and on the Aran Islands. The Great Yellow Bumblebee (*B. distinguendus*) is perhaps the most famous, as it has declined rapidly over most of Britain and Ireland, but the Red-shanked Carder Bee (*B. ruderarius*) is also notable (the Shrill (*B. sylvarum*) and Moss (*B. muscorum*) carder bees were discussed in Chapter 7). Later in the year, cuckoo-bees appear – one should look out particularly for the Field Cuckoo-bee (*B. campestris*) and Red-tailed Cuckoo-bee (*B. rupestris*). The last is a southern species and one of the few bumblebees that is spreading significantly.

The Small Blue Butterfly (*Cupido minimus*) is generally associated with sand dunes throughout Ireland as it feeds on Kidney Vetch (*Anthyllis vulneraria*), but because this foodplant occurs all over the Burren, the insect is much more widespread here. The greatest numbers are seen on the coast and on the Aran Islands, but it would not be a surprise to encounter some on the limestone pavement. Small is certainly the size of the butterfly, but grey or dark brown is more the colour than the blue suggested by its common name. The caterpillar feeds in the flower clusters of the vetch, overwintering as a larva and then pupating for only one to two weeks. The adults fly in May and June.

BIRDLIFE

The coastal birds in the Burren and on the Aran Islands change throughout the year in line with other coasts, although there is more of an emphasis here on species that breed in western Scotland and Iceland. No year passes without a constant passage of Whimbrel (*Numenius phaeopus*) in the spring, large numbers of divers (*Gavia* spp.) on the sea and Purple Sandpipers (*Calidris maritima*) on the rocky coasts.

Large flocks of waders and wildfowl do not occur, but small numbers of every species are seen, albeit rather spread out. The Aran Islands have the greatest diversity of nesting species, as there are several waders as well as seabirds. Inishmore still supports a few breeding Common Sandpipers (*Actitis hypoleucos*) along the north coast, while the machair and coast at the eastern end

ABOVE: **FIG 345.** Great Northern Diver (*Gavia immer*) off the Flaggy Shore, May. The bird is in full breeding plumage and probably on its way to its northern breeding grounds in Iceland. (Fiona Guinness)

FIG 346. Whimbrel (*Numenius phaeopus*) in Ballyvaghan Bay, mid-May. It is known colloquially as the May Bird because of the regularity of its northward passage to Iceland, given away by the piping notes of its flight call. (Fiona Guinness)

FIG 347.
Oystercatchers
(*Haematopus ostralegus*)
are seen everywhere
along the coast but nest
only in small numbers.
(David Cabot)

are a major site for other waders. Up to 20 pairs of Ringed Plover (*Charadrius hiaticula*) and Lapwing (*Vanellus vanellus*) nest on the area immediately beside An Trá Mhór according to a survey in 2010, along with a few Oystercatchers (*Haematopus ostralegus*) (Suddaby, 2010). Many more are likely on the peninsula to the north, which was not covered by that survey. This is a major concentration for the Burren area, especially for Lapwing, which otherwise nests as singles or small groups at a few turloughs. The birds must benefit from a lack of ground predators on Inishmore, if not of Hooded Crows (*Corvus cornix*), although they are surprisingly tolerant of the small airport and games pitch nearby. Ringed Plover also nest on the other Aran Islands and on the north coast of the mainland, along with one or two Oystercatchers.

Terns are often associated with sandy beaches, and nesting has occurred on all three Aran Islands as well as in Ballyvaghan Bay (Lysaght, 2002). In some years, Inishmaan contains just the right mix of stones, sand and vegetation on its upper beach for terns to nest – colonies of 300 pairs of Arctic Tern (*Sterna paradisaea*) and 20 of Little Tern (*Sternula albifrons*) have been recorded there in the past. Terns move around a great deal and have a multitude of alternative sites around west Galway, so smaller numbers are more usual. Little Terns are perhaps the more regular species, with a few pairs also on Inisheer, and more around An Trá Mhór on Inishmore.

The cliffs of Inishmore and Inishmaan have sizeable seabird colonies, which have to be viewed from the sea rather than the cliff top. The Guillemot (*Uria aalge*) is most numerous – about 800 pairs nest here, along with 40 pairs of Razorbill (*Alca torda*). The Black Guillemot (*Cepphus grylle*) is more widespread and breeds

FIG 348. Cover for nesting terns on the upper beach on Inishmaan, September. (Roger Goodwillie)

FIG 349. Unlike the larger tern species, the Little Tern (*Sternula albifrons*) is not given to such strong attacks on predators that venture into the colony. (John Fox)

on all the islands, as well as on the mainland near Black Head and at Aughinish. A total of 120 pairs is likely. This species is by far the most visible of the auks as it feeds in inshore waters and even in winter does not go further than 2 kilometres or so out to sea. The Shag (*Phalocrocorax aristotelis*) is also seen on every visit to the coast, even though the population numbers only 40 pairs. Like the Black Guillemot, it feeds in inshore waters. It may also nest among boulders at the cliff base, in caves or on ledges, but it has the oddest breeding habits of any. Eggs may be found in almost every month of the year, and the timing of the first egg in a colony under observation has varied in different years from the 1 March to 16

FIG 350. Black Guillemots (*Cepphus grylle*) nest in small numbers on the mainland and on the Aran Islands. (John Fox)

May (Mitchell *et al.*, 2004). In some years, three-quarters of eligible adults fail to breed – which adds to their visibility. They can live for 20–30 years as adults, but at the same time are susceptible to mass mortality when onshore winds prevent feeding for a week or more.

The Fulmar (*Fulmarus glacialis*) has a flourishing population of about 300 pairs on Inishmore, while gulls – apart from the Kittiwake (*Rissa tridactyla*) – are rather poorly represented. About 500 pairs of Kittiwake nested on Inishmore in the recent past and the colonisation of the cliffs at Poulsallagh suggests that the population is increasing or at least stable, in contrast to major declines elsewhere in Britain and Ireland. The Kittiwake is certainly seen over the sea throughout the area during the breeding season, whereas on the shore itself the Common Gull (*Larus canus*) is the most frequent gull, especially in autumn. This species has nested sporadically in Inishmore, but occurs all along the north Burren coast as it has nesting colonies throughout Co. Galway. Inishmore also supports relatively good numbers of Peregrine (*Falco peregrinus*) and Chough (*Pyrrhocorax pyrrhocorax*), although the latter may be declining somewhat from its peak in the 1990s, when a dozen birds could be seen together in late summer on the island and on Inisheer (Gray *et al.*, 2003). Odd pairs also breed in buildings on the north Burren coast.

FIG 351. Kittiwakes (*Rissa tridactyla*) on the cliffs at Poulsallagh. (Roger Goodwillie)

FIG 352. Kittiwake (*Rissa tridactyla*) defending its nest and chick. (John Fox)

FIG 353. Chough (*Pyrrhocorax pyrrhocorax*), the crow of the western seaboard. It is frequent on the Aran Islands and often seen in the coastal Burren. (John Fox)

There are no extensive mudflats on the north coast that could accommodate large flocks of migrating or wintering shorebirds. However, Ballyvaghan Bay and the lagoons to the east support Wigeon (*Anas penelope*), Teal (*A. crecca*) and Brent Geese (*Branta bernicla*) in winter, along with the common waders. The lagoons also provide a secure area for moulting Mute Swan (*Cygnus olor*) and Mallard (*A. platyrhynchos*) after breeding. Divers of all species and sea ducks feed off all this coast for the winter months until they migrate north to breed in April and May. There seems to be a build-up in late spring, perhaps so that young birds can pair up. The two smaller species – the Red-throated (*Gavia stellata*) and Black-throated (*G. arctica*) divers – remain in the relative shelter of Galway Bay (inside Black Head), but the Great Northern Diver (*G. immer*) is also seen off the west coast. Black Head is a lookout point for ocean-going birds such as shearwaters, petrels and skuas, which are otherwise seen only from the ferries heading to and from the Aran Islands.

MAMMALS

Not many mammals have adopted the inshore habitat – only seals are fully at home there, and are often seen in the kelp zones on the shores of the Aran Islands in summer, although less so around the mainland Burren. There is a substantial breeding population of Grey Seal (*Halichoeris grypus*) and Harbour Seal (*Phoca vitulina*) in Co. Galway and some of the Aran animals are visitors from these populations. However, a survey carried out at moulting time (August) recorded about 60 Harbour Seals in Ballyvaghan Bay and 40 in sheltered parts of Inishmore (Cronin *et al.*, 2004). Grey Seals were also counted at this time, with 11 animals found around the Aran Islands. Pups are occasionally born at the western end of Inishmore and in Ballyvaghan Bay, but these are not traditional or large breeding sites.

Otters (*Lutra lutra*) are frequently seen all around the coast and on the Aran Islands, partly because they occur at higher density in the sea than in fresh waters and also because they are active by day. The home range of an animal is typically of 4–5 kilometres of coastline (different female territories overlap somewhat), whereas on rivers a stretch or 20–30 kilometres is more usual (Scottish Natural Heritage, 2017). The animals are perhaps easiest to see in the inlets along the north Burren coast, but can be met with anywhere. A study of Otter spraints on Inishmore showed that they feed largely on inshore fish – species of rockling (Gadidae) and wrasse (Labridae) (Kingston *et al.*, 1999). The European Eel (*Anguilla anguilla*) was also taken by the animals, probably from the brackish lakes behind

FIG 354. Adult female Grey Seal (*Halichoeris grypus*) showing the vibrissae (whiskers) with which it detects prey movements in the water – even blind seals can survive quite well. (David Cabot)

FIG 355. Grey Seal (*Halichoeris grypus*) pup in October. The females come ashore to give birth. (David Cabot)

the shore, while crustaceans, molluscs and sea urchins also formed a small part of their diet. The Purple Sea Urchin was recorded in 9 per cent of the spraints analysed and was taken between November and January.

The American Mink (*Neovison vison*) arrived on the Burren coast within the last 10 years but has not yet been recorded on the Aran Islands (Lysaght & Marnell, 2016). Its competitive relationship with the Otter has not been studied in the Burren, although research elsewhere has found that the Otter's diet remains unchanged in the presence of American Mink but that the latter feed more on terrestrial prey if Otters are in the area. One can only feel concern for the terns and waders that nest on the Aran Islands once the mink does manage to make the trip there.

CHAPTER 12

The Future

To the unsuspecting tourist, whisked by coach along the coastal road from Ballyvaghan to Poulsallagh, much of the countryside appears too bleak, rocky and inhospitable for any sort of farming. Yet the Burren is an agricultural landscape that has been farmed continuously since the first settlers began clearing the forest cover in the neolithic period. Today, there are several hundred farms within the Burren area as defined for this book (p. 9). Most of the farmers live and work there, and are crucial for the Burren's future as an area of unique landscape and ecological interest.

What is in store for the Burren's landscape and biodiversity is very much intertwined with the future direction of farming policies and other activities such as tourism. The current Burren Programme commenced in 2016 and is supported by the Rural Development Programme, the Department of Agriculture, Food and the Marine (DAFM), and the National Parks and Wildlife Service (NPWS). The Burrenbeo Trust organises active community involvement, provision of information and education activities, which complement the farming policies and support increased awareness of the area's unique heritage. The trust is also spearheading a Burren charter that aims to set out a community vision for the area.

Tourism in the Burren has been accelerated and stimulated by the promotion of the Wild Atlantic Way, which includes the coastal road around Black Head. Its impact, along with other tourism-related issues, is currently being addressed by the GeoparkLIFE project. Any future expansion of the Burren National Park, coupled with more vigilant, but judicious, land management, has the potential to enhance biodiversity protection. In April 2010, the Irish government submitted a proposal to list the Burren as a UNESCO World Heritage Site, together with five

FIG 356. What better place than Black Head to ponder the future of the Burren?
(Fiona Guinness)

other Irish projects (World Heritage Ireland, 2010). It set the Burren's size as some 72,000 hectares – almost double the area of our definition. The implications of this proposal are as yet unclear.

In this final chapter, we look first at the history and evolution of farming in the Burren, before reflecting on the various agricultural and other activities that are working today to protect and manage the area's heritage for the future.

PAST AGRICULTURAL PRACTICES

The first farmers

As we saw in Chapter 4, the early neolithic farmers in the Burren kept cattle, sheep, goats and pigs in the period after 4000 BCE. Some crops were grown, while hunting provided supplementary food. Small areas of the woodland that covered most of the upland landscape would have been cleared with stone axes. This exposed some soil suitable for crops, and created grassland around the edges to feed grazing animals. It seems from pollen data that the woodland clearances were on a minor scale, transient and intermittent. Once the benefits of clearance had been taken, farmers moved on to new areas in a shifting agriculture. Part of their culture was to build stone or megalithic monuments as burial chambers or as focal points for religious ceremonies. There are three major types of these megaliths in the Burren: court tombs (of which four are known in the Burren), portal tombs (two known) and wedge tombs (75 known). Today, these stone memorials and hundreds of later cairns stand starkly on the exposed limestone uplands, testimony to a once thriving population of farmers. It has often been claimed that the density of these monuments in the Burren is the highest found in any part of Ireland, which led Tim Robinson, the writer and cartographer, to state in 1999 that the Burren is 'a vast memorial to bygone cultures'.

FIG 357. A wedge tomb, with suggestive evidence that it was covered with earth soon after it was constructed about 3,000 years ago. (Fiona Guinness)

Excavations of the famous Poulnabrone portal tomb provided some information on the nature of farming around 3500 BCE. A miniature polished stone axe, arrowheads and scrapers were discovered in the tomb among the remains of 22 people. Dental information from the skeletons showed that these people had had a coarse diet, suggestive of stoneground cereals.

As the neolithic period progressed, farmers became more settled and organised. The first stone walls were built and walled settlements were constructed, the remains of which can still be seen in many locations today. All archaeological evidence indicates that the Burren uplands were densely populated and intensively farmed some 4,500 years ago.

Upland farming activity continued into the Bronze Age (2,000–500 BCE), and the clearance of woodlands progressed more earnestly. More effective bronze axes had replaced the earlier stone ones, but burning of woodland was also introduced, as evidenced by the discovery of Hazel (*Corylus avellana*) charcoal radiocarbon dated to this period. More grass and weed pollen appeared, indicating an expansion of grasslands. These farmers also added their own monuments on the uplands, building cairns both large and small. The cairns have burial cists inside, but they are not as elegant as those created by the earlier neolithic people. Nevertheless, they show continued reverence for the dead and departed. Many of them remain unexcavated today, holding their hidden secrets within.

FIG 358. An upland Bronze Age cairn surrounded by a dense sward of Mountain Avens (*Dryas octopetala*). May. (Fiona Guinness)

FIG 359. Abandoned farm enclosures and stone walls. (David Cabot)

The Bronze Age also left a legacy of more than 300 ancient cooking sites or *fulacht fiadh*, small horseshoe-shaped mounds of burnt stones, usually located in wet or marshy ground (Gosling, 1991). The stones were heated in a nearby fire and then placed in a water-filed trough, which brought the water to near boiling point. Joints of sheep, cattle or deer, assumed to have been wrapped in leafy vegetation, were placed in the water and left for slow cooking. This process has been successfully replicated today. *Fulacht fiadh* may also have been used for other activities such as bathing.

Woodland clearance continued and the soils, now unprotected by trees, were open to rain-wash erosion. A weather change to wetter conditions may have accelerated the erosion of the soil cover on the limestone pavement (p. 101). The soils were washed away, first into grikes and then downwards to the underground streams and waters, revealing the barren, but nevertheless attractive, upland landscapes of today. Some of the soil can now be seen in the Doolin Cave, below an impressive stalactite.

Farming activity was apparently subdued during the Iron Age (approximately 500 BCE–400 CE), relaxing some of the previous pressure on the land. The reasons for the lull are not fully understood, but are probably related to the deteriorating weather conditions and perhaps the loss of soil in some localities. There was some regeneration of Hazel woodlands, indicated by a resurgence of Hazel pollen in Iron Age peat horizons (Feeser & O'Connell, 2009).

Increased farming and the importance of cattle

Farming activity was renewed and better organised in the early Christian period, from about 500 CE onwards. It was driven by new skills and a more regular demand for food products from settled communities and ecclesiastical establishments. One of these was Corcomroe Abbey, established in 1200 CE by Cistercian monks and strategically located between upland grazing sites and lowland fields used for tillage. The monks introduced better crop husbandry and milk processing.

FIG 360. Corcomroe Abbey, strategically placed between the uplands for winterage and the richer lowland soil, which is ideal for cows and milk production as well as for crops. (Fiona Guinness)

FIG 361. Cows and calves grazing on a mound of glacial till south of Bell Harbour in May. (Fiona Guinness)

By this stage, cattle and sheep were the most important farm animals. More pasturage was needed for the expanding numbers of animals, especially cows, while milk, cheese and butter production benefitted from the enhanced expertise provided by the monks. The farmers of the time constructed at least 500 ring forts, or *cahers*, which are scattered throughout the Burren but mostly confined to the upland areas. These functioned as family farmsteads, and provided protection for cattle and other animals from predators such as Grey Wolves (*Canis lupus*), as well as from marauding people.

The most dramatic of the stone forts in our area is Dún Aonghasa (Dun Aengus) on Inishmore, a semicircle perched on the edge of a cliff above the Atlantic. Inishmore and Inishmaan have two stone forts each, but on the mainland they are rivalled by Cathair Chomáin, set on a 30 metre-high cliff edge in the high Burren. It looks out over a valley that is now filled by Hazel scrub, although it would have been clear when the structure was built, in about 800 CE. It was constructed on top of earlier settlements dating to the period 400–500 CE, and consists of a triple ring of stone walls, although the outermost two are now incomplete. The Third Harvard Archaeological Expedition to Ireland excavated the Cathair Chomáin site in 1934, finding 12 stone buildings inside the innermost and most substantial wall, which is some 8.5 metres thick and 4.3 metres tall at its highest point. The fort was thought to have been the residence of 40–50 inhabitants at most, possibly representing an extended family. It was probably not occupied later than 1000 CE (O'Neill Hencken, 1938).

Cathair Chomáin was clearly a significant settlement for the area. Numerous spindle whorls, used in spinning wool, were recovered, together with pin beaters, used for beating down the weft in weaving. Archaeologist Paul Gosling suggested that this may have been the site of a wool industry where shearing, wool production and weaving were undertaken on a relatively large scale (Gosling, 1991). Such an activity would indicate the economic importance of sheep at the time.

Animal bones recovered during the Harvard excavations came from cattle, sheep, pig, deer, goat, horse and Red Deer (*Cervus elaphus*). The most significant remains were of cattle, comprising 97 per cent by weight of all animal bones (sheep and goat bones together made up 1 per cent, pig 1 per cent, and the remaining 1 per cent comprised horse and Red Deer bones). The high proportion of cattle bones emphasises the importance of these animals for the farmers of the period, and they would have grazed the surrounding lands. Tillage crops were probably confined to the deeper drift soils, mainly in the lowlands and particularly in the extensive valleys extending southwards from Bell Harbour and Ballyvaghan.

FIG 362. The inner and outer walls of Cathair Chomáin, built 1,200 years ago. (David Cabot)

BELOW: FIG 363. Panoramic view of the inner wall of Cathair Chomáin, looking north. The open side is protected by a steep cliff. (Fiona Guinness)

FIG 364. Ewe on an upland field in early May. (Fiona Guinness)

FIG 365. Gleninagh Castle, a sixteenth-century tower house built for the O'Loughlin family, who were still there in the 1840s. The Irish tower house was used for residence and defence against raiding and marauding expeditions from other clans. (David Cabot)

The Burren's reputation as an area of rich agricultural produce during the medieval period (fifth to fifteenth centuries CE) was well known, especially for cattle. Knowledge of these resources brought many raiding and marauding expeditions into the area, when cattle and other goods were plundered. The impression of the Burren to one outside observer in 1317 was of a 'hilly grey expanse of jagged points and slippery steeps, nevertheless overflowing with milk and yielding luscious grass' (O'Grady, 1929).

The expansion of estates throughout Ireland in the seventeenth century, when Cromwellian officers and others were given land by the British authorities, led to the confiscation and redistribution of land and the replacement of the abbeys and monasteries as landowners. Many of the new landlords became absentees and left the management of the estates to agents. Walls were built to mark boundaries and divide holdings. Sheep replaced cattle as the major livestock, and vast flocks were reported in 1700–1880. Some 250,000 sheep were exported from the Burren in 1846–48, just after the Great Famine, to prevent them being stolen by starving tenants (Kirby, 1981).

Thus, we see that from early neolithic times there has been an unbroken tradition of farming in the Burren, characterised initially by cattle and some tillage, and later by sheep. Latterly, cattle have again taken the lead in the form of dry stock in the twentieth century and suckler cattle in the twenty-first century. These are now supplemented by alternative farm enterprises.

Modern changes in farming practices

Economic forces, market demands and farming technology have dramatically changed the nature of farming in the recent past, driven especially by the European Union (EU) Common Agricultural Policy (CAP) in the 1970s and 1980s. An increasing emphasis was put on productivity and efficiency, and the use of intensive methods became necessary for farms to remain viable. The Burren's unique ecosystems and fragile ecology were threatened by a variety of practices. First, there was a decline in winter grazing on the uplands, leading to the spread of Hazel scrub and the loss of floral diversity. Second, supplementary feeding increased in the uplands, causing nutrient enrichment and invasion by lowland and commoner plant species. Third, limestone pavement and stony areas were converted to pasture by machinery, and the spreading of fertilisers led again to the replacement of some of the unique plant communities by grasses. And finally, both agricultural and new domestic buildings increased the discharge of wastes to the groundwater, threatening water quality for all.

While EU agricultural policies initially drove farm productivity to unrealistic heights, it has since evolved towards environmental sustainability and accountability. Today, the financial support for farmers emphasises environmental protection, delivery of environmental services, and the protection and enhancement of biodiversity in particular. This new direction, actively adopted by Burren farmers and supported by many local initiatives, provides the way forward to securing and managing of the Burren's unique landscape and ecosystems.

Without financial support from EU programmes, Burren farmers would find it impossible to remain economically independent. In the 1990s, most farmers lacked the opportunity to generate off-farm income and their only alternative was to leave the land altogether and move elsewhere. Nearly 20 years ago, a survey of a relatively small sample of 65 farmers was carried out by Brendan Dunford (at the time of writing, programme manager of the Burren Programme; see below). The farmers were living in the Ballyvaghan Rural District (covering about 75 per cent of the Burren's area), and the survey provided early information on farm family demography and economy, farm composition, general upland farm management issues, and attitudes to the Rural Environment Protection Scheme (REPS), the EU agricultural support programme that was in force at the time (Dunford, 2002). Nearly 43 per cent of the respondents considered that farming alone could not provide an acceptable living for their families. An additional 45 per cent felt that their farm was barely able to provide enough for them.

Clearly, the economic circumstances of the farms had to change, otherwise there would be a flight from the land. The consequence of abandoning farmland was 'rewilding' by scrub and woodland, obliterating much of the landscape. The unique habitats and ecosystems of the Burren, as we know them today, were threatened and, in places, would have disappeared. Fortunately, this progression was recognised and measures were taken to reverse it, thanks initially to the imaginative EU-funded LIFE programme. This has led to a range of local initiatives, designed to protect and manage the landscape. In this process, the farmers, as principal custodians of the landscape and its biodiversity, play the critical role. These initiatives are described below, but first some background information on one of the old traditions of Burren farming, winterage.

Winterage

A recurrent practice for generations of Burren farmers has been the tradition of winterage, which is the opposite of the mainland European practice of transhumance. During transhumance, cattle are moved up into the hills and mountains in spring for summer grazing and then, in autumn, are brought down to the lowland pastures for winter. In winterage in the Burren, in contrast, cattle are walked up in October on ancient drover tracks and pathways into the hills, where they graze extensively on the vegetation. In spring, they are then brought back to the lowlands, to graze the fields that have been rested during the winter. When the cattle are up in the hills, many cows give birth to their calves. Winter grazing removes much of young scrub and taller plants that would otherwise obstruct the following year's new growth. This allows the survival of the smaller species, including many of the Burren's iconic and characteristic flowers.

This movement of cattle is also known as booleying and dates to the early medieval period, or even earlier. The tradition may seem strange to outsiders, as the uplands appear bereft of vegetation. On closer inspection, however, it is surprising how much forage is present when the vegetation in hollows in the rock, grikes and on the terraces of the hills is combined. By coincidence, winterage became an early and sophisticated management tool of the uplands, keeping woody species under control and ensuring survival and even spread of the flora.

One of the factors benefitting winter plant growth in the Burren hills is the 'storage heater' effect of limestone. The rock retains some of its summer heat and, combined with its fast drainage, provides a relatively dry and warm environment

FIG 366. Old trackway leading up to the winterage on Gortaclare Mountain. Many of these routes have become green roads for hiking. (Fiona Guinness)

FIG 367. Nipped Hazel (*Corylus avellana*) scrub on uplands above the Glen of Clab, March. (Roger Goodwillie)

FIG 368. A patch of deeper soil in the hills, used for grazing. Early May. (Fiona Guinness)

for resting cattle, something akin to the effects of our underfloor heating. During bad winter weather, cattle will shelter in hollows or behind specially constructed stone walls, some designed in the form of a 'T' or a large 'X' to give protection from wind blowing from any direction.

Many more cattle than sheep are seen on the uplands today, but feral Goats (*Capra hircus*) also have a considerable impact on the vegetation. These animals are thought to have descended from goats brought to the area during the neolithic

FIG 369. Special sheep passes such as this example in the Glen of Clab are built through stone walls in some upland areas, their height restricting the movement of larger animals. (David Cabot)

FIG 370. One of numerous shelter walls on the upland winterage areas, built to provide protection for animals or shepherds in inclement weather. (Fiona Guinness)

period, around 5,000 years ago. The population has a complementary role in controlling the spread of woody species, as the animals eat more scrub than grass and take bark from Ash (*Fraxinus excelsior*) and willow trees. The spectre of completely prostrate Yew (*Taxus baccata*) on limestone pavement, compared with its almost lusty growth on inaccessible cliffs, is testimony to the proficiency of grazing goats. A small number of domesticated goat herds are also kept for the production of cheese and other products.

FIG 371. Old Irish type of feral goat, one of many keen consumers of woody plants in the Burren. (Fiona Guinness)

FIG 372. Luxuriant growth of Yew (*Taxus baccata*) on a limestone cliff, out of the reach of goats. (Fiona Guinness)

FIG 373. Herd of modern domesticated goat breeds, kept for milking and cheese making. (Fiona Guinness)

FIG 374. Goat's cheeses maturing in a curing house, made from the flock of goats on the road in the previous photograph. (Fiona Guinness)

The number of goats currently in the Burren is uncertain, but is of the order of 1,500 animals. Only two national censuses have been attempted, neither with satisfactory results (Werner, 2010). However, in May 2016 we counted approximately 600–700 animals in one herd alone, grazing on the south-eastern side of Gortaclare Mountain. While on the subject of goats, concern has been expressed about the survival of the small number of animals still exhibiting 'ancient Irish' goat characteristics. These and other issues are considered in a draft management plan for the feral goats, published by the BurrenLIFE Project (Werner, 2010).

The availability of water for grazing animals in upland hills is an issue during the winter. Historically, special water collectors were constructed to provide a supply and these are still important on the

Aran Islands, where animals are brought to the 'back of the island' in winter. On the mainland today, some water is saved from springs on the hills and the rest is pumped up into drinking troughs and tanks.

FIG 375. Goat herd on the southern side of Gortaclare Mountain. (Fiona Guinness)

FIG 376. Abandoned farmstead near Glencolumbkille. (Fiona Guinness)

SPECIES INTRODUCTIONS

In previous chapters we met many species introduced to the Burren, from the
Wireweed (*Sargassum muticum*) and other seaweeds seen on the coasts, to White
Stonecrop (*Sedum album*) and Wall Lettuce (*Mycelis muralis*) on the rocks, not
to mention the Rabbit (*Oryctolagus cuniculus*), Slow-worm (*Anguis fragilis*) and
American Mink (*Neovison vison*). In addition, there are reasonable suspicions that
the Purple Milk-vetch (*Astragalus danicus*), Common Whitebeam (*Sorbus aria*) and
Land Winkle (*Pomatias elegans*) are also alien introductions.

Some of these organisms were brought in by people who, in some misguided
way, thought they were doing nature a favour, but others have arrived accidentally,
carried on ships or in their cargo, on vehicle tyres or footwear, or latterly, on
nursery plants and flowerpots. All introductions interfere with the natural
ecology of an area in one way or another, by displacing native species through
shading or competition, or by eating them – but they are a fact of life in our
travel-addicted society. Given time, all the temperate regions of the world will

FIG 377. The Pinnacle Well or Tobercornan, near Ballyvaghan. Built in 1860, it covers a
small spring well that is considered holy. Fairy Foxglove (*Erinus alpinus*), planted in the 1940s,
now grows on the limestone walls and rubble-stone roof. (Fiona Guinness)

have the same flora and fauna, and there will be a wholesale loss of biodiversity as less vigorous species decline and disappear, being replaced by others that have evolved elsewhere to grow more strongly. Often the newcomers arrive without their natural pests, so there may be no biological controls available. If we want to prevent or at least curtail this slow-moving calamity, we have to choose which species are really damaging and which we can live with. And then, also, which we can do something about controlling.

Aquatic introductions are the most difficult to control; Wireweed spores and larvae of the acorn barnacle *Elminius modestus* travel on the ocean currents, while the free-swimming larvae of the Zebra Mussel (*Dreissena polymorpha*) are spreading this species throughout aquatic habitats in Ireland, undeterred by control measures – it is currently recorded in the Fergus system (Lough Cullaun and Lough Skeardeen, adjacent to Lough Bunny). Terrestrial plants are easier to see and, depending on their dispersal methods, easier to monitor and control. A walk on Abbey Hill will reveal Entire-leaved Cotoneaster (*Cotoneaster integrifolius*) in abundance, Alexanders (*Smyrnium olusatrum*) dominates many views on the Aran Islands, and pink forms of Columbine (*Aquilegia vulgaris*) and the Fairy Foxglove (*Erinus alpinus*) always excite attention.

FIG 378. Fairy Foxglove (*Erinus alpinus*), also known as Alpine blossom. It is a neat and attractive alpine species, native of central and southern Europe. (Fiona Guinness)

FIG 379. Red Valerian (*Centranthus ruber*) on shattered limestone pavement at Gortnacloch. (Fiona Guinness)

A plant that is dispersed by wind and suited to growing in many types of habitat is the most difficult to eradicate, and none more so than Red Valerian (*Centranthus ruber*). Not only will this plant grow well on any patch of exposed limestone in full sun, in any mortared wall and on any roadside, but it also has minute windborne seeds that remain viable for at least seven years (Lavorel *et al.*, 1991). There is little doubt that, left to itself, it will follow the example of the Wall Lettuce in colonising the whole area. However, the end result will be much more damaging in that it will outcompete much of the native flora. One can imagine the valerian supplanting species such as Irish Saxifrage (*Saxifraga rosacea*) and Wild Madder (*Rubia peregrina*), and even Hoary Rock-rose (*Helianthemum oelandicum*) in time to come. It is a plant with few natural enemies and one whose spread will be further encouraged by a warming climate since it is native to the Mediterranean. A visit to the stony desert of Gortnacloch (Fig. 379) at the foot of Slievecarran is a sobering reminder of what an introduced plant can do if left to itself.

PROTECTING THE BURREN

The Burrenbeo Trust

The Burrenbeo Trust was established in 2001 as an independent registered charity with the principal objective of creating a greater appreciation of the Burren as a living landscape of international importance. Other goals were to engage in research and dissemination of information, and to promote the holistic and self-sustaining conservation and development of the Burren

FIG 380. Kinvarra in 2000, looking north-west to Galway Bay. The 'estuary' is partly the outflow of a large underground river from the Gort region, east of the Burren. (David Cabot)

FIG 381. Burren National Park Information Point, Corrofin. (Fiona Guinness)

(Walsh, 2009). The trust, based in Kinvarra, is extremely active in promoting 'the connection of all of us to our places and our role in caring for them'. It achieves this by organising 40 different programmes each year, and it also carries out periodic farmer surveys to determine current opinions and implementation.

FIG 382. National Parks and Wildlife Service Information Centre at Corrofin, an important information point offering helpful advice for naturalists visiting the Burren. (Fiona Guinness)

FIG 383. Burrenbeo Trust headquarters in Kinvarra, another important source of information as well as a key centre for the organisation of local initiatives in support of the Burren Programme. (Fiona Guinness)

One of the recent surveys, undertaken in association with the Irish Farmers' Association, sought information on the needs of Burren farm families and was published as *The Voice on the Ground: A Survey of the Needs of Burren Farm Families* (Walsh, 2009). One of the key conclusions concerned significant changes in farming practices:

> *Among the most significant changes that have taken place in farming in the Burren has been the change in farming enterprises with significant numbers of farmers having gone into sucklers, and substantial numbers having got out of dairy and sheep. The building of slatted sheds and the housing of animals over winter has also proved to be a significant change for the farmers, and indeed for the wider landscape as upland areas that used to be grazed in the winter are now abandoned and slowly being taken over by scrub.*

Another key conclusion concerned the role of farming and farmers in the Burren:

> *The farm families surveyed believed that farming is central to the conservation of the unique landscape of the Burren. They saw farming and farmers as having a key role in controlling scrub and conserving the plant life of the Burren. They identified themselves as being particularly knowledgeable about the wildlife and cultural history of the area but identified that they would like to learn more about not only the wildlife and cultural heritage of the area but also the archaeology, geology and plant life. Family and friends were identified as the most important source of knowledge of the Burren suggesting that there is a repository of local knowledge within the farming community in Burren that may have to potential to be shared with others.*

The BurrenLIFE Project (2005–10)

An innovative environmental management scheme for the Burren was developed from a Ph.D. research project by Brendan Dunford. His thesis, titled 'The impact of agricultural practices on the natural heritage of the Burren', was published in 2002 as *Farming and the Burren* (Dunford, 2002). This report highlighted the important role that farming plays in supporting the rich biodiversity and cultural heritage of the Burren, and also the worrying breakdown in traditional farming systems and the habitats dependent on them. Funding was then sought from the EU LIFE programme to develop 'A blueprint for the sustainable agricultural management of the Burren' (Burren Programme, 2017). The result was the BurrenLIFE project (2005–10), the first major farming-for-conservation project in Ireland and one of the very few EU projects that have placed farmers at the centre of the conservation agenda. Over the five years it ran, it successfully piloted a range of sustainable agricultural practices on 20 farms covering some 2,500 hectares. A detailed research and monitoring programme then confirmed that these actions had had a positive impact on the priority habitats on these farms.

The project identified a number of key factors and actions that benefitted conservation and farming, including the following:

- Scrub encroachment reduces biodiversity. To address this, the programme cleared scrub from 100 hectares of priority habitats.
- To reduce the housing of livestock over winter, the project extended the winter grazing on traditional winterages by 25 per cent.
- The scheme introduced a special supplementary feed, in a bid to encourage farmers to cut down on the use of silage.
- The project improved water facilities on the winterages by installing pumps and tanks on 18 farms.
- 15,000 metres of internal stone walls were restored, using local labour, to enable better management of grazing.

The culmination of this research has been the development of a blueprint for sustainable agriculture in the Burren, containing a suite of practical, costed actions aimed at safeguarding the natural heritage. The project paved the way for a new programme to tackle the most pressing issues affecting the region. It was chosen by judges as one of the top five nature conservation projects in Europe of the past 25 years assisted by the EU-funded LIFE programme, and ended up as a joint winner. In fact, the BurrenLIFE Project was the only Irish scheme to make the shortlist, which also included projects from Bulgaria, Spain, Belgium and Slovenia. Judges cited the BurrenLIFE project for its 'phenomenal impact on a unique landscape... this Irish project pioneered a novel approach to farming and conservation' (Burren Programme, 2017). Its success led to the parallel development of the AranLIFE Project in 2013, which is ongoing.

Burren Farming for Conservation Programme (2010–15)

Arising from the BurrenLIFE project, the Burren Farming for Conservation Programme was launched in 2010 by its funders, the DAFM and the NPWS. It operated from 2010 to 2015, and worked with 160 farmers over an area of 15,000 hectares of prime Burren habitat. It built directly on the lessons learnt during BurrenLIFE and supported and incentivised farmers to maintain and enhance the habitats, effectively tackling many of the issues identified in the original research project a decade previously.

The programme had a farmer-led approach to farming and conservation, which saw farmers paid for work undertaken but, more importantly, for the delivery of environmental results. The programme divided its annual farmer payments almost equally between payments for actions (e.g. scrub clearance) and payments for output (e.g. well-managed grassland with abundant wildflowers). Each designated farm field within the scheme was scored annually using a 'health check' for the grasslands and other habitats, which was specifically developed for the Burren region. The higher the score, the greater the payment made to farmers. Simply put, the farmers who delivered the highest environmental benefits were rewarded the most. In this way, conservation becomes as much a product for the farmer as the livestock. Total payments under the six-year programme were €6 million for farm-level actions and environmental performance rewards.

FIG 384. Clearance and removal of Hazel (*Corylus avellana*) scrub and rocks for new pasture. (David Cabot)

FIG 385. Typical Burren stone wall with a vertical emphasis in the stonework. Many walls that had fallen or been knocked down have been restored under the BurrenLIFE Project. (Fiona Guinness)

The Burren Programme (2016–)

The current Burren Programme commenced in 2016 and at the time of writing is working with more than 300 farmers, with a target of some 400-plus farmers by 2018. It developed as a natural extension of the earlier Burren Farming for Conservation Programme and will continue to work closely with all the interested parties – farmers, farm advisers, the EU, the DAFM and the NPWS. The aim is to implement the measures that help manage and protect the Burren.

These programmes and projects show that the environmental and management issues required to protect, maintain and enhance the Burren's ecological features are better understood now than ever before. Success with achieving the objectives depends upon continued moral and financial support from the government for the programme, creative administration of the programme and, most important of all, the understanding and cooperation of Burren farmers.

Burren National Park

The Burren Programme has many conservation objectives and is also being carried out on an area of state-owned land in the south-eastern corner of the area – the Burren National Park. This land was bought by the government and, together with a small parcel of land donated by An Taisce (the National Trust for Ireland), constitutes an area of 2,013 hectares. The park was established in 1991 and is managed by the NPWS. At present, the land is leased to farmers participating in the Burren Programme. Although relatively small, the park embraces some of the most visually attractive areas within the Burren, with Mullagh More – one of the more ecologically interesting areas within the region – as its centrepiece.

FIG 386. Construction of new roadway above the Caher River valley for access to the uplands, May 2017. (Fiona Guinness)

FIG 387. The Burren National Park, although small in size compared with other Irish national parks, is attracting increasing numbers of visitors each year. (Fiona Guinness)

The park is not a single landholding but consists of five areas that are separated by sparsely settled countryside and linked by a rural road network. The Mullagh More area is the largest, accounting for nearly two-thirds of the total area. It includes Mullagh More itself and Slieve Roe, the most southerly of the line of limestone hills that mark the eastern edge of the high Burren. The Keelhilla area is made up of the Slieve Carran Nature Reserve and adjacent lands. This area contains the site of St Colman MacDuagh's hermitage. The Cathair Chomáin area embraces one of the most impressive medieval monuments in the Burren (p. 387). The Burren National Park landholdings are:

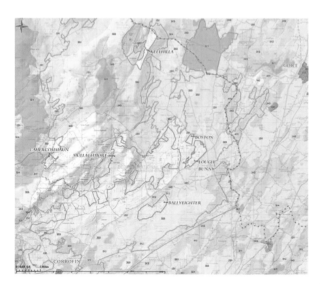

FIG 388. Landholdings comprising the Burren National Park – the areas within the red lines. From Brendan McGrath & Associates (2017).

- Mullagh More, 1,321 hectares (66 per cent of the total park area).
- Keelhilla, 371 hectares (18 per cent).
- Cathair Chomáin, 45 hectares (2 per cent).
- Lough Bunny and Boston, 130 hectares (6 per cent).
- Ballyeighter, 146 hectares (7 per cent).

The park contains examples of all the major habitats: limestone pavement, calcareous grassland, Hazel scrub, Ash–Hazel woodland, turloughs, lakes, petrifying springs, cliffs and fens. Nearly all the iconic Burren species occur here: plants, butterflies and moths, and other invertebrates. The park is managed and staffed by park rangers. There are seven well-marked walking trails, together with information boards.

The proposal to establish an interpretative centre and car park on the edge of the national park in 1991 unleashed a sometimes vehement controversy over the suitability of such a development. The proposed centre was close to ecologically sensitive habitats in the form of turloughs and a limestone lake, and also to Mullagh More, which then assumed an almost sacred status. At the time, the author and philosopher John O'Donohue (1956–2008) said that the Burren Action Group's opposition to the centre 'was motivated by an old-fashioned,

FIG 389. Part of the Burren National Park, with Lough Gealáin on the left and Mullagh More rising on the right. (David Cabot)

FIG 390. Burren National Park trail and information board. There are seven well-designed walking trails in the park. (Fiona Guinness)

almost innocent love of a mountain' (Anon., 2001). The developer was the Office of Public Works (OPW), a government agency that undertakes public works such as the building of schools and police stations, and the maintenance of public and heritage buildings. As such, the OPW was not required to obtain planning permission under the Local Government (Planning and Development) Acts 1963–69. Opponents to the project eventually won their case and the project was cancelled, despite a modified proposal being submitted. The controversy had a long-lasting outcome, in that the OPW is now required to seek planning permission for all of its development works.

Increasing numbers of people are visiting the park yearly, now estimated as upwards of 10,000. This leads to access issues and parking problems, with cars often scattered on the sides of the narrow roads. There is still no car park, nor road signs to the park. The National Park Site Assessment Report for the area (Hoctor, 2014) concluded that:

> The lack of official and adequate parking facilities within the Gortlecka–Crag Road area of the Burren National Park is causing severe congestion during the busy summer period. This is causing difficulties for residents and is a serious safety issue at times. At present parking occurs in several areas along the grass verge. This is impacting on the roadside vegetation and the limestone pavement especially in areas along the Crag Road.

In 2012, the NPWS lodged plans with Clare County Council for a 27-space car park in the same area where the original one was proposed as part of the interpretative centre. It argued that this was necessary to reduce the frequency of parking along the road verges, to cater for existing visitor usage, and to improve road safety, traffic congestion and access problems, particularly for local communities. The Burren Action Group, which objected to the original project in 1991, condemned the proposal as 'piecemeal' and 'premature', and the NPWS withdrew the application in March 2013.

Most national parks throughout Europe provide visitor access and other facilities at park entrances, as was originally proposed by the OPW for the Burren. Even here, with low numbers of visitors, access and other facilities need to be addressed soon, otherwise the existing impacts may become intolerable. Uncontrolled access may also lead to degradation of some aspects of the park.

Supporting the national park is the Burren Centre in Kilfenora, claimed to be the first interpretative centre in Ireland and originally partially funded by the Carnegie UK Trust. The centre provides information on the park and a regular screening of the film *A Walk Through Time*, covering the Burren's history and its flora and fauna. In addition, the NPWS runs the Burren National Park Information Point on the ground floor of the Clare Heritage Centre in Corrofin, which is open to the public from April until the end of September. It includes an excellent educational display, and a well-briefed information officer provides additional details on the park. During the summer, a free shuttle bus service runs from the centre to the park.

Special Areas of Conservation (SACs)

While the national park is totally protected and inviolable by virtue of its state ownership, it forms only a very small part (3.6 per cent) of our defined mainland area (41,900 hectares) of the Burren. Much larger areas have been designated by the government as Special Areas of Conservation (SACs) under the EU Habitats Directive. Our area includes four Burren SACs (National Parks and Wildlife Service, n.d.):

- East Burren Complex (18,808 hectares).
- Black Head–Poulsallagh Complex (land area 5,464 hectares; marine 2,341 hectares).
- Ballyvaghan Turlough (21.73 hectares). This turlough, discussed on p. 30, is much modified by agricultural development.
- Moneen Mountain (6,107.45 hectares).

In addition to the above, some 300 hectares of the Flaggy Shore on the northern edge of the Burren are included in the Galway Bay Complex SAC (National Parks and Wildlife Service, n.d.). The total area of terrestrial land included in these four SACs plus the Flaggy Shore section is 30,701.18 hectares, or 73.3 per cent of the

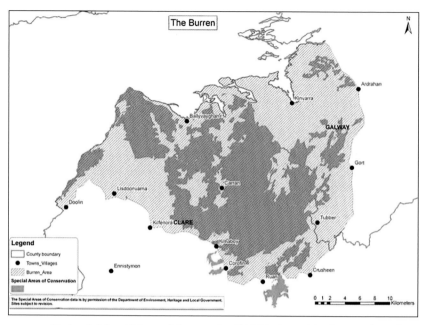

FIG 391. Designated Special Areas of Conservation (SACs) in the mainland Burren (note that marine SACs are not mapped). (National Parks and Wildlife Service)

Burren's mainland defined area. In the Aran Islands, a further 3,201 hectares of land (70 per cent) and 12,809 hectares of sea have been designated as SACs.

SACs are habitat and wildlife conservation areas considered important on a European as well as an Irish level. Their selection is justified by certain qualifying features – habitats and species that are listed in the directive and that must be protected within these areas. Conservation objectives and site plans have to be set for these features and are essential when assessing the impacts of other plans and projects that are proposed for the included land. No site management plans have been published for any of the SACs according to the relevant websites as of October 2017. While SACs provide some control on adverse impacts to habitats and species, the degree of protection is not as tight or absolute as it is within the Burren National Park.

Slieve Carran Nature Reserve

The Slieve Carran Nature Reserve was established in 1986 and covers an area of 145.5 hectares. It lies within the East Burren Complex SAC (see above) and is managed by the NPWS. It is a most interesting area, with much of the region's

characteristic flora spread on extensive limestone pavements. It includes an Ash woodland at the base of dramatic cliffs, and the remains of a significant early medieval heritage site, St Colman's oratory. A pair of Golden Eagles (*Aquila chrysaetos*) once nested on the cliffs but had abandoned the site by the end of the nineteenth century. Today, Ravens (*Corvus corax*) and Peregrine Falcons (*Falco peregrinus*) patrol the cliffs, while Kestrels (*F. tinnunculus*) are also in the area. Feral Goats sometimes venture onto the cliff ledges. This is one of the sites in the Burren where introduced Slow-worms (see pp. 53–4) can be observed. The reptiles have also been seen more widely across the east Burren, from Tubber to Carran (Brendon

FIG 392. Peregrine Falcon (*Falco peregrinus*). (David Cabot)

FIG 393. Slievecarran, with Eagle's Rock on the cliffs in the distance. The area has rich plant communities on thin soil overlying limestone pavement. (Fiona Guinness)

FIG 394. St Colman Macduagh's Cave, at the base of Slievecarran, surrounded by a prolific growth of Soft Shield-fern (*Polystichum setiferum*). (Fiona Guinness)

Dunford, pers. comm., 2017). Viviparous Lizards (*Zootoca vivipara*) often sun themselves on rocks here.

There are numerous butterflies in the area, attracted by the diverse flora. At Eagle's Rock below the cliffs, St Colman MacDuagh (c.560–632 CE), son of the Irish chieftain Duac, built a small wooden chapel towards the end of the sixth century. He chose this remote area to live as a hermit, devoting seven years of his life to prayer and spiritual contemplation. The ruins of a later stone church are visible today, surrounded by prolific growth of Ramsons (*Allium ursinum*) and ferns. There is also a 'holy' well, and a cave where McDuagh may have slept.

Special Protection Areas (SPAs)

There are three SPA designations in the Burren concerned with birdlife:

- Inishmore, Aran Islands – for cliff-nesting seabirds, which scarcely occur on the mainland Burren.
- Corrofin Wetlands, incorporating Inchiquin Lough, Lough Atedaun, Lough Cullaun and their associated calcareous wetlands – important for migrant and wintering wildfowl and waders.

• Inner Galway Bay, from Gleninagh east to Kinvarra – but only up to high-
water mark – for shorebirds.

TOURISM AND CONSERVATION

Global Geopark status was awarded to the combined Burren and Cliffs of Moher
area in 2011 for its outstanding geological and cultural heritage. It thus joins
a network of more than 100 Geoparks worldwide. The Geopark was further
designated as a UNESCO site at the UNESCO General Conference in Paris in
November 2015 (UNESCO, 2017).

The GeoparkLIFE project, an innovative tourism-for-conservation project,
was set up in 2012 by the Geopark 'to seek a collaborative balance between the
tourism interests and the conservation needs of the Geopark region' (Burren and
Cliffs of Moher Geopark, 2018). EU LIFE funding of €1.1 million was awarded
to finance the project's work programme, with matching grants provided by
the project's stakeholders: Clare County Council, Geological Survey of Ireland,
Fáilte Ireland, NPWS, National Monuments Service, OPW, Heritage Council,
National University of Ireland Galway and University College Dublin. The project
began in October 2012 and was completed in December 2017. It was designed to
achieve several results by the end of its life: to strengthen tourism enterprises in
conservation, to improve archaeological and natural sites, and to enhance skills
for conservation management.

FIG 395. Tour bus stopping at Poulsallagh, one of the favourite Burren visitor sites. Up to five buses may been seen here at times, with most visitors wandering over the limestone pavement towards the sea. (David Cabot)

FIG 396. Party of tourists setting up their luncheon with all the comforts on a small grass patch between limestone outcrops, Poulsallagh. (Fiona Guinness)

The GeoparkLIFE project has published an impressive range of 14 research reports, providing all the necessary details for the development of general policy and site-specific actions. It is unlikely that any other area in Ireland has been studied so intensively as the Burren with regard to the impact of tourism and the need to protect and manage the natural and built resources. The critical issue facing Geopark Burren now is to secure the implementation of the various recommendations.

NEW ENTERPRISES

Today, there are many flourishing economic and cultural enterprises in the Burren that derive their inspiration from the local natural resources. Activities

FIG 397. The Burren Outdoor Education Centre was established by the County Clare Vocational Education Committee in 1982. It was one of the first purpose-built outdoor education centres in the country and since its establishment has been used by tens of thousands of visitors. The centre is an education partner and activity provider with the Burren and Cliffs of Moher Geopark, as well as being a member of the Burren Ecotourism Network. (Fiona Guinness)

FIG 398. The Burren Perfumery was one of the earliest non-farming activities in the Burren. Founded 40 years ago, it is a small operation that makes cosmetics and perfumes inspired by the surrounding landscape and flora. (Fiona Guinness)

range from a perfumery (which, correctly, does not use flowers from the Burren) to cheesemaking, a chocolatier, a falconry, an outdoor pursuits centre, a national park, national monuments, a cave centre, an art college, field study centres and many more. The photographs here provide only a selection of these activities.

FIG 399. Hazel Mountain Chocolate, in Bell Harbour, claims be Ireland's first stoneground, bean-to-bar chocolate maker and chocolatier, and has a public viewing area. (Fiona Guinness)

FIG 400. Entrance to the Burren College of Art at Newtown Castle. (Fiona Guinness)

FIG 401. Newtown Castle is a sixteenth-century tower house located close to the village of Ballyvaghan, and is part of the Burren College of Art. It is cylindrical in its upper part but rises from a square base, a unique feature for an Irish tower house. (Fiona Guinness)

FIG 402. The Burren Nature Sanctuary, near Ballyvaghan, offers a range of activities with an emphasis on younger children. (Fiona Guinness)

FIG 403. The National University of Ireland Galway Field Station at Carran has excellent facilities for students and researchers, and is placed strategicaly in the Burren. The university has another field study station for marine biological research near Lough Murree. (Fiona Guinness)

FIG 404. The Bird of Prey Centre at Aillwee Cave has a collection of raptors and daily flight demonstrations during the summer. (Fiona Guinness)

FIG 405. The Clare Heritage Museum in Corrofin, housed in what was once St Catherine's Church of Ireland. This was built in the eighteenth century by Catherine Keightly, a cousin of Queen Mary II and Queen Anne. (Fiona Guinness)

FIG 406. Dunguaire Castle is a sixteenth-century tower house near Kinvarra. The name derives from the Dun of King Guaire, the legendary king of Connacht. The castle's 23 metre-high tower and its defensive wall have been restored, and it is now a major tourist attraction. (Fiona Guinness)

WHERE DO WE GO FROM HERE?

Encouraging results have been achieved in establishing the Burren Programme and providing sustainable farming that is compatible with the outstandingly rich ecological and historical heritage of the region. Farming activities and the natural Burren together form a dynamic system that has been evolving for thousands of years and will continue to do so. Responding to, and managing, future agricultural change presents creative opportunities for all participants. We cannot 'freeze' the Burren as it is now; the challenge is to manage the dynamic process with appropriate support from government, while maintaining clear environmental objectives. In this situation, the Burrenbeo Trust has a vital role to play with the development of its charter, together with providing upwards of 40 annual educational, information and community programmes.

Tourism has become a major activity in the Burren, with more than an estimated million visitors to the region every year. To the outside observer, this growth does appear to be placing increased environmental pressures on the area, although as yet this is very localised. The GeoparkLIFE programme has researched visitor usage in the Burren, the frequency of visits to key sites and visitor satisfaction, and identified issues that need to be addressed as well as many more matters (Millward Brown, 2014). These concerns are being actively examined by the project in conjunction with local and community-based organisations. GeoparkLIFE Programme has also commissioned several other important reports, including *Sustainable Tourism and Conservation Management: Mapping Policy*, prepared by Gabriel Cooney and Joanne Gaffrey. Let us hope that the development of the Burren as a tourist destination, boosted by the inducement of Fáilte Ireland to visitors to travel the Wild Atlantic Way as 'a once-in-a-lifetime experience', will not become a nightmare scenario. The consequences of the Burren's possible designation as a UNESCO World Heritage Site are unknown. Such a listing may encourage further tourism growth but at the same time may require further conservation measures.

Acknowledging the pressures exerted on the Burren by farming and tourism, it becomes essential to measure what is happening to the environment, monitoring its condition and knowing how to modify exploitation if this becomes necessary. A periodic state of the environment report for the region, in which standard ecological measurements are repeated and assessed, would be the most valuable outcome. This could be coordinated by one of the existing bodies such as the Burrenbeo Trust, in association with an academic institution. There would be inputs from national bodies such as the Environmental Protection Agency, the NPWS and the Meteorological Service, but only a peripheral role for those agencies dedicated to one exploitative land use or another. Such monitoring might include:

- Water quality of the Caher River and groundwater representing the major drainage catchments of the Burren.
- Quantitative vegetation analyses in a representative range of the major Burren habitats.
- Overall extent and height of scrub.
- Frequency and abundance of indicator plants and ferns based on fixed sampling areas.
- Frequency and abundance of butterflies, the larger moths and Odonata at fixed points.
- Breeding bird surveys in specified areas.
- Measurement of the spread of introduced species.

Without periodic reports on the quality of the natural environment, there is no objective information as to whether things are improving or deteriorating over time. Student projects are valuable to elucidate particular circumstances and species, but their methods differ from one another and they cannot give the necessary overall view. There is too much at stake to depend upon anecdotal accounts of environmental change.

The monitoring system will be important to evaluate the success of farming programmes, as well as measuring potential impacts from increasing tourism. A large and unknown driver of ecological change is climate change, which may well influence environmental quality in the Burren more than any other impact.

References

Akeroyd, J. R. & Parnell, J. A. N. (1981). *Bulletin of the Irish Biogeographical Society*, **5**, 25–26.

Allen, D., O'Donnell, M., Nelson, B. *et al.* (2016). *Ireland Red List No. 9: Macromoths (Lepidoptera).* National Parks and Wildlife Service, Dublin.

Allott, N., Coxon, C. & Cunha Pereira, H. (2015). Turlough algae. In Turlough hydrology, ecology and conservation (ed. E. Waldren). Unpublished report. National Parks and Wildlife Service, Dublin.

Andresen, H., Bakker, J. P., Brongers, M., Heydemann, B. & Irmler, U. (1990). Long-term changes of salt marsh communities by cattle grazing. *Vegetatio*, **89**, 137–148.

Anon. (1867). Obituary of Frederick J. Foot, M.A., F.R.G.S.I. *Geological Magazine or Monthly Journal of Geology*, **4**, 95–96.

Anon. (1884). *The Naturalist*, **9**, 194–195.

Anon. (1885). Obituary of T. H. Corry. *Journal of Botany, British and Foreign*, **21**, 813–814.

Anon. (2001). Back to nature in the Burren. *Irish Times*, 9 June.

Arnold, G. W. (1962). Factors within plant associations affecting behaviour and performance of grazing animals. In *Grazing in Terrestrial and Marine Environments* (ed. D. J. Crisp). Blackwell, Oxford.

Arukwe, A. & Langeland, A. (2013). Mitochondrial DNA inference between European populations of *Tanymastix stagnalis* and their glacial survival in Scandinavia. *Ecology and Evolution*, **3**, 3868–3878. doi: 10.1002/ece3.756.

Asher, J., Warren, M., Fox, R., Harding, P., Jeffcoate, G. & Jeffcoate, S. (2001). *The Millennium Atlas of Butterflies in Britain and Ireland.* Oxford University Press, Oxford.

Asplund, J. & Gauslaa, Y. (2008). Mollusc grazing limits growth and early development of the old forest lichen *Lobaria pulmonaria* in broadleaved deciduous forests. *Oecologia*, **155**, 93–99.

Babington, C. C. (1859). Hints towards a Cybele Hibernica. *Natural History Review*, **6**, 533–537.

Baker, C. A. & Kentish, H. E. (1913). Cave exploring in County Clare. *Yorkshire Ramblers' Club Journal*, **4**, 128–133.

Ballantyne, C. K. & Ó Cofaigh, C. (2017). The last Irish ice sheet: extent and chronology. In *Advances in Irish Quaternary Studies* (eds P. Coxon, S. McCarron & F. Mitchell). Atlantis Press, Paris. doi: 10.2991/978-94-6239-219-9.

Balmer, D. E., Gillings, S., Caffrey, B. J., Swann, R. I., Downie, I. S., & Fuller, R. J. (2013). *Bird Atlas 2007–2011: The Breeding and Wintering Birds of Britain and Ireland.* BTO Books, Thetford.

Bilton, D. T. & Lott, D. A. (1991). Further records of aquatic Coleoptera from Ireland. *Irish Naturalists' Journal,* **23**, 389–397.

Biological Records Centre (BRC) and Botanical Society of Britain and Ireland (BSBI) (2017). *Online Atlas of the British and Irish Flora.* www.brc.ac.uk/plantatlas/index.

Birchall, E. (1866). Catalogue of the Lepidoptera of Ireland. *Proceedings of the Natural History Society of Dublin,* **5**, 57–85.

Böcher, J., Kristensen, N. P., Pape, T. & Vilhelmsen, L. (eds) (2015). *The Greenland Entomofauna: An Identification Manual of Insects, Spiders and their Allies.* Brill, Leiden.

Bond, K. G. M. & O'Connor, J. P. (2012). Additions, deletions and corrections to *An Annotated Checklist of the Irish Butterflies and Moths (Lepidoptera)* with a concise checklist of Irish species and *Elachista biatomella* (Stainton 1848) new to Ireland. *Bulletin of the Irish Biogeographical Society,* **36**, 60–179.

Boston, H. L. & Adams, M. S. (1987). Productivity, growth and photosynthesis of two small 'isoetid' plants, *Littorella uniflora* and *Isoetes macrospora. Journal of Ecology,* **75**, 333–350.

Bradley, J. D. (1952). Microlepidoptera collected in the Burren. *Entomologist's Gazette,* **3**, 185–192.

Bradley, J. D. (1953). Microlepidoptera collected in the Burren, Co. Clare, in 1952, including a plume moth new to the British list. *Entomologist's Gazette,* **4**, 135–140.

Bradley, J. D. & Pelham-Clinton, E. C. (1967). The Lepidoptera of the Burren, Co. Clare, W. Ireland. *Entomologist's Gazette,* **18**, 116–153.

Brantjes, N. B. M. (1981). Ant, bee and fly pollination in *Epipactis palustris* (L.) Crantz (Orchidaceae). *Acta Botanica Neerlandica,* **39**, 59–68.

Braun-Blanquet, J. (1932). *Plant Sociology.* (Transl. G. D. Fuller and H. S. Conrad). McGraw-Hill, New York.

Breen, J. & O'Brien, A. (1995). Species richness in oldfield limestone grassland. In *Irish Grasslands – Their Biology and Management* (eds D. W. Jeffrey, M. Jones & J. H. McAdam). Royal Irish Academy, Dublin.

Brendan McGrath & Associates (2017) *Burren National Park Management Plan 2017–2030 – Draft.* www.burrennationalpark.ie/images/downloads/1ExecutiveSummary.pdf. Accessed March 2018.

Brian, M. V. (1977). Ants. Collins New Naturalist Library 59. William Collins, Sons and Co., Glasgow.

Bunce, C. (2010). Cave of the Wild Horses. *Irish Speleology,* **19**, 3–5.

Burren and Cliffs of Moher Geopark (2018). *GeoparkLIFE.* www.burrengeopark.ie/geopark-life.

Burren Programme (2017). *Burren Programme.* http://burrenprogramme.com. Accessed January 2018.

Burrenbeo Trust (2005). *Burren Library Ecology.* http://web.archive.org/web/20061118165758/http://www.burrenbeo.com/learning-library-ecology.aspx. Accessed February 2018.

Butterfly Conservation (2017). *Butterfly Conservation.* https://butterfly-conservation.org. Accessed October 2017.

Byrne, A., Moorkens, E. A., Anderson, R., Killeen, I. J. & Regan, E. C. (2009). *Ireland Red List No. 2 – Non-marine Molluscs*. National Parks and Wildlife Service, Dublin.

Cabot, D. (1965). The Green Lizard *Lacerta viridis* in Ireland. *Irish Naturalists' Journal*, **15**, 111.

Cabot, D. (1999). *Ireland*. HarperCollins, London.

Cameron, D. D., Coats, A. M. & Seel, W. E. (2006). Differential resistance among host and non-host species underlies the variable success of the hemi-parasitic plant. *Annals of Botany*, **98**, 1289–1299.

Carter, C. (1846). Botanical ramble in Ireland. *The Phytologist*, **2**, 512–514.

Cassina, F., Dalton, C., de Eyto, E., & Sparber, K. (2013). The palaeolimnology of Lough Murree, a brackish lake in the Burren, Ireland. *Biology and Environment: Proceedings of the Royal Irish Academy*, **113**, 1–17.

Ceiridwen J. E., Suchard, M. A., Lemey, P. *et al.* (2011). Hybridization and an Irish origin for the modern Polar Bear. *Current Biology*, **21**, 1251–1258.

Cham, S., Nelson, B., Parr, A., Prentice, S., Smallshire, D. & Taylor, P. (2014). *Atlas of Dragonflies in Britain and Ireland*. Biological Records Centre, Wallingford.

Clare Birdwatching (2018). *ClareBirdwatching.com*. www.clarebirdwatching.com. Accessed January 2018.

Clark, C. D., Hughes, A. L. C., Greenwood, S. L, Jordan, C. & Sejrup, H. P. (2012). Pattern and timing of retreat of the last British–Irish ice sheet. *Quaternary Science Reviews*, **44**, 112–146.

Clark P. U., Dyke A. S., Shakun J. D. *et al.* (2009). The last glacial maximum. *Science*, **325**, 710–714.

Classey, F. W. & Robinson, H. S. (1951). Burren – 1950. *Entomologist's Gazette*, **2**, 87–94.

Cockayne, E. A. (1954). The Irish subspecies of *Calamia virens* L. (Lep. Caradrinidae). *Entomologist's Gazette*, **5**, 155–156.

Cohen, K. M. & Gibbard, P. (2011). Global chronostratigraphical correlation table for the last 2.7 million years. Subcommission on Quaternary Stratigraphy (International Commission on Stratigraphy), Cambridge.

Colhoun, E. A., Dickson, J. H., McCabe, A. M. & Shotton, F. W. (1972). A middle Midlandian freshwater series at Derryvree, Maguiresbridge, County Fermanagh, Northern Ireland. *Proceedings of the Royal Society of London. Series B, Biological Sciences*, **180**, 273–292.

Cooper, F., Stone, R. E., McEvoy, P., Wilkins, T. & Reid, N. (2012). *The Conservation Status of Juniper Formations in Ireland*. Irish Wildlife Manuals No. 63. National Parks and Wildlife Service, Dublin.

Cope, T. & Grey, A. (2009). *Grasses of the British Isles*. Botanical Society of the British Isles, London.

Copini, P., den Ouden, J., Robert, E. M. R., Tardif, J. C., Loesberg, W. A., Goudzwaard, L. & Sass-Klaassen, U. (2016). Flood-ring formation and root development in response to experimental flooding of young *Quercus robur* trees. *Frontiers in Plant Science*, 14 June. doi: 10.3389/ fpls.2016.00775.

Coppins A. M. & Coppins, B. J. (2010). *Atlantic Hazel*. Scottish Natural Heritage, Inverness.

Corbel, J. (1957). *Les Karsts du Nord-ouest de l'Europe*. Memoires et Documents 12. Institut des Études Rhodaniens de l'Université de Lyons, Lyons.

Corry, T. H. (1880). Notes of a botanical ramble in the county of Clare, Ireland. *Proceedings and Reports of the Belfast Natural History and Philosophical Society,* **1879–1880**, 167–207.

Corry, T. H. (1882). *A Wreath of Wildflowers.* Privately printed, Belfast.

Cotton, D. (2015). *Pleistocene Epoch ('Ice Ages') (2.6 Ma to 11,700 BP).* http://staffweb. itsligo.ie/staff/dcotton/Pleistocene_ Epoch_2.6_Ma_to_11,700_BP.html. Accessed January 2018.

Cox, R., Zentner, D. B., Kirchner, B. J. & Cook, M. S. (2012). Boulder ridges on the Aran Islands (Ireland): recent movements caused by storm waves, not tsunamis. *Journal of Geology,* **120**, 249–272.

Coxon, P. (1988). Remnant periglacial features on the summit of Truskmore, Counties Sligo and Leitrim, Ireland. *Zeitschrift für Geomorphologie,* **71**, 81–91.

Coxon, P. & Waldren, S. (1995). The floristic record of Ireland's Pleistocene temperate stages. In *Island Britain: A Quaternary Perspective* (ed. R. C. Preece). Geological Society Special Publication No. 96. Geological Society, London.

Cronin, M., Duck, C., Ó Cadhla, O., Nairn, R., Strong, D. & O' Keeffe, C. (2004). *Harbour Seal Population Assessment in the Republic of Ireland: August 2003.* Irish Wildlife Manuals No. 11. National Parks and Wildlife Service, Dublin.

Croot, D. G. & Sims, P. C. (1996). Early stages of till genesis: an exposure from Fanore, Co. Clare, Ireland. *Boreas,* **25**, 37–46.

Cunha Pereira, H. (2011). Hydrochemistry and algal communities of turloughs (karstic seasonal lakes). Ph.D. thesis, University of Dublin.

Curtis, T. & Thompson, R. (2009). *The Orchids of Ireland.* National Museums, Northern Ireland, Belfast.

Curtis, T. G. F. (1981). A further station for *Mercurialis perennis* L. in the Burren with comments on its status there. *Irish Naturalists' Journal,* **20**, 184–185.

Curtis, T. G. F. & McGough, H. N. (1984). An unusual habitat for *Dryas octopetala. Bulletin of the Irish Biogeographical Society,* **7**, 11–13.

Curtis, T. G. F. & McGough, H. N. (1987). The occurrence of *Limosella aquatica* L. in limestone solution hollows at Fisherstreet, Co. Clare (H9). *Irish Naturalists' Journal,* **22**, 248–249.

Dale, A. & Elkington, T. T. (1984). Distribution of clones of *Galium boreale* L. on Widdybank Fell, Teesdale. *New Phytologist,* **96**, 317–330.

D'Arcy, G. (2016). *The Breathing Burren.* Collins Press, Cork.

Deenihan, A. (2011). Bumblebees in prime landscapes with special reference to the Aran Island bumblebee (Hymenoptera: Apidae). Ph.D. thesis, University of Limerick.

Desmond, R. (1994). *Dictionary of British and Irish Botanists and Horticulturalists.* Taylor & Francis and the Natural History Museum, London.

de Witte, L. C., Armbruster, G. F. J., Gielly, L., Taberlet, P. & Stöcklin, J. (2012). AFLP markers reveal high clonal diversity and extreme longevity in four arctic–alpine species. *Molecular Ecology,* **21**, 1081–1097.

Dickinson, C. H., Pearson, M. C. & Webb, D. A. (1964). Some micro-habitats of the Burren, their micro-environments and vegetation. *Proceedings of the Royal Irish Academy,* **63B**, 291–302.

Doddy, P. & Roden, C. M. (in press). The fertile rock: productivity and erosion in limestone solution hollows of the Burren, Co. Clare. *Biology and Environment: Proceedings of the Royal Irish Academy.*

Doddy, P. & Roden, C. M. (2014). The nature of the black deposit occurring in solution hollows on the limestone pavement of the Burren, Co. Clare. *Biology and Environment: Proceedings of the Royal Irish Academy,* **114**B, 71–77.

Donnelly, A., Caffarra, A. and O'Neill, B. F. (2011). A review of climate-driven mismatches between inter-dependent phenophases in terrestrial, aquatic and agricultural ecosystems. *International Journal of Biometeorology,* **55**, 805–817.

Dowd, M. (2016). A remarkable cave discovery. *Archaeology Ireland,* **30**(2), 21–25.

Doyle, G. J. (1985). A further occurrence of *Neotinea maculata* in woodland. *Irish Naturalist's Journal,* **21**, 502–503.

Doyle, G. J. (1993). *Cuscuta epithymum* (L.) L. (Convolvulaceae), its hosts and associated vegetation in a limestone pavement habitat in the Burren lowlands in County Clare (H9), western Ireland. *Biology and Environment: Proceedings of the Royal Irish Academy,* **93**B, 61–67.

Drew, D. (1986). *Aillwee Cave and the Caves of the Burren.* Irish Heritage Series No. 43. Eason, Dublin.

Drew, D. P. (1982). Environmental archaeology and karstic terrains: the example of the Burren, Co. Clare, Ireland. *BAR International Series,* **146**, 115–127.

Drew, D.P. & Cohen, J. M. (1980). Geomorphology and sediments of Aillwee Cave, Co. Clare, Ireland. *Proceedings of the University of Bristol Spelæological Society,* **15**, 227–249.

Druce, G. C. (1909). Notes on Irish plants. *Irish Naturalist,* **18**, 209–213.

Druce, G. C. (1916). *Ajuga pyramidalis* × *reptans.* Report of the Botanical Society and Exchange Club of the British Isles, **4**, 207.

Dublin Naturalists' Field Club (2017). *Butterflies of Ireland.* www.butterflyireland.com. Accessed October 2017.

Dudman, A. A. & Richards, A. J. (1997). *Dandelions of Great Britain and Ireland.* BSBI Handbook No. 9. Botanical Society of the British Isles, London.

Duffy, K. J., Scopece, G., Cozzolino, S., Fay, M. F., Smith, R. J. & Stout, J. C. (2009). Ecology and genetic diversity of the dense-flowered orchid, *Neotinea maculata,* at the centre and edge of its range. *Annals of Botany,* **104**, 507 –516.

Dunford, B. (2002). *Farming and the Burren.* Dublin, Teagasc.

Dunford, B. & Feehan, J. (2001). Agricultural practices and natural heritage: a case study of the Burren uplands, Co. Clare. *Tearmann: Irish Journal of Agri-environmental Research,* **1**, 19–34.

Dunning, J. W. (1884). *Proceedings of the Entomological Society of London,* **1884**, xlii.

Edwards, R. J. & Brooks, A. J. (2008).The island of Ireland: drowning the myth of an Irish land-bridge? In *Mind the Gap: Postglacial Colonisation of Ireland* (eds J. J. Davenport, D. P. Sleeman & P. C. Woodman). Special Supplement to the *Irish Naturalists' Journal* No. 29, Dublin.

El Haak, M. A., Mujahid, M. M. & Wegmann, K. (1997). Ecophysiological study on *Euphorbia paralias* under soil salinity and sea water spray treatments. *Journal of Arid Environments,* **35**, 459–471.

Elkington, T. T. (1972). Variation in *Gentiana verna* L. *New Phytologist,* **71**, 1203–1211.

Elkington, T. T. (1989). Cytotaxonomic variation in *Potentilla fruticosa* L. *New Phytologist*, **68**, 151–160.

Elkington, T. T. & Woodell, S. R. J. (1963). Biological flora of the British Isles: *Potentilla fruticosa* L. *Journal of Ecology*, **51**, 769–781.

Elliott, B. & Skinner, B. (1995). Some preliminary notes on *Odontognophos dumetata* Treitschke ssp. *hibernica* Forder (Lep.: Geometridae). *Entomologists Record and Journal of Variation*, **107**, 1–3.

Eriksson, O. & Bremer, B. (1993). Genet dynamics of the clonal plant *Rubus saxatilis*. *Journal of Ecology*, **81**, 533–542.

Farran, C. (1847). Panegyric on Mr. M'Alla and other Irish botanists. *The Phytologist*, **2**, 742–746.

Farrington, A. (1965). The last glaciation of the Burren. *Proceedings of the Royal Irish Academy*, **64B**, 33–39.

Feeser, I. & O'Connell, M. (2009). Fresh insights into long-term changes in flora, vegetation, land use and soil erosion in the karstic environment of the Burren, western Ireland. *Journal of Ecology*, **97**, 1083–1100.

Feeser, I. and O'Connell, M. (2010). Late Holocene land-use and vegetation dynamics in an upland karst region based on pollen and coprophilous fungal spore analyses: an example from the Burren, western Ireland. *Vegetation History and Archaeobotany*, **19**, 409–426. doi: 10.1007/s00334-009-0235-5.

Finch, T. F. (1966). Slieve Elva, Co. Clare – a nunatak. *Irish Naturalists' Journal*, **15**, 133–136.

Finch, T. F. (1971). *Soils of County Clare*. Bulletin No. 23 of the National Soil Survey. An Foras Taluntais, Dublin.

Fischer, S. F., Poschlod, P. & Beinlich, B. (1996). Experimental studies on the dispersal of plants and animals on sheep in calcareous grassland. *Journal of Applied Ecology*, **33**, 1206–1222.

Florian, P., Schiestl, F. P. & Cozzolino, S. (2008). Evolution of sexual mimicry in the orchid subtribe orchidinae: the role of preadaptations in the attraction of male bees as pollinators. *BMC Evolutionary Biology*, **8**, 27.

Foged, N. (1977). *Freshwater diatoms in Ireland*. Bibliotecha Phycologica No. 34. J. Cramer, Stuttgart.

Foot, D. E. J. (1862). On the distribution of plants in Burren, County of Clare. *Transactions of the Royal Irish Academy*, **24** (Science Part III), 143–160.

Foot, F. J. (1860). *Explanations to Accompany Sheets 131 and 132 of the Maps of the Geological Survey of Ireland Illustrating Parts of the Counties of Clare and Galway*. Memoirs of the Geological Survey of Ireland, Dublin.

Foot, F. J. (1863). *Explanations to Accompany Sheets 114, 122 and 123 of the Maps of the Geological Survey of Ireland Illustrating Parts of the Counties of Clare and Galway*. Memoirs of the Geological Survey of Ireland, Dublin.

Foot, F. J. (1864). On the distribution of the plants in Burren, County Clare. *Transactions of the Royal Irish Academy*, **24** (Science Part III), 143–160.

Foot, F. J. & Baily, W. H. (1860). *Explanations to Accompany Sheets 140 and 141 of the Maps of the Geological Survey of Ireland Illustrating Parts of the Counties of Clare and Kerry, with Palaeontological Notes by W. H. Baily*. Memoirs of the Geological Survey of Ireland, Dublin.

Forbes, E. (1846). On the connexion between the distribution of the existing fauna and flora of the British Isles, and the geological changes which have affected their areas, especially during the epoch of the Northern Drift. *Great Britain Geological Survey Memoir*, **1**, 336–432.

Ford, E. B. (1955). *Moths*. Collins New Naturalist Library 30. Collins, London.

Forder, P. (1993). *Odonthognophos* [*sic*] *dumetata* Treitschke (Lepidoptera: Geometridae) new to the British Isles with a description of a new form *hibernica* Forder ssp.nov. *Entomologists Record and Journal of Variation*, **105**, 201–202.

Fossilworks (2017). *Fossilworks*. www.fossilworks.org. Accessed July 2017.

Foster, G. N., Nelson, B. H. & O Connor, Á. (2009). *Ireland Red List No. 1 – Water Beetles*. National Parks and Wildlife Service, Dublin.

Francis, R., Finlay, D. & Read, D. J. (1986). Vesicular-arbuscular mycorrhiza in natural vegetation systems. (iv) Transfer of nutrients in inter- and intra-specific combinations of host plants. *New Phytol*, **102**, 103–111.

Froberg, L., Baur, A. & Baurj, B. (1993). Differential herbivore damage to calcicolous lichens by snails. *Lichenologist*, **25**, 83–95.

Gallagher, R. N. & Fairley, J. S. (1979). A population study of fieldmice *Apodemus sylvaticus* in the Burren. *Proceedings of the Royal Irish Academy*, **79B**, 123–147.

Gleeson, E., Donnelly, A., McGrath, R., ní Bhroin, A., F. O'Neill, B. F. & Semmler, T. (2013). Assessing the influence of a range of spring meteorological parameters on tree phenology. *Biology and Environment: Proceedings of the Royal Irish Academy*, **113B**, 47–56.

Glenza, C., Schlaepfer, R., Iorgulescu, I. & Kienast, F. (2006). Flooding tolerance of central European tree and shrub species. *Forest Ecology and Management*, **235**, 1–13.

Gloser, V. & Gloser, J. (1999). Production processes in a grass *Calamagrostis epigejos* grown at different soil nitrogen supply. In *Grassland Ecology V. Proceedings of the Fifth Ecological Conference, Banská Bystrica, Slovakia, 23–25 November 1999* (eds D. Ferienčíková, N. Gásorčík, L. Ondrášek, E. Uhliarová and M. Zimková). Grassland and Mountain Agriculture Research Institute, Banská Bystrica.

Good, J. A. (1998). *The Potential Role of Ecological Corridors for Habitat Conservation in Ireland: A Review*. Irish Wildlife Manuals No. 2. National Parks and Wildlife Service, Dublin.

Good, J. A. (2004). Lake-shore fens and reedbeds as a habitat for Staphylinidae and Carabidae (Coleoptera) in the east Burren area, Co. Clare, Ireland. *Bulletin of the Irish Biogeographical Society*, **28**, 163–188.

Gosling, P. (1991). The Burren in early historic times. In *The Book of the Burren* (eds J. W. O'Connell & A. Korf). Tir Eolis Press, Kinvarra.

Grainger, J. N. R. (1976). Further records for the fairy shrimp *Tanymastix stagnalis* (L.). *Irish Naturalists' Journal*, **18**, 326.

Gray, N., Thomas, G., Trewby, M & Newton, S. F. (2003). The status and distribution of Choughs *Pyrrhocorax pyrrhocorax* in the Republic of Ireland 2002/03. *Irish Birds*, **7**, 147–156.

Greene, J. (1854). A list of Lepidoptera hitherto taken in Ireland as far as the end of the Geometrae. *Natural History Review*, **1**, 165–196, 238–244.

Guiry, M. D. (2018). *Sargassum muticum* Wireweed. www.seaweed.ie/sargassum. Accessed January 2018.

Haflidason, H., King, E. L. & Kristensen, D. K. (1997). Marine geological/

geophysical cruise report on the western Irish margin: Donegal Bay, Clew Bay, Galway Bay, Irish Shelf and Rockall Trough. University of Bergen, Bergen.

Hanrahan, S. A. & Sheehy Skeffington, M. (2015). *Arctostaphylos* heath community ecology in the Burren, western Ireland. *Ecological Questions*, **21**, 9–12.

Harding, J. M. (2008). *Discovering Irish Butterflies and their Habitats.* Jesmond Harding, Dublin.

Harrington, T. J. (2003). Relationships between macrofungi and vegetation in the Burren. *Biology and Environment: Proceedings of the Royal Irish Academy*, **103B**, 147–159.

Harrington, T. J. & Mitchell, D. T. (2005). Ectomycorrhizas associated with a relict population of *Dryas octopetala* in the Burren, western Ireland. I. Distribution of ectomycorrhizas in relation to vegetation and soil characteristics. *Mycorrhiza*, **15**, 425–433

Hart, H. C. (1875). *A List of Plants Found in the Islands of Aran, Galway Bay.* Hodges, Foster & Co., Dublin.

Hatier, J.-H. B. & Gould, K. S. (2008). Anthocyanin function in vegetative organs. In *Anthocyanins: Biosynthesis, Functions, and Applications* (eds K. Gould, K. M. Davies & C. Winefield). Springer-Verlag, New York.

Healy, B. & Oliver, G. A. (1996). A survey of Irish coastal lagoons: aquatic fauna. Unpublished report for Dúchas, the Heritage Service. Dublin.

Heery, S. & Madden, B. (1997). A summer concentration of Little Grebes *Tachybabtus ruficollis* in south County Galway. *Irish Birds*, **6**, 53–54.

Heslop-Harrison, J. (1949). *Pinguicula grandiflora* Lam. in North Clare. *Irish Naturalists' Journal*, **9**, 311.

Heslop-Harrison, J. (1960). A note on the temperature and vapour pressure deficit under drought conditions in some microhabitats of the Burren limestone, Co. Clare. *Proceedings of the Royal Irish Academy*, **61B**, 109–114.

Heslop-Harrison, J., Wilkins, D. A. & Greene, S. W. (1961). *Arenaria norvegica* Gunn in Co. Clare. *Irish Naturalists' Journal*, **13**, 267–268.

Heuertz, M., Fineschi, S., Anzidei, M. et al. (2004). Chloroplast DNA variation and postglacial recolonization of common ash (*Fraxinus excelsior* L.) in Europe. *Molecular Ecology*, **13**, 3437–3452.

Hoctor, Z. (2014). *Burren National Park Site Assessment Report.* www.burrengeopark. ie/wp-content/uploads/2016/05/Burren-National-Park-Site-Assessment-Report. pdf.

Hogan, A. R. (1855). Catalogue of Irish microlepidoptera. *Natural History Review*, **2**, 109–115.

How, W. (1650). *Phytologia Britannica, Natales Exhibens Indigenarum Stirpium Sponte Emergentium.* Richard Cotes, Oxford.

Howard-Williams, E. (2013). A phylogeographic study of *Arenaria ciliata* and *Arenaria norvegica* in Ireland and Europe. Ph.D. thesis, National University of Ireland, Maynooth.

Humboldt, A. von H. & Bonpland, A. (1805). *Essai sur la Géographie des Plantes: Accompagné d'un Tableau physique des Régions équinoxiales, Fondé sur des Mesures exécutées, Depuis le Dixième Degré de Latitude boréale jusqu'au Dixième Degré de Latitude australe, Pendant les Années 1799, 1800, 1801, 1802 et 1803.* Chez Levrault, Schoell et co., Paris. [Transl. S. Romanowski, with an introduction by S. T. Jackson (2009). *Essay on the Geography of Plants.* University of Chicago Press, Chicago, IL.]

Ingrouille, M. J. & Stace, C. A. (1985). Pattern of variation of agamospermous Limonium (Plumbaginaceae) in the British Isles. *Nordic Journal of Botany*, **5**, 113–125.

Ivimey-Cook, R. B. (1965). The vegetation of solution cups in the limestone of the Burren, Co. Clare. *Journal of Ecology*, **53**, 437–445.

Ivimey-Cook, R. B. & Proctor, M. C. F. (1966). The plant communities of the Burren, Co. Clare. *Proceedings of the Royal Irish Academy*, **64B**, 211–301.

Jacquemyn, H. & Hutchings, M. J. (2010). Biological flora of the British Isles: *Spiranthes spiralis* (L.) Chevall. *Journal of Ecology*, **98**, 1253–1267.

Jacquemyn, H., Brys, R., Honnay, O. & Hutchings, M. J. (2009). Biological flora of the British Isles: *Orchis mascula* (L.) L. *Journal of Ecology*, **97**, 360–377.

Jeffrey, D. W. (2003). Grasslands and heath: a review and hypothesis to explain the distribution of Burren plant communities. *Biology and Environment: Proceedings of the Royal Irish Academy*, **103B**, 111–123.

Jeličič, L. & O'Connell, M. (1992). History of vegetation and land use from 3200 B.P. to the present in the north-west Burren, a karstic region of western Ireland. *Vegetation History and Archaeobotany*, **1**, 119–140.

Jensen N. B., Zagrobelny M., Hjernø K. *et al.* (2011). Convergent evolution in biosynthesis of cyanogenic defence compounds in plants and insects. *Nature Communications*, **2**. doi: 10.1038/ncomms1271.

Jinks, R. L., Parratt, M. & Morgan, G. (2012). Preferences of granivorous rodents for seeds of 12 temperate tree and shrub species used in direct sowing. *Journal of Forest Ecology and Management*, **278**, 71–79.

Joose, E. N. G. (1976). Littoral apterygotes (Collembola and Thysanura). In *Marine Insects* (ed. L. Cheng). North-Holland Publishing Company, Amsterdam and Oxford.

Jordaens, K., Platts, E. & Backeljau, T. (2001). Genetic and morphological variations in the Land Winkle *Pomatia elegans* (Müller) (Caenogastropoda: Pomatiasidae). *Journal of Molluscan Studies*, **67**,145–152. https://doi.org/10.1093/mollus/67.2.145.

Kame, K. & Fong, P. (2000). A fluctuating salinity regime mitigates the negative effects of reduced salinity on the estuarine macroalga, *Enteromorpha intestinalis* (L.) Link. *Journal of Experimental Marine Biology and Ecology*, **254**, 53–69.

Kelly-Quinn, M., Bradley, C., Murray, D. *et al.* (2003). Physico-chemicla characteristics of the Caher River. *Biology and Environment: Proceedings of the Royal Irish Academy*, **103B**, 187–196.

K'Eogh, J. (1735). *Botanalogia Universalis Hibernica*. George Harrison, Cork.

Kinahan, G. H. (1875). *Valleys and their Relation to Fissures, Fractures and Faults*. Trübner, London.

Kingston, S., O'Connell, M. & Fairley, J. S. (1999). Diet of Otters *Lutra lutra* on Inishmore, Aran Islands, west coast of Ireland. *Biology and Environment: Proceedings of the Royal Irish Academy*, **99B**, 173–182.

Kirby, E. N. (1981). An ecological and phytosociological study of *Corylus avellana* (L.) in the Burren, Western Ireland. Ph.D. thesis, University College Galway.

Knowles, M. C. (1913). The maritime and marine lichens of Howth. *Scientific*

Proceedings of the Royal Dublin Society, **14**, 79–143.

Labarde, J. & Thompson, A. (2009). Post-dispersal fate of hazel nuts and consequences for the management and conservation of scrub-grassland mosaics. *Biological Conservation,* **142**, 974–981.

Lamb, J. G. D. (1966). The occurrence of *Gentiana verna* L. on different soil types in the Burren. *Irish Naturalists' Journal,* **15**, 187–191.

Lambert, S. J. & Davy, A. J. (2011). Water quality as a threat to aquatic plants: discriminating between the effects of nitrate, phosphate, boron and heavy metals on charophytes. *New Phytologist,* **189**, 1051–1059.

Langangen, A. (2005). Charophytes collected in Cos Clare (H9) and south-east Galway (H15) in 2003. *Irish Naturalists' Journal,* **28**, 151–158.

Lansbury, I. (1965). Notes on the Hemiptera, Coleoptera, Diptera and other invertebrates of the Burren, Co. Clare and Inishmore, Aran Islands. *Proceedings of the Royal Irish Academy,* **64B**, 89–115.

Lauga, B., Cagnon, C., D'Amico, F., Karama, S. & Mouchès, C. (2005). Phylogeography of the white-throated dipper *Cinclus cinclus* in Europe. *Journal of Ornithology,* **146**, 257–262.

Lavorel, S., Lebreton, J. D., Debussche, M. & Lepart, J. (1991). Nested spatial patterns in seed bank and vegetation of Mediterranean old-fields. *Journal of Vegetation Science,* **2**, 367–376. doi: 10.2307/3235929.

Lawrey, J. D. (1986). Biological role of lichen substances. *The Bryologist,* **89**, 111–122.

Leake, J. P. (2005). Plants parasitic on fungi: unearthing the fungi in myco-heterotrophs and debunking the 'saprophytic' plant myth. *Mycologist,* **193**, 113–122.

Lee, A. (1998). The calcicole–calcifuge problem revisited. In *Advances in Botanical Research. Vol. 21* (ed. J. A. Callow). Elsevier Science Publishing, San Diego, CA.

Lee, J. (2013). *Yorkshire Dales.* Collins New Naturalist Library 130. William Collins, London.

Lhwyd, E. (1707). *Archaeologia Britannica.* Privately published, Oxford.

Lhwyd, E. (1712). Some farther observations relating to the antiquities and natural history of Ireland. In a letter from the late Mr. Edw. Lhwyd, Keeper of the Ashmolean Museum in Oxford, to Dr. Tancred Robinson, F.R.S. *Philosophical Transactions of the Royal Society of London,* **27**(336), 524–526.

Light, M. H. S & MacConaill, M. (2006). Appearance and disappearance of a weedy orchid, *Epipactis helleborine. Folia Geobotanica,* **41**, 77–93.

Long, M. P. (2011). Plant and snail communities in three habitat types in a limestone landscape in the west of Ireland, and the effects of exclusion of large grazing animals. Ph.D. thesis, Department of Botany, Trinity College Dublin.

Longley, M. (2000). *The Weather in Japan.* Jonathan Cape, London.

Lousley, J. E. (1950). *Wild Flowers of Chalk and Limestone.* Collins New Naturalist Library 16. William Collins, London.

Lucas, C. (1740). A description of the cave at Kilcorney, in the barony of Burren. *Philosophical Transactions of the Royal Society of London,* **41**, 360–364.

Ludlow, E. (1722). *The Memoirs of Edmund Ludlow Esq., Lieutenant-General of the Horse, Commander in Chief of the Forces* *of Ireland.* 2nd edn. W. Mears & F. Clay, London.

Lynch, A. (2014). *Poulnabrone: An Early Neolithic Portal Tomb in Ireland.* Wordwell Publications, Dublin.

Lyne, G. J. & Mitchell, M. E. (2014). *The Travels of Joseph Woods, Architect and Naturalist 1809.* https://celt.ucc.ie/published/E800005-003.html. CELT Corpus of Electronic Texts, University College Cork. Electronic edition compiled and proofed by Beatrix Färber.

Lyons, M. (2015). The flora and conservation status of Irish petrifying springs. Ph.D. thesis, Trinity College Dublin.

Lysaght, L. (2002). *An Atlas of Breeding Birds of the Burren and the Aran Islands.* BirdWatch Ireland, Dublin.

Lysaght, L. & Marnell, F. (2016). *Atlas of mammals in Ireland 2010–2015.* National Biodiversity Centre, Waterford.

McAllister, H. A. & Rutherford, A. (1990). *Hedera helix* L. and *H. hibernica* (Kirchner) Bean (Araliaceae) in the British Isles. *Watsonia,* **18**, 7–15.

McAney, K., O'Mahony, C., Kelleher, C., Taylor, A. & Biggane, S. (2013). *The Lesser Horseshoe Bat in Ireland: Surveys by the Vincent Wildlife Trust.* Irish Naturalists' Journal, Belfast.

McCabe, A. M. (1987). Quaternary deposits and glacial stratigraphy in Ireland. *Quarterly Science Reviews,* **6**, 259– 299.

McCarthy, D. M. & Mitchell, M. E. (1988). *Lichens of the Burren Hills and the Aran Islands.* Officina Typographica, Galway.

McCarthy, P. M. (1981). An ecological study of saxicolous lichen communities at three locations in coastal areas of south-western and western Ireland. Ph.D. thesis, National University of Ireland.

McCarthy, T. K. (1977). The Slow-worm *Anguis fragilis* L: a reptile new to the Irish fauna. *Irish Naturalists' Journal,* **19**, 49.

McGeever, A. H. & Mitchell, F. J. G. (2016). Re-defining the natural range of Scots Pine (*Pinus sylvestris* L.): a newly discovered microrefugium in western Ireland. *Journal of Biogeography,* **43**, 2199–2208.

McNamara, M. E. (2009). *The Geology of the Burren Region, Co. Clare, Ireland.* www.geoneed.org/wp-content/uploads/downloads/2011/06/Burren-Technical-geological-review.pdf. Part of WP2 of the Northern Environmental Development Project. EU Northern Periphery Programme 2007–13.

MacGowran, B. (1985). Phytosociological and ecological studies on turloughs in the west of Ireland. Ph.D. thesis, Department of Botany, University College Galway.

Mackay, J. T. (1836). *Flora Hibernica.* William Curry Jun and Co., Dublin.

Malloch, A. J. C. (1976). An annotated bibliography of the Burren. *Journal of Ecology,* **64**, 1093–1105.

Manaeva, E. S., Naumova, E. I., Kostina, N. V., Umarov, M. M. & Dobrovolskaya, T. G. (2012). Functional features of microbial communities in the digestive tract of field voles (*Microtus rossia meridionalis* and *Clethrionomys glareolus*). *Biology Bulletin,* **39**, 346–350.

Marschall, M. & Proctor, M. C. F. (2004). Are bryophytes shade plants? Photosynthetic light responses and proportions of chlorophyll a, chlorophyll b and total carotenoids. *Annals of Botany,* **94**, 593–603.

Matthews, J. R. (1955). *Origin and Distribution of the British Flora.* Hutchinson, London.

Mayland-Quellhorst, E., Föller, J. & Wissemann, V. (2012). Biological flora of the British Isles: *Rosa spinosissima* L. *Journal of Ecology*, **100**, 561–576.

Meehan, R. T. (2016). Glacial geomorphology of the last Irish ice sheet. In *Advances in Irish Quaternary Studies* (eds P. Coxon, S. McCarron & F. Mitchell). Atlantis Press, Paris. doi: 10.2991/978-94-6239-219-9.

Meeuse, B. J. D. (1989). *Handbook of Flowering. Vol. 6* (ed. A. H. Halevy). CRC Press, Boca Raton, FL.

Met Éireann (2018). *Climate of Ireland*. https://www.met.ie/climate/climate-of-ireland.asp. Accessed January 2018.

Meulebrouck, K., Ameloot, E., Van Assche, J. A. & Verheyen, K. (2008). Germination ecology of the holoparasite *Cuscuta epithymum*. *Seed Science Research*, **18**, 25–34.

Millward Brown (2014). *Burren and Cliffs of Moher Geopark Life Programme Visitor Survey*. www.burrengeopark.ie/wp-content/uploads/2015/03/Milward-Brown-Visitor-Survey.pdf. Accessed January 2018.

Mitchell, F. (1976). *The Irish Landscape*. Collins, London.

Mitchell, F. & Ryan, M. (1997). *Reading the Irish Landscape*. Town House & Country House, Dublin.

Mitchell, P. I., Newton, S. F., Ratcliffe, N. & Dunn, T. E. (eds.) (2004). *Seabird Populations of Britain and Ireland: Results of the Seabird 2000 Census (1998–2002)*. Poyser, London.

Mole, S. R. C. (2010). Changes in relative abundance of the Western Green Lizard *Lacerta bilineata* and the Common Wall Lizard *Podarcis muralis* introduced onto Boscombe Cliffs, Dorset, UK. *Herpetological Bulletin*, **114**, 24–29.

Moles, R. (1982). A study of some bird communities of the Burren, Co. Clare. *Irish Naturalists' Journal*, **20**, 419–423.

Molloy, K. & O'Connell, M. (2014). Post-glaciation plant colonisation of Ireland; fresh insights from An Loch Mór, Inis Oírr, western Ireland. In *Mind the Gap II: New Insights into the Irish Postglacial* (ed. D. P. Sleeman). Irish Naturalists' Journal Occasional Publication No. 14. Irish Naturalists' Journal, Dublin.

Monaghan, N. T. (2017). Irish Quaternary vertebrates. In *Advances in Irish Quaternary Studies* (eds P. Coxon, S. McCarron & F. Mitchell). Atlantis Press, Paris. doi: 10.2991/978-94-6239-219-9_9.

Montgomery, W. I., Lundy, M. G. & Reid, N. (2012). Invasional meltdown: evidence for unexpected consequences and cumulative impacts of multispecies invasions. *Biological Invasions*, **14**, 1111–1125.

Moore, D. & More, A. G. (1866). *Contributions Towards a Cybele Hibernica*. Hodges, Smith and Co., Dublin.

Moore, D. (1864). *Neotinia intacta* Reichb. a recent addition to the British flora. *Journal of Botany*, **2**, 228–229.

Moorkens, E. A. & Killeen, I. J. (2011). *Monitoring and Condition Assessment of Populations of* Vertigo geyeri, Vertigo angustior *and* Vertigo moulinsiana *in Ireland*. Irish Wildlife Manuals No. 55. National Parks and Wildlife Service, Dublin.

More, A. G. (1860). Localities for some plants observed in Ireland, with remarks on the geographical distribution of others. *Proceedings of the Dublin University Zoological and Botanical Association*, **2**, 54–65.

More, A. G. (1865). Notes on the discovery of *Neotinea intacta*, Reich. in Ireland.

Transactions of the Botanical Society of Edinburgh, **8**, 265–266.

Morris, M. G. (1967). Weevils (Coleoptera, Curculionoidea) and other insects collected in north-west Clare, with special reference to the Burren region. *Proceedings of the Royal Irish Academy*, **65B**, 349–371.

Morris, M. G. (1974). Auchenorhyncha (Hemiptera) of the Burren, with special reference to species-associations of the grasslands. *Proceedings of the Royal Irish Academy*, **74B**, 7–30.

Morris, M. G. (1985). *Bagous brevis* Gyllenhal. New to Ireland from the Burren, Co. Clare, with a brief review of the Bagoini (Coleoptera; Curculionidae). *Irish Naturalists' Journal*, **21**, 401–403.

MothsIreland (2017) *Moths Ireland: Mapping Ireland's Moths*. www.mothsireland.com. Accessed August 2016.

Müller, E., Cooper, E. J. & Alsos, I. G. (2011). Germinability of arctic plants is high in perceived optimal conditions but low in the field. *Botany*, **89**, 337–348.

Nash, D., Boyd, T. & Hardiman, D. (2012). *Ireland's Butterflies – A Review*. Dublin Naturalists' Field Club, Dublin.

National Parks and Wildlife Service (n.d.). *Special Areas of Conservations (SAC)*. www.npws.ie/protected-sites/sac. Accessed January 2018.

National Parks and Wildlife Service (2015). Inishmore Island SAC (site code: 213) – conservation objectives supporting document: coastal lagoons. Version 1, January 2015. Unpublished report. National Parks and Wildlife Service, Dublin.

Naughton, O., Johnston, P. & Gill, L. (2015). Hydrology. In Turlough hydrology, ecology and conservation (ed. E. Waldren). Unpublished report.

National Parks and Wildlife Service, Dublin.

Nelson, B. (2000). Dragonflies of the Burren and surrounding areas. *Atropos*, **11**, 9.

Nelson, B., Hughes, M. & Bond, K. (2011). The distribution of *Leptidea sinapis* (Linnaeus, 1758) and *L. reali* Reissinger, 1989 (Lepidoptera: Pieridae) in Ireland. *Entomologist's Gazette*, **62**, 213–233.

Nelson, B., Ronayne, C. & Thompson, R. (2011). *Ireland Red List No. 6: Damselflies and Dragonflies (Odonata)*. National Parks and Wildlife Service, Dublin.

Nelson, C. (2016). *The Wild Plants of the Burren and the Aran Islands*. Collins Press, Cork.

Nelson, E. C. (1979). Records of the Irish flora published before 1726. *Bulletin of the Irish Biogeographical Society*, **3**, 51–79.

Nelson, E. C. (1991). *The Burren*. Boethius Press and Conservancy of the Burren, Aberystwyth, Wales, and Kilkenny, Ireland.

Nelson, E. C. (1997). *The Burren, a Companion to the Wildflowers of an Irish Limestone Wilderness*. 2nd edn. Conservancy of the Burren, Dublin. Reprint, Samton.

Ni Bhriain, B., Gormally, M. & Sheehy Skeffington, M. (2003). Changes in land-use practices at two turloughs on the east Burren limestones, Co. Galway with reference to nature conservation. *Biology and Environment: Proceedings of the Royal Irish Academy*, **103B**, 169–176.

O'Callaghan, E., Foster, G. N., Bilton, D. T. & Reynolds, J. D. (2009). *Ochthebius nilssoni* Hebauer new for Ireland (Coleoptera: Hydraenidae), including a key to Irish *Ochthebius* and *Enicocerus*. *Irish Naturalists' Journal*, **30**, 19–23.

O'Connell, M. (2013). The Burren, north Clare – an exceptional landscape, a place apart. In *Secrets of the Irish*

Landscape (eds M. Jebb & C. Crowley). Cork University Press, Cork.

O'Donovan, G. (1987). An ecosystem study of grasslands in the Burren National Park, Co. Clare. Ph.D. thesis, Trinity College Dublin.

O'Grady, S. H. (ed.) (1929). *Caithreim Thoirdhealbhaigh: The Triumphs of Turlough*. Irish Texts Society No. 26. Simpkin, Marshall Ltd., London.

Oliver, G. A. (2007). Conservation status report: coastal lagoons (1150). Unpublished report. National Parks and Wildlife Service, Dublin.

O'Mahony, D. C. & Turner, P. (2006). *National Pine Marten Survey of Ireland 2006*. COFORD, Dublin.

O'Mahony, T. (1858). Notes of a botanical excursion in Clare. *Back Proceedings of the Dublin Natural History Society*, **1**, 30–34.

O'Neill Hencken, H. (1938). *Cahercommaun: A Stone Fort in County Clare*. Extra volume of the Royal Society of Antiquaries, Dublin.

Ordnance Survey Ireland (1995). *Discovery Series 51*. 1st edn. 1:50,000 map. Ordnance Survey Ireland, Dublin.

Ordnance Survey Ireland (1998). *Discovery Series 52*. 1st edn. 1:50,000 map. Ordnance Survey Ireland, Dublin.

Osborne, B., Black, K., Lanigan, G., Perks, M. & Clabby, G. (2003). Survival on the exposed limestone pavement of the Burren: photosynthesis and water relations of three co-occurring plant species. *Biology and Environment: Proceedings of the Royal Irish Academy*, **103B**, 125–137.

Pangua, E., Belmonte, R. & Pajarón, S. (2009). Germination and reproductive biology in salty conditions of *Asplenium marinum* (Aspleniaceae), a European coastal fern. *Flora*, **204**, 673–684

Parker, A. G., Goudie, A. S., Anderson, D. E., Robinson, M. A. & Bonsalle, C. (2002). A review of the mid-Holocene elm decline in the British Isles. *Progress in Physical Geography*, **26**, 1–45.

Parr, S., O'Donovan, G., Ward, S. & Finn, J. A. (2009). Vegetation analysis of upland Burren grasslands of conservation interest. *Biology and Environment: Proceedings of the Royal Academy*, **109B**, 11–33.

Pearce, R., Giuggioli, L. & Rands, S. (2017). Bumblebees can discriminate between scent-marks deposited by conspecifics. *Scientific Reports*, **7**. doi: 10.1038/srep43872.

Penck, M., Waldren, S., Allott, N. *et al.* (2015). Integration of work packages: turlough ecological functioning. In Turlough hydrology, ecology and conservation (ed. E. Waldren). Unpublished report. National Parks and Wildlife Service, Dublin.

Peric, B. (ed.). (2012). *The Škocjan National Park*. Park Škocjanske, Koper.

Perrin, P. M., Kelly, D. L. & Mitchell, F. J. G. (2006). Long-term deer exclusion in yew-wood and oakwood habitats in southwest Ireland: natural regeneration and stand dynamics. *Forest Ecology and Management*, **236**, 356–367.

Perring, F. H. (1995). David A. Webb memorial issue of *Watsonia*. *Watsonia*, **21**, 1.

Petty, W. (1685). *Hiberniae Delineato*. University of Ireland, Dublin.

Phillips, R. A. (1923). The Pearl-bordered Fritillary in Ireland. *Irish Naturalists' Journal*, **32**, 91–92.

Platts, E., Bailey, S., McGrath, D. & McGeough, G. (2003). A survey of the Land Winkle *Pomatias elegans* (Müller, 1774) in the Burren, Co. Clare. *Biology* and Environment: Proceedings of the Royal Irish Academy, **103B**, 197–201.

Platts, E. A. (1977). The Land Winkle *Pomatias elegans* (Müller) confirmed as an Irish species. *Irish Naturalists' Journal*, **19**, 10–12.

Plunkett Dillon, E. C. (1985). The field boundaries of the Burren, County Clare. Ph.D. thesis, Trinity College Dublin.

Porst, G. & Irvine K. (2009). Distinctiveness of macroinvertebrate communities in turloughs (temporary ponds) and their response to environmental variables. *Aquatic Conservation: Marine and Freshwater Ecosystems*, **19**, 456–465.

Porter, M. & Foley, M. (2017). *Violas of Britain and Ireland*. BSBI Handbook No. 17. Botanical Society of Britain and Ireland, Bristol.

Potts, M. (2000). Nostoc. In *The Ecology of Cyanobacteria: their Diversity in Time and Space* (eds B. A. Whitton & M. Potts). Kluwer Academic Publishers, Dordrecht.

Praeger, R. L. (1895). Notes on the flora of Aranmore. *Irish Naturalist*, **4**, 249–252.

Praeger, R. L. (1903). *Pinguicula grandiflora* in Clare. *Irish Naturalist*, **12**, 269.

Praeger, R. L. (1909). *A Tourist's Flora of the West of Ireland*. Hodges, Figgis & Co., Dublin.

Praeger, R. L. (1932). The flora of the turloughs: a preliminary note. *Royal Irish Academy*, **41B**, 37–45.

Praeger, R. L. (1934). *The Botanist in Ireland*. Hodges, Figgis & Co., Dublin.

Praeger, R. L. (1939). A further contribution to the flora of Ireland. *Proceedings of the Royal Irish Academy*, **45B**, 231–254.

Praeger, R. L. (1949). *Some Irish Naturalists: A Biographical Notebook*. Dundalgen Press, Dundalk.

Preston, C. D. (1995). *Pondweeds of Great Britain and Ireland*. BSBI Handbook No. 8. Botanical Society of the British Isles, London.

Preston, C. D., Pearman, D. A. & Dines, T. D. (eds) (2002). *New Atlas of the British and Irish Flora*. Oxford University Press, Oxford.

Proctor, M. C. F. (1956). Biological flora of the British Isles: *Helianthemum* Mill. *Journal of Ecology*, **44**, 675–692.

Proctor, M. C. F. (2010). Environmental and vegetational relationships of lakes, fens and turloughs in the Burren. *Biology and Environment: Proceedings of the Royal Irish Academy*, **110B**, 17–34.

Pybus, C., Pybus, M. J. and Ragneborn-Tough, L. (2003). Phytoplankton and charophytes of Lough Bunny, Co. Clare. *Biology and Environment: Proceedings of the Royal Irish Academy*, **103B**, 177–185.

Reed, R. H. & Russell, G. (1979). Adaptation to salinity stress in populations of *Enteromorpha intestinalis* (L.) Link. *Estuarine and Coastal Marine Science*, **8**, 251–258.

Regan, E. C. & Meagher, O. (2011). The distribution and habitat of the Pearl-borderd Fritillary butterfly *Boloria euphrosyne* (Linnaeus) (Lepidoptera: Nymphalidae) in Ireland. *Bulletin of the Irish Biogeographical Society*, **35**, 150–161.

Regan, E. C., Lovatt, J. K. & Wilson, C. J. (2010). Natural fluctuations in numbers of the Holly Blue butterfly (*Celastrina argiolus* (L.) in Ireland. *Irish Naturalists' Journal*, **31**, 123–126.

Reichenbach, H. G. (1865). *Neotinea intacta*, Rehb.fil, the new Irish orchid. *Journal of Botany*, **3**,1–5.

Reynolds, J. D. (2016). Invertebrates of Irish turloughs. In *Invertebrates in Freshwater Wetlands* (eds D. Batzer & D. Boix).

Springer International Publishing, Switzerland.

Richards, O. W. (1961). The fauna of an area of limestone pavement on the Burren, Co. Clare. *Proceedings of the Royal Irish Academy*, **62B**, 1–7.

Robinson, G. S., Ackery, P. R., Kitching, I. J., Beccaloni, G. W. & Hernández, L. M. (2010). HOSTS – a Database of the World's Lepidopteran Hostplants. www. nhm.ac.uk/our-science/data/hostplants. Accessed January 2018.

Robinson, T. (1999). *The Burren: A Two-inch Map of the Uplands of North-west Clare.* Folding Landscapes, Roundstone.

Roche, J. (2010). The vegetation ecology and native status of Scots Pine (*Pinus sylvestris* L.) in Ireland. Ph.D. thesis, Trinity College Dublin.

Roche, J. R., Waldren, R., Stefanini, B. & Mitchell, F. J. G. (2010). Palaeoecological evidence for the continued survival of Scots Pine (*Pinus sylvestris*) through the late Holocene in the Burren, western Ireland. In The vegetation ecology and native status of Scots pine (*Pinus sylvestris* L.) in Ireland (J. Roche). Ph.D. thesis, Trinity College Dublin.

Roche, N., Aughney, T., Marnell, F. & Lundy, M. (2014). *Irish Bats in the 21st Century.* Bat Conservation Ireland, Virginia.

Rödel, H. G. & Starkloff, A. (2014). Social environment and weather during early life influence gastro-intestinal parasite loads in a group-living mammal. *Oecologia*, **176**, 389–398.

Roden, C. & Murphy, P. (2013). *A Survey of the Benthic Macrophytes of Three Hard-water Lakes: Lough Bunny, Lough Carra and Lough Owel.* Irish Wildlife Manuals No. 70. National Parks and Wildlife Service, Dublin.

Rodwell, J. S. (ed.) (1991). *British Plant Communities. Vol. 2: Mires and Heaths.* Cambridge University Press, Cambridge.

Rohling, E. J., Grant, K., Bolshaw, M. *et al.* (2009). Antarctic temperature and global sea level closely coupled over the past five glacial cycles. *Nature Geoscience*, **2**, 500–504. doi: 10.1038/ngeo557.

Rook, A. J., Dumont, B., Isselstein, J. *et al.* (2004). Matching type of livestock to desired biodiversity outcomes in pastures – a review. *Biological Conservation*, **119**, 137–150.

Russell F. M., Russell F. M. & Mizell, R. A. (2002). Trolling: a novel trapping method for *Chrysops* spp. (Diptera: Tabanidae). *Florida Entomologist*, **85**, 356–366.

Ryland, J. S. & Nelson-Smith, A. (1975). Littoral and benthic investigations on the west coast of Ireland: IV. (Section A: faunistic and ecological studies.) *Proceedings of the Royal Irish Academy*, **75B**, 245–266.

Scannell, M. J. P. (1988). *Ajuga* × *hampeana* Braun & Vatke in the Burren: P.B. O'Kelly vindicated. *Irish Naturalists' Journal*, **22**, 488–490.

Scharff, R. F., Ussher, R. J., Cole, A. J., Newton, E. T., Dixon, A. F. and Westropp, T. J. (1906). The exploration of the caves of Co. Clare. *Transactions of the Royal Irish Academy*, **33B**, 1–76.

Scottish Natural Heritage (2017). *Otter.* www.nature.scot/plants-animals-and-fungi/mammals/land-mammals/otter. Accessed January 2018.

Sejrup, H. P., Hjelstuen, B. O., & Dahlgren, K. I. T. (2005). Pleistocene glacial history of the NW European continental margin. *Marine and Petroleum Geology*, **22**, 1111–1129.

Sharkey, N., Waldren, S., Kimberley, S., González, A., Murphy, M. & O'Rourke, A. (2015). Turlough vegetation: description, mapping and ecology. In *Turlough Hydrology, Ecology and Conservation* (ed. E. Waldren). Unpublished report. National Parks and Wildlife Service, Dublin.

Sheehy Skeffington, M. J. & Curtis, T. G. F. (2000). The Atlantic element in Irish salt marshes. In *Biodiversity: The Irish Dimension* (ed. B. S. Rushton). Royal Irish Academy, Dublin.

Sheehy Skeffington, M. J. & Jeffrey, D. W. (1985). Growth performance of an inland population of *Plantago maritima* in response to nitrogen and salinity. *Vegetation*, **61**, 265–272.

Sheehy Skeffington, M. J. & Wymer, E. D. (1991). Irish salt marshes – an outline review. In *A Guide to the Sand Dunes of Ireland* (ed. M. B. Quigley). Compiled for the Third Congress of the European Union for Dune Conservation and Coastal Management, Galway, Ireland, June 1991.

Shields, L. & Fitzgerald, D. (1989). The Night of the Big Wind in Ireland, 6–7 January 1839. *Irish Geography*, **22**, 31–43.

Simms, M. (2006). *Exploring the Limestone Landscapes of the Burren and the Gort Lowlands*. Mike Simms, Belfast.

Simms, M. J. (2003). The origin of enigmatic, tubular, lakeshore Karren: a mechanism for rapid dissolution of limestone in carbonate-saturated waters. *Physical Geography*, **23**, 1–20.

Sjøtun, K., Heesch, S., Lluch, J. R. *et al.* (2017). Unravelling the complexity of salt marsh '*Fucus cottonii*' forms (Phaeophyceae, Fucales). *European Journal of Phycology*, **52**, 360–370.

Smiddy, P. & O'Halloran, J. (1998). Breeding biology of the Grey Wagtail *Motacilla cinerea* in southwest Ireland. *Bird Study*, **45**, 331–336

Smiddy, P., O'Halloran, J., O'Mahony, B. & Taylor, A. J. (1995). The breeding biology of the Dipper *Cinclus cinclus* in southwest Ireland. *Bird Study*, **42**, 76–81.

Smith, M. E., Douhan, G. W. &. Rizzo, D. M. (2007). Intra-specific and intra-sporocarp ITS variation of ectomycorrhizal fungi as assessed by rDNA sequencing of sporocarps and pooled ectomycorrhizal roots from a *Quercus* woodland. *Mycorrhiza*, **18**,15–22.

Soane, G. A. (1980). Food selection by the Rabbit. Ph.D. thesis, University of Wales, Bangor.

Sparling, J. H. (1968). Biological flora of the British Isles: *Schoenus nigricans* L. (*Chaetospora nigricans* Kunth). *Journal of Ecology*, **56**, 883–899.

Speight, M. C. D. (1975). The husbandry of *Schoenus nigricans* (Cyperaceae) for thatching in Co. Galway. *Irish Naturalists' Journal*, **18**, 192–193.

Speight, M. C. D. (1985). The extinction of indigenous *Pinus sylvestris* in Ireland: faunal data. *Irish Naturalists' Journal*, **21**, 449–453.

Stace, C. (2010). *New Flora of the British Isles*. Cambridge University Press, Cambridge.

Stanton, W. I. (1986). Snail holes (helixigenic cavities) in hard limestone – an aid to the interpretation of karst landforms. *Proceedings of the University of Bristol Spelæological Society*, **17**, 218–226.

Stevens, C. J., Wilson, J. & McAllister, H. A. (2012). Biological flora of the British Isles: *Campanula rotundifolia*. *Journal of Ecology*, **100**, 821–839.

Suddaby, D., Nelson, T. & Veldman, J. (2010). *Resurvey of Breeding Wader Populations of Machair and Associated Wet Grasslands in North-west Ireland*. Irish

Wildlife Manuals No. 44. National Parks and Wildlife Service, Dublin.

Sweeting, M. M. (1966). The weathering of limestones, with particular reference to the Carboniferous limestones of Northern England. In *Essays in Geomorphology* (ed. G. H. Dury). Heinemann Educational Books, London.

Synge, F. M. (1969). The Würm ice limit in the west of Ireland. In *Quaternary Geology and Climate* (ed. H. E. Wright). National Academy of Sciences Publication 1701. National Academy of Sciences, Washington, DC.

Takahashi, Y., Kagawa, K., Svensson, E. I. & Kawata, M. (2014). Evolution of increased phenotypic diversity enhances population performance by reducing sexual harassment in damselflies. *Nature Communications*, 5. doi:10.1038/ncomms5468.

Tansley, A. G. (1939). *The British Islands and Their Vegetation*. Cambridge University Press, Cambridge.

Thomas, P. A. (2016). Biological flora of the British Isles: *Fraxinus excelsior*. *Journal of Ecology*, 104, 1158–1209.

Thomas, P. A., El-Barghathi, M. & Polwart, A. (2010). Biological flora of the British Isles: *Euonymus europaeus* L. *Journal of Ecology*, 99, 345–365.

Thompson, W. (1840). Catalogue of the land and freshwater mollusca of Ireland. *Annals and Magazine of Natural History*, 6, 16–34, 109–126, 194–206.

Thompson, W. (1842). *Cyclostoma elegans* Lam. an Irish shell. *Annals and Magazine of Natural History*, 8, 228.

Threlkeld, C. (1727). *Synopsis Stirpium Hibernicum*. Powell, Davys, Norris & Worrall, Dublin.

Titz, W. (1972). Evolution of the *Arabis hirsuta* group in central Europe. *Taxon*, 21, 121–128.

Toftegaard, T., Iason, G. R., Alexander, J., Rosendahl, S. & Taylor, A. (2010). The threatened plant intermediate wintergreen (*Pyrola media*) associates with a wide range of biotrophic fungi in native Scottish pine woods. *Biodiversity and Conservation*, 19, 3963–3971.

Tolan-Smith, C. (2008). Mesolithic Britain. In *Mesolithic Europe* (eds C. Bailey & P. Spikins). Cambridge University Press, Cambridge.

Tratman, E. K. (ed.) (1969). *The Caves of North-west Clare, Ireland*. David and Charles, Newton Abbot.

Turgeon, J., Stoks, R., Thum, R. A., Brown, J. M. & McPeek, M. A. (2005). Simultaneous Quaternary radiations of three damselfly clades across the Holarctic. *American Naturalist*, 165, E78–E107. https://doi.org/10.1086/428682.

UNESCO (2017). *UNESCO World Heritage Centre*. www.whc.unesco.org. Accessed March 2017.

Urquhart, D. (comp.) (2015). *Essex Moth Group Annual Newsletter 2015*. www.essexfieldclub.org.uk/resource/essex_moth_group_newsletter_2015.pdf.

Valvasor, J. V. (1689). *Die Ehre deß Hertzogthums Crain: das ist, Wahre, gründliche, und recht eigendliche Belegen- und Beschaffenheit dieses Römisch-Keyserlichen herrlichen Erblandes*. Erasmum Francisci, Laibach (Ljubljana).

Vernon, J. D. R. (1972). Feeding habitats and food of the Black-headed and Common gulls. Part 2 – food. *Bird Study*, 19, 173–186.

von Euler, T. (2006–07). *Sex Related Colour Polymorphism in* Antennaria dioica. Plants and Ecology, Department of

Botany, Stockholm University. www.
su.se/polopoly-fs/1.71162.1326706393!/
PlantsEcology_2007_6.pdf.

Waldren, S. (ed.). (2015). Turlough hydrology,
ecology and conservation. Unpublished
report. National Parks and Wildlife
Service, Dublin.

Waldren, S., Lynn, D. & Murphy, S. (2006).
The turlough form of Ranunculus repens
L. (Creeping Buttercup). In Botanical
Links in the Atlantic Arc (eds S. J. Leach,
C. N. Page, Y. Peytoureau & M. N.
Sanford). Conference Report No. 24
English Nature/Botanical Society of the
British Isles.

Walker, K. J. (2015). Ajuga pyramidalis L.
Pyramidal Bugle. https://bsbi.org/
species-accou. Accessed January 2018.

Walker, K. J., Howard-Williams, E. &
Meade, C. (2013). The distribution and
ecology of Arenaria norvegica Gunn. in
Ireland. Irish Naturalists' Journal, 32,
1–12.

Walsh, K. (2009) The Voice on the Ground:
A Survey of the Needs of Burren Farm
Families. Burrenbeo Trust and Burren
Irish Farmers Association, Dunboyne.

Wardle, P. (1961). Biological flora of the
British Isles: Fraxinus excelsior L. Journal
of Ecology, 49, 739–751.

Warren, W. P. & Ashley, G. M. (1994).
Origins of the ice-contact stratified
ridges (eskers) of Ireland. Journal of
Sedimentary Research, A64, 433–449.

Watson, J. E., Brooks, S. J., Whitehouse,
N. J., Reimer, P. J., Birks, H. J. B. &
Turney, C. (2010). Chironomid-inferred
late-glacial summer air temperatures
from Lough Nadourcan, Co. Donegal,
Ireland. Journal Quaternary Science, 25,
1200–1210.

Watts, W. A. (1963). Late-glacial pollen zones
in western Ireland. Irish Geography, 4,
367–376.

Watts, W. A. (1984). The Holocene vegetation
of the Burren, western Ireland. In Lake
Sediments and Environmental History (eds
E. Y. Haworth & J. W. G. Land). Leicester
University Press, Leicester.

Watts, W. A. & Ross, R. (1959). Interglacial
deposits at Kilbeg and Newton, Co.
Waterford. Proceedings of the Royal Irish
Academy, 60B, 79–134.

Webb, D. A. (1952). Narrative of the Ninth
I.P.E. In Die Pflanzenwelt Irlands
(The Flora and Vegetation of Ireland).
Ergebnisse der 9. Internationalen
Pflanzengeographischen Exkursion
durch Irland 1949 (ed. W. Lüdi).
Veröffentlichungen des Geobotanischen
Institutes Rübel in Zürich No. 25. Verlag
Hans Huber, Bern.

Webb, D. A. (1962). Noteworthy plants
of the Burren: a catalogue raisonné.
Proceedings of the Royal Irish Academy
62B, 117–134.

Webb, D. A. (1980). The Flora of the Aran
Island. Journal of Life Sciences Royal
Dublin Society 2, 51–83.

Webb, D. A. (1983). The flora of Ireland
in its European context. Journal of
Life Sciences. Royal Dublin Society, 1983,
143–160.

Webb, D. A. & Scannell, M. J. P. (1983). Flora
of Connemara and the Burren. Cambridge
University Press, Cambridge.

Weidema, I. R., Magnussen, L. S. &
Phillipp, M. (2000). Gene flow and
mode of pollination in a dry grassland
species Filipendula vulgaris (Rosaceae).
Heredity, 84, 311–320.

Welker, J. M., Molau U., Parsons, A. N.
Robinson, C. H. & Wookey, P. A.
(1997). Responses of Dryas octopetala to
ITEX environmental manipulations:
a synthesis with circumpolar
comparisons. Global Change Biology, 3,
61–73.

Werner, R. (ed.) (2010). *A Management Strategy for Maintaining a Feral Goat Population on the Burren that is Sustainable and does not Conflict with Farming Interests, Alongside a Strategy for Preserving the Remnant of the old Irish Goat Breed that is Retained Within it; Being the Outcomes of a Conference Organised by the BurrenLife Project.* BurrenLIFE, Carran.

Westbury, D. B. (2004). Biological flora of the British Isles: *Rhinanthus minor* L. *Journal of Ecology*, **92**, 906–927.

Whitla, F. (1851). On *Orobanche alba*. The *Phytologist*, **4**, 168.

Whittier, P. (1981). Spore germination and young gametophyte development of *Botrychium* and *Ophioglossum* in axenic culture. *American Fern Journal*, **71**, 13–19.

Wilkman, K. (2010). *Nilssons bagge – ett irländskt dragplåster.* News post, 7 October, Umeå Universitet, Umeå. www.umu.se/om-universitetet/aktuellt/arkiv/aktum/aldre-aktum/aktum-2010--nr-4/nilssons-bagge---ett-irlandskt-dragplaster. Accessed September 2016.

Williams, D. M. & Hall, A. M. (2004). Cliff-top megaclast deposits of Ireland, a record of extreme waves in the North Atlantic – storms or tsunamis? *Marine Geology*, **206**, 101–117.

Williams, P. W. (1966). Limestone pavements, with special reference to

Western Ireland. *Transactions of the Institute of British Geographers* **40**, 155–172.

Wilson, S. & Fernández, F. (2013). *National Survey of Limestone Pavement and Associated Habitats in Ireland.* Irish Wildlife Manuals No. 73. National Parks and Wildlife Service, Dublin.

Winther, J. L. & Friedman, W. E. (2007). Arbuscular mycorrhizal symbionts in *Botrychium* (Ophioglossaceae). *American Journal of Botany*, **94**, 1248–1255.

World Heritage Ireland (2010). *Tentative List Submission Format: The Burren.* www.worldheritageireland.ie/fileadmin/user_upload/documents/tentative_list/Burren_TL_form.pdf. Accessed January 2018.

Wright, W. S. (1951). A moth new to British and Irish lists in Co. Clare. *Irish Naturalists' Journal*, **10**, 135.

Wyse Jackson, M., FitzPatrick, U., Cole, E. *et al.* (2016). *Ireland Red List No. 10: Vascular Plants.* National Parks and Wildlife Service, Dublin.

Yalden, W. (1986). Diet, food availability and habitat selection of breeding Common Sandpipers *Actitis hypoleucos. Ibis*, **128**, 23–36.

Young, R. (1976). *Tanymastix stagnalis* (Linn.) in County Galway, new to Britain and Ireland. *Proceedings of the Royal Irish Academy*, **76B**, 369–378.

Index

GENERAL INDEX